A WALK IN THE PARK

A WALK IN THE PARK

BY

ROBIN CLARKE

HARROW AND HESTON
Open Access Publishers
Australia, New York & Philadelphia

A Read-Me.Org Imprint

ISBN: 978-0-911577-64-8 (Paperback)

ISBN: 978-0-911577-65-5 (Digital)

Library of Congress Control Number: 2022939423

Cover design by Read-Me.Org

TABLE OF CONTENTS

1. Prologue

We had just arrived in the Hotel Santa Cruz from the airport after the UK Foreign Office had selected me to help with one of their Overseas Aid projects.

And here I sit in what is going to be our bed-sit until we find a more permanent and much larger accommodation, but it does boast a tiny patio overlooking the street below complete with the roar of dozens of motorbikes, the hooting of impatient drivers in their vehicles and friendly greetings, lots of "Hola, gringo" from the crowds of passers-by.

Cheri went off to scout the possibilities of finding all those things our fair sex requires for every moment of the day.

What we didn´t realize was the whole night was a close copy of the day´s noise. We didn´t get any sleep. Immediately after breakfast, my three cups of coffee and many cigarettes I requested transfer to somewhere much quieter, and it certainly was! Now I was on the top floor at the back of the building with a bigger patio and a fantastic view towards the foothills of the Andean Mountains. Cheri had gone to bed leaving me to look at an amazing rainstorm sweeping down the higher ground straight towards me.

But wake up you dumb-cluck, don´t you realize what you are seeing? Rain for the City´s water supply and, even more beneficial, uncontaminated fresh air.

Just at that moment one of the hotel maids came to make the beds, so I asked her if she knew the name of the mountains I was looking at. "Yes" she said that is Amboró. Is there anybody you know that I could talk to who would know a lot about Amboró? "Yes", she said, you have to go and talk to Professor Kempff Mercado. If you like, one of the hotels messenger boys will take you to visit him in the morning.

And that is what I did, not realizing that I had just committed myself to a life changing decision and the repercussions that were to make me many more enemies than friends!

1

****

When we left England, a friend who had come to see us off pressed into my hands an old book, Exploration Fawcett, the account of Lt.-Col. P.H. Fawcett's travels in and around Bolivia and later Brazil, during 1906-1925. My flight from Miami to Santa Cruz de la Sierra, in the eastern lowlands of Bolivia, had not been monotonous, for whatever else the book was, it was not dull. Fascinating, incredible, perceptive, unbelievable, preposterous. Like many others who had read the book before me, my judgement was filled with a variety of conflicting emotions.

As I sat staring down on the rain forest far below, I thought to myself, one day I would like to go and find out. The book, a strange mixture of unbelievable cruelty and debauchery, describes in vivid detail the man's perilous adventures in a world between. Between the cold scientific realism of the developed world (as it was then) and the twilight of the unexplored Amazon rain forest as it remains today. Another world indeed, where fact and fiction walk side by side; where legends of almost mythical proportions have since been well authenticated. A world of superlatives: longest, largest, smallest, oldest, highest, deepest, fastest, strangest, wildest, and remotest. Yes, these and much, much more. Animals, birds, people, straight from the pages of a bestiary: Pandora's Box, the lid not quite firmly closed, seeping odours of incredulity. So, finally, what are we to think? Upon which of the many accounts can cold reason rest? The mind boggles, becomes blank, reason sinks to the chest and leaves us with a feeling of hopelessness. Who knows?

****

Yesterday it had rained. Today it was hot and the ground was now dusty. Everybody was sweating in the stifling atmosphere of the bus. In the last town we had waited more than an hour for the bus to fill up. The driver refused to set off until every seat was occupied. Ten minutes after leaving the town behind, so many others had boarded there was not even standing room now. A pig tied to the roof rack was complaining bitterly, probably at the stench of humanity below. The bus had left the market town of Montero only half an hour ago and I was already feeling irritable. South American travel books regularly refer to the hazards and discomforts of bus travel on the continent. But these authors are complaining about the flotas, Inter-city Coaches that cater for the better off and the tourists.

Our bus was an old Scania, vintage '53, held together with a variety of Heath-Robinson devices and strips of rubber. I was in the back corner, the 'gringo seat', known by the locals as the one to avoid since it is adjacent to the exhaust pipe, collects all the dust and takes the worst pounding from the countless potholes and ruts in the road.

The most alarming part of the trip was the crossing of the River Piraí. True, the actual river was crossed by a modern steel bridge many hundreds of feet long, but a good part of the river had gone off on a new course and now flowed across the road. Beside the flood a huddle of expectant tractor drivers waited to tow vehicles across. Most of the tractors, bought with farm subsidies, should have been gainfully employed raising the agricultural production of the country, but their owners, thinking otherwise, were now touting their expensive services in this new enterprise. The bus slowed to a crawl and sent forth a cloud of black smoke, accompanied by an ear-throbbing belch as it entered the water. Friendly tractor drivers, secretly hoping to distract him, jeered our chauffeur on. They knew his reckless attempt to pioneer a new phase of unaided crossings would, if successful, doom them back to the farm as word spread that the river was, now, not so high. Not to be flustered, our enterprising driver took on board a young lad who was to be our pilot, I doubt he was more than twelve years old.

The riverbed was sandy and deep pools quickly formed in the most unexpected places. Indeed, we had not gone far before we passed an overturned lorry that, like some prehistoric beast, lay in such a pool. Its driver sat dejectedly on the side of the cab munching a long piece of fried yuca, the staple root crop of the region. His load of bagged rice, crates of empty bottles and bundles of hobbled hens were spread in sorry disarray on an adjacent sand bank. Unconcerned, two other men sat on the back of the half-submerged vehicle fishing for the bony "Sabalo" that were plentiful in most of the rivers. Our driver gave vent to his own bout of jeering as we passed them. Distracted by his puerile enjoyment he very nearly collided with an oncoming tractor.

Though this was officially the dry season, marked by the passage of "San Juan" on the 24th of June the eastern lowlands of Bolivia are subject to rain throughout the year and humidity remains high except on a few rare days. It comes as a surprise to many a stranger to see the rivers overflowing at this time of the year. Even the smallest streams rarely dry up and few are the days or nights free of the ubiquitous mosquitos that emanate in clouds from the ever-present swamps close by.

****

Nine months ago, I arrived in Bolivia to take up my post as Crop Protection Officer/Research Entomologist to the British Tropical Agriculture Mission (BTAM). As a zoologist by training one of my objectives had been to out-Bates Bates, the famous British explorer-naturalist of the last century. Now, somewhere not too far to the west, but still hidden by the partly forested plain, lay the mountains of Amboró and my destination. These mountains form part of the "Fajas sub-Andinas" ("The Andean foothills") and my interest in them lay in a proposed National Park for the area. Profesor Noel Kempff Mercado, Director of the zoo in Santa Cruz, invited me to go and explore Amboró on behalf of the Academia Nacional de Ciencias. He wanted me to collect enough information to justify the establishment of the National Park.

I had been waiting several months for an end to the rains in order to make my first trip. I had studied maps of the area that showed the proposed Park to be divided into two distinct zones, both heavily forested. Most of it was wild mountainous country cut into steep-sided ravines by the numerous rivers and streams that drained it. The rest, relatively flat rolling country, was known to contain some peasant farmers communities. Cerro Amboró (that I was politely informed, but incorrectly, meant "Big Chief" in the Guarani language), is a rocky prominence that gives the area its name. Visible from the city of Santa Cruz it is a landmark that had attracted me since the day of my arrival.

At long last the rains had abated and travel was possible again. I checked out of my office for a welcome few days off. Amboró offers the more adventurous the nearest untouched tract of wilderness to the big city. If I had had any sense, I would have thought about it more carefully; the words untouched, near city, should have been warning enough, but "Mad dogs and Englishmen…".

We left the river in much the same way as a dog will, the water-soaked bus giving its back-end a thorough shake as it accelerated down the rutted track. My shoes got wet when the water welled up through a hole in the floor. I would have raised my legs, but the buses are designed for maximum profit and the space between the seats for the local people, few of whom are over 5'6". As it was, my knees were immovably jammed against the seat back in front of me.

****

We were now approaching my intended staging post, the small Jesuit reduction of Buena Vista where 300 families lived. The pig was still complaining as the bus farted its way down the first hill we had met since leaving Montero. The trip had taken four hours, twice as long as scheduled. I thanked God that in a minute I would return to the sane part of the world I hoped still existed. We entered the central plaza in a cloud of dust and flying gravel somehow thrown up by the bus's almost bald tyres. As I was stepping down the attendant stuffed a fistful of pesos into my hand, the change for my fare I had completely forgotten about. I appeared to be the only one getting off.

The little town stood astride the top of one of the many laterite-coloured hills characteristic of the region. Not the slightest breeze relieved the soporific midday heat, nor did the faintest breath of air stir the shredded banana leaves. Nothing moved, not even the shrill scream of an Ovenbird disturbed the town as it dozed under its thick blanket of humidity. Had I been able to climb to the top of a tall microwave tower dominating the village, I would have looked down into a child's cardboard model, so still, so medieval did the town appear. No one walked under the "Jorori" and "Toborochi" trees offering such delicious shade to the town square. No horses pulled at the decumbent grass. The village dogs, having given up their search for scraps, now lay on their flanks fast asleep in the cool of the piazza which ran around three sides of the square. Everything appeared to be waiting, all activity suspended, not in anticipation of things to come, because nothing ever happened here, nothing ever came except the rain. No column of ants disturbed the dust, no fly-provoked slap, even the "marahuí", tiny biting flies had fled to the shade of the forest. Far above, had anyone had the energy to look up, a single Turkey Vulture wheeled effortlessly under one dark cloud. The cloud, itself, too weary to move across the sun and supply some relief to the exhausted town.

All this I saw in a flash, as if for one fleeting moment I had absorbed the detail of an old postcard. I was not to know, then, how events were to move along their fated path. That this dead little village was to become my home and was to see a number of disturbing events of which I was to be the cause.

But today I was a stranger and a thirsty one. I headed for the nearest bar in search of a cold beer. I entered an open door with a small sign beside it: "Alojamiento Se vende Comida Cerveza Hielo" ("Rooms Meals Beer Ice"). There was nobody around. I clapped my hands, the socially accepted form of attracting attention. A well-developed doe-eyed girl, who seemed to have something seriously wrong with her ambulation, like she couldn't lift her feet, slouched into the room. She returned after a few moments with a cold bottle of "Ducal" (a refreshing local lager) and a chipped glass a yoghurt would have found cramping. I couldn't hide my disdain for the glass and the faulty Spanish of my request for a larger one "No hay otra más grande, Señorita?" elicited a tired shrug but no replacement. Since the glass was dirty as well, I drank from the bottle, first wiping off the rust left by the cap. The gassy beer and my well-shaken innards conspired to emulate the bus and forced a sudden belch from my throat that took even me by surprise. I heard the girl, who had given me a scornful look and slouched out of the room, altercating in a shrill voice with somebody in the back. I was new to the country, and only years later did I learn to appreciate these people and their ways, what was and what was not taboo. Disdain or sarcasm between strangers is the epitome of bad manners. Drinking from the bottle, thereby signifying an unwillingness to share it, is mean. Belching is unacceptable unless one is clearly too drunk to be considered human. It seemed I had not made a good start in this little town.

One of the things for which they all have respect is money; and wishing to leave the rest of the day free for more important things, I requested that supper be prepared for me and booked a room for the night. The doe-eyed girl gave me a friendly leer as I left. Having satisfied my first two needs I went in search of a guide to lead me to the mountains. The last proved to be impossible. "Did not the Señor realise that this was ploughing time, clearing time, thatching time?" Every sort of time, but not time for guides. After more fruitless enquiries I concluded that I would have to find somebody on my way towards the mountains in the morning.

Now, in the late afternoon, with the sun slipping behind the Andes to the west, I felt tired, yet full of unjustifiable elation. I had slipped out of town by way of one of the sandy roads that lead eventually to the River Surutú two miles away. I was perched on a ridge of red soil, below which the land had slipped away in a minor landslide, probably, to judge from the lack of regrowth, during the last rains. In front of me the land fell in a series of hills and hollows to a broad level valley running towards the south; here and there, where its surface reflected the last rays of the sun, the river flashed messages to me.

The first impression was one of unbroken forest, but the more I looked the more I perceived how little forest was left. Much of the vegetation was secondary in nature, marked by indicator trees such as Balsa and the ubiquitous "ambaiba" (a species of Cercropia) that the locals esteem for the sugary taste of their condom-like fruit; and which provide them with an antiseptic for minor cuts and scrapes. Some extensive pasture had been established in the bottom lands and I could make out further erosion scars, the bright red soil exposed like open wounds on an animal's flank. Taller forest still dominated many of the ridge tops, some steeper slopes and the wetlands skirting the river.

The far side of the Surutú looked more densely forested, and I eagerly scanned it with my field glasses for any hint of betrayal, for the river formed the northeastern boundary of the proposed National Park. With a map on my knee and the land spread out at my feet, I realised that my little excursion might turn into a major expedition, for the distance involved now looked a good deal farther than I had imagined. The proposed park (about the same size as some small countries) was a triangular area of more than 800 square miles as measured on the map, much more including the wrinkles.

I could see a secondary ridgeline on the other side of the river that I reckoned to be about six miles away. It hid a substantial area of dead ground between itself and the foot of the Andes. The map indicated this to be about six miles wide. The wall of the mountain quickly rose in a jumble of minor ones to a major ridgeline forming the horizon in front of me. The map showed this section of the ridge to be about twenty miles long.

I planned to go up one of the many rivers that tumbled down from this ridge, and with considerable meandering cross the lowlands to join the Surutú. I calculated that this approach would nearly double the distance (say twenty miles) to the foot of the Andes. Then there would be the extra distance of my zigzag path (in the rivers themselves) and the total would be considerably more. I imagined, with luck, I could make it in one day. That was as far as I had thought it out but, I intended to go as far into the mountains as I thought was possible. Now I saw what a feasible objective: a huge cliff of bare white rock. It rose majestically to the back of another cliff (this one's crest hidden by an intervening ridge). I liked the idea of heading for something specific and easily recognisable once I got there. And, I thought, those cliffs might be home to all sorts of animals and birds I wouldn't see during the rest of my trip.

I had heard vague rumours about large Macaws and the Cock-of-the-Rock being found in Amboró: birds that might be attracted to nest in such places. It would have been futile to prophesy how long this part of the adventure would take but, I decided then and there, come hell or (and I shouldn't have mentioned it) "high water", that's where I was going. With a definite plan to guide me, my growing apprehension abated and my feeling of elation returned. I now had time to relax and pay attention to that part of the Serrania ("the mountains") lying further to the south.

I could see Cerro Amboró projecting from a series of ridges around it: the main one ran towards the north to pass above my cliffs. From here the name 'Big Chief' took on real meaning, for there he was reclining on the top of the serrania; and Cerro Amboró now painted pink by the setting sun looked just like, "well I durst say!" Staring at these huge ominous rocks and deep sinister-looking canyons, and faced by my determination to enter alone if necessary: the folly of it all was the worm in the apple of my complaisant calm, and suddenly I felt terrified that I had bitten off more than I could chew. That I would go was sure, for a deep-down urge, one that had sent many explorers to their end, had settled in my breast. I call the feeling "Fawcett's Folly". If the world were flat, this feeling would drive the men it affected to the very lip, where they would be found hanging over the edge examining the underside. For certain Fawcett, that incredible man, must have set off on many of his, what? Mad? No: ("Fated adventures" – "Yes") driven, as he was, by this same feeling. And what was to be my fate?

****

When I was about five years old (living on a coffee farm in Tanganyika) a little old woman (said to be the wife of the local Witch Doctor) took my small head into her hands, peered into my eyes and told me my fortune. "Bwana Mdogo" ("Little Boss") she started, "You will never fail your lessons" [wrong]. "You will always be happy" [rightish]. "You will never be rich" [right]. "You will die by the crocodile", gulp!

****

Since there are no reports of man-eating crocodiles in these shallow rivers, I took some consolation in remembering this curious incident, and set off for my room with a light heart. The stars were by now fully out; so absorbed had I been with my thoughts I'd hardly noticed that darkness had fallen. Immediately above I could plainly see Scorpio, or could it be a crocodile? No, forget it, this trip was going to be a doddle.

I passed a fitful night, more bothered by indigestible food and a back-breaking bed than by tarantulas, man-eating lizards, or evil-intentioned ne'er-do-wells. I was woken well before sunup by the landlords sinewy "Gallo fino" ("Fighting cock") that had taken to screaming insults at the one next door. It was five-thirty. I spent an hour preparing my pack whilst creating unnecessary noises to signify I was awake and wanted my coffee. To no avail, all the doors around the inner patio remained stubbornly shut. I wandered into the back garden that boasted a magnificent Tamarind tree laden with hard grey pods. The Buenavisteñians soak the pulp in water to prepare a pleasantly refreshing cordial.

I saw two species of hummingbirds: a pair of Fork-tailed Wood-nymphs and a Reddish Hermit feeding on the nectar of some small red and yellow flowers I did not recognise. Below a dripping tap a pair of Rufous Horneros (or "Ovenbirds" as we call them) were collecting beak-fulls of mud to add to their nest on a level branch of the Tamarind. The nests, that weigh several kilograms when completely dry, take on the consistency of concrete.

****

Some years later I calculated that the birds bring close to 18,000 beak-fulls of mud to make it. The brood chamber is cut off from the outside by a vestibule that leaves just enough space for the horneros to squeeze through. Clever little blighters.

I am sure the design has evolved because of their frequent need to repel would-be tenants: like the ubiquitous Parrotlets, as well as protection for their eggs and young from the predations of other birds: like Chestnut-eared Araçaris with their long probing bills.

****

This pair was busily working on the inside, putting the finishing touches to the vestibule, normally the last of the walls to be finished.

Every now and again, determined to help me obtain my coffee, they set up the most fearful racket.

I kicked an old beer can around: shades of Butch Cassidy and the Sundance Kid. Was that bit in Bolivia? You remember, they have just arrived, Redford says to Newman, something like "So this is Bolivia," but I seem to remember he kicked a pile of horse shit.

Whatever, the tin did it. The slouching girl of yesterday finally emerged through an oil-starved door, gave a tremendous yawn during which one large breast fell through a rip in her nightgown. She stared at me as though in recognition of my innate insolence and re-entered her room. I am not sure what it is in me, but a performance like that leaves me feeling intimidated, almost a little scared by its cold apathy. So, I waited patiently for another half hour during which I could hear much water-sluicing and pot-bashing coming from the kitchen. More importantly I saw and smelt smoke and in the morning, to me, this means Coffee! I was not wrong. With the coffee came two hard little buns probably baked in the northern Sahara. These delicacies I sacrificed to the chickens, slurped down three muddy cups of coffee, paid my bill, and left.

It was a magnificent morning. The sun striking the mountains with all the glory of a tropical sunrise, bestowed a red tint to my cliffs. "A rose-red city half as old as time." Who knows? Maybe I will find another "Petra" below those cliffs. Perhaps even Fawcett himself will greet me with, "Taken your time, what?", or some such English off-handedness. Such are the nature of my muddled thoughts in the early morning.

Last night, even though tired, I did not go straight to my room. There were two oldish men drinking in the tiny bar and I fell into conversation with them, albeit only the fatter one ever said anything. His companion, very tall and thin, grinned or looked serious as the situation demanded. The pair reminded me of Hilda Baker and her stooge: "Aye, she knows you know", as they used to appear on TV some time back in the sixties. The small man had been very definite about the best way to approach my objective. Both had done a good deal of hunting in the mountains, mainly for the Wild pig and Tapir that they assured me were abundant. After a rather tiresome half hour wasted in explaining my disinclination to search for gold and my real inclination to see some natural history, they proved to be useful sources of information. Neither wished to accompany me since they were too busy, "logging time". But, with the aid of the map they did show me, more or less, the way to go. I was surprised they were so adept at map reading: and with my inadvertent gringo habit to patronise the locals, I mentioned it. Without the slightest trace of annoyance the smaller man explained that they were in the army and I was talking to Captain Eduardo Aguirre and his batman Jose Alvarez. Furthermore, I was now invited to join them for supper: a very bland Spaghetti Bolognese. As it turned out Eduardo was my landlord, the sloppy girl his older daughter. He was an "Excombatiente": a veteran of the Chaco War that, here, carries about the same prestige and meaning as would a veteran of the "Dunkirk fiasco" in England.

According to the captain's instructions, I was to take the first vehicle that came down the Huaytu road, where I was now waiting. I was to get off at "El Terminal" ("The Finish"): I hoped it wouldn't be! From there I had to walk down to the Surutú River, cross over and follow the path to the River Chonta (via Agua Blanca, a village I would find on the way), and from there ask again.

Here comes the first vehicle. Now if you were to take this particular one up, say, the M1 Motorway in England, not only would you end up in jail, but also cause a multi-car pile-up, so astonishing the sight. It was a lorry, "was" being the operative word. It had no cab as such; in fact, it had no bodywork of any kind. The engine was completely open to the elements, as was the only seat: the driver's (that was made from a wooden crate bolted to the chassis). Behind, where the cab should have been, was a flatbed of crudely hewn planks, along the sides of which were a number of crude posts rudely let into holes: posts that swayed and wobbled as the lorry approached. The tyres were mostly down to the canvas: the one in good condition only serving to emphasise the extreme dilapidation of the others. One of these had a patch sewn to its side that accounted for the rhythmic flapping-noise as the vehicle approached. There were no lights, indicators, horn, or (as I was told after we had set off), brakes! Once we got moving the engine sounded surprisingly sweet and I was not surprised to be told it was a 1951 Italian Fiat.

Sitting in the driver's seat was Captain Aguirre beaming at me like a Cheshire cat, the sun winking off the carefully worked gold bordering his teeth. He was a chubby man with a round "cherub-like face", the face set off by the compulsory droopy moustache and pierced by black fathomless eyes, that now he was grinning so delightfully, almost disappeared into a fold of skin running from ear to ear. The whole affect: the delightful Captain, his ridiculous truck with its tired old body and youthful purring engine, the dusty pot-holed road, the sweet early morning air ahead, behind the cloud of red dust thrown up by our passing combined to give me a moment of inexpressible joy.

The Captain was very proud of his vehicle that "Never gives me any serious trouble and would get me anywhere, regardless of the conditions." I was surprised we made it up the first hill. After that we made a shallow descent to the level ground of the valley bottom. Near the river the road bent abruptly away and we regained slightly higher ground. Any chance of views were quickly cut off by the thick vegetation pressing in on either side but, for one fleeting moment the whole of the Serrania came into view, now much closer and more sinister looking than it had from my perch outside town. When we reached El Terminal, so called because here the gravelled road ended, I had expected to get off for the three-mile walk down to the river, but the Captain motioned me to stay where I was while two or three campesinos paid him for their ride. We then set off down the bumpy track to the river, passed through a sticky patch of mud with consummate ease, took a minor diversion around a fallen tree: and there we were, beside the Surutú. The good Captain refused to accept the customary fare, embraced me warmly, and with a final wink and a "Buena suerte. Buen viaje" ("Good Luck. Have a good trip") disappeared the way we had come. I knew the tone of that good luck wish, and that impish wink, I am sure he still thought I was looking for gold, "You can't fool me," he had said.

When the sound of his truck died away and civilisation appeared to have gone with him, I turned to pay attention to my immediate surroundings. A long line of "Oropendolas" (Crow-like birds with yellow bills and tails) flew overhead in the same direction I was going. A Morpho butterfly, its wings large enough to hide my face, flew past not a yard from me. Its electric blue wings flashing in the sunlight, and as it jinked away looked like a windblown leaf from a tree on a magical planet. A flock of "Canary-winged Parakeets" were chattering busily in a nearby treetop; and directly overhead I heard the harsh screams of "Chestnut-fronted Macaws", and the answering call of others at some distance. A "Moustached Wren", a master ventriloquist, scolded me from, I don't know where, for no matter which way I turned it was always behind me. I recognised the call of a male Antbird and the less vocal response of its mate somewhere further ahead. I knew too little about the calls of these interesting birds to be able to identify the species.

For now, I had another, more immediate, problem: the river. Like most relatively placid ones, and this was no exception, the water is often muddy and its depth difficult to assess. Should I strip off and prepare myself for a swim? After all, there was always Fawcett's technique of cutting a stout pole some eight feet in length, sitting on one end with your pack tied to the other, and then sort of... "Splosh!"

I decided to leave my pack on the sandy beach and see how far I could get by wading. But what a fool! I had let my attention wander and was now firmly up to my thighs in thixotropic soil, or to the layman: Quicksand! A moment of real panic, then I realised I wasn't going to sink further. After an absurdly graceless struggle, I managed to regain the beach. Why I had not thought to ask about the river when I could have, the Lord only knows. Nobody said anything about it being dangerous, but you don't get told the answers unless you ask the questions.

I was feeling annoyed and flustered; the river crossing was obviously quite impossible and here my journey appeared to have come to an ignominious end. I sat on the beach and calmed myself with a cigarette, summoning up the courage to give it another go, watching a man as he made his way across the river, how dumb! The man was already more than halfway across and the water hardly covered his knees; and his knees, he probably wasn't more than five feet tall. I was so relieved I somewhat startled him with an over-enthusiastic "Buenos días" and a few cigarettes to speed him on his way. In fact he went off with such haste, I think he took me for an escaped lunatic. I made certain I looked as nonchalant as possible when I stepped into the river because; I could see he had now paused at a bend in the road to eye me from a safe distance. I was sure if I started floundering about again I might get myself arrested. I pushed to the back of my mind that intimidating quicksand, concentrated on the line I thought he had taken, and drove forward. The river was about one hundred yards wide. The first part was shallow and firm as I had witnessed, but further on the footing was loose and the sand formed numerous pools into which I frequently plunged to my waist. At one point I stepped on something that moved. I tried to forget those horrific accounts I had read of Pirañas and Freshwater Rays.

The rays are said to abound in South American Rivers, and this looked like just the right place. Even though I'd kept my gym shoes on, I knew they wouldn't protect me from the wicked dorsal spine that every ray, if you believe some books, dreams of driving into the feet of intrepid explorers. The only cure for the resulting paralysis was the amputation of the whole limb. Later, the locals told me that no rays lie in wait in the Surutú, but further downstream, where it joins the Rio Yapacaní instances had occurred. One assured me the best cure was to keep the foot under water for a whole day. Another recommended immersion of the leg in cow's milk, or better still, menstrual blood. I hope I never have to try it.

Well, I had got across the river but, not without loss of grace. As I took my sand-filled shoes off I noticed that in the middle, and slightly upstream from the point I had taken, a line of flimsy sticks. I instantly realised that they marked the way I should have gone. While I washed my shoes and socks, I mused upon a similar, but even sillier incident in Ethiopia.

****

The leader of our Grain Storage team decided he would take us on a field trip to demonstrate extension techniques to the local peasant farmers. "We are," he informed us as he pulled up the jeep, "going to visit this village as a training exercise." The village lay about 300 yards off the road on the other side of a stream that tumbled through a small, but deep ravine. "Follow me," he said, and turning about set off: and with some difficulty and soaking wet, we arrived in the dusty space left between the village's dozen houses. There a group of farmers was now awaiting us.

Our leader explained to them who we were and what we had come for: how important it was for them to store their grain in the right way; the factors responsible for the degradation of their wheat and sorghum; and that he was here with his team to introduce them to modern storage techniques, that if applied correctly, would make them money.

Stepping back with a self-satisfied grin, he then asked if any of them had any questions. There was a long embarrassing silence. Not to be put off our leader insisted that someone must have a question for him: "Are you so wise that you know everything? Don't you want to know anything?"

"Yes," said one old man, pushing to the front. "Why didn't you use the bridge?"

****

My foolish floundering in the river and my subsequent musing had lost me a precious hour. By the time I was hastening down the pathway towards Agua Blanca, the sun was well up and the morning was getting hot. Most of the wildlife had already adjourned for its mid-morning siesta, but a whole tribe of Smooth-billed Anis were preening themselves on a fallen tree. I have noticed that during the dry season, or winter as it is here, the birds are busy until about 10 a.m. and again in the evening, from about 4 p.m. to dusk. On summer mornings they tend to withdraw earlier but: are active for an hour during midday and then retire to rest until the late afternoon. There are also a number of species, apart from owls, which are active after dark. The Goatsuckers or Whip-poor-wills, their curious relatives the Potoos, Chachalacas, Wood-rails, Tinamous and some Woodcreepers all disturb the quiet of the night with their cries. My favourites are the Tinamous with their sad, flute-like calls, and the Common Potoo whose local name Guajojo is a fair likeness to its melancholy cry: "gua ho, ho-ho-hohohoho" that starts on a high note and ends with a hollow, ironical laugh.

I soon got into a walking rhythm, easing the pain caused by the weight of my pack with a short cigarette-break every hour or so. I passed through a variety of scenery: from open fields and cultivations to boggy overgrown woodland rich in palms and tacuara, the local name for this spiteful bamboo with its vicious thorns. Some small patches of tall forest sprang up here and there. I saw no animals apart from domestic ones and heard no sounds except for dogs barking. I didn't stop to examine any of the insects, of which there were many, so determined was I to put some distance behind me.

I passed quite a few houses, mainly simple one-roomed affairs, their adobe walls and palm leaf thatches in various states of disrepair. Most of them appeared to have a single entrance, no more than one or two very small windows and a raised sleeping platform at the centre. To one side a dilapidated shed held the kitchen where, next to the walls an assortment of goods and chattels were scattered on the floor. Pigs, chickens, and children ran in and out of the door. Most of the grownups were absent, probably busy tending their livestock or weeding their plantations of citrus and cooking bananas. From out of some of these huts ran mean looking dogs, their ribs as clearly marked as a shoe scrapers. They were always wary of anyone who carried a stick and ready to flee and yelping in anticipation of the pain if you pretended to throw a stone at them. Some just sat around looking as apprehensive as their masters: who in these parts appeared to be mainly migrant Collas from the altiplano, the broad, thin-soiled valley of highland Bolivia.

These people, darker and even smaller than the Cambas are intensely disliked by the lowlanders. They call them "paisanos" meaning fellow countrymen, but accompanied by grimaces and an upward nod of the head to indicate their disdain or, even worse call them "cholos". This name has taken on the extra connotation of inferior quality, even though the Collas may be the true descendants of the Incas. They speak their own language, Aymara or Quechua, according to their tribal group. I read somewhere (probably in good ol' Gould) that Quechua is the World's most logical language: very computable and as such would be ideally suited for inter-galactic communications. Maybe, but I assure you, Stephen Jay, trying to communicate with them here on earth is much more complicated. They have a completely separate culture and those who adopt religion are normally Evangelicals whereas the Cambas tend to be Catholics.

With a shortage of fertile land in the altiplano many families have had to migrate in search of a living. Introduced to the lowlands through seasonal labour schemes (picking cotton, harvesting sugar cane, pineapples, rice, or soya beans) many return to begin their own farms: colonising huge areas of Bolivia even, as I now saw, in federal reserves like Amboró. Some of the land is inevitably given over to coca for which they have a great liking. Most of the men I met on the path had one cheek distended by a dossal of masticated leaves that; every once in a while, they suck on to release the stimulating alkaloids.

****

Here, while I stop and have another cigarette, is as good a moment as any to briefly discuss the disagreements that have led to the coca leaf wars, wars that have become Bolivia's Gordian knot. Let me summarise the position of the two antagonists: Native Tradition v Government Modernisation.

Native Tradition: Our coca plant contains small amounts of cocaine; chewing the leaf, especially when mixed with lime, releases the alkaloid that invigorates our bodies while suppressing our appetites. It is also a cure for mountain sickness. It won't surprise you to learn that the cultivation of coca and its sale are thriving businesses among us, Altiplano Natives. Our ancestors believed, and many of us still do, that useful plants were gifts from divine beings. Coca, too, has its divine mother, "Mama Coca". It's got nothing to do with us if half the Yank population misuses our divine right. Sacred, healthy, and profitable, no one is going to take it away from us.

Government Modernisation: Cocaine is an addictive drug causing depression and indolence, and eventually serious mental disorders. Whereas we don't deny chewing coca leaves seems to be a healthy happy activity, the leaves do contain cocaine, and its long-term effects are not known. Maybe this explains your ornery character. It's not your sacred plant — the Incas were the only people allowed to use it — and they're all dead. Whereas it is true the coca economy is a thriving one, it is almost impossible to control its exploitation by the Drug Mafia — and subsequently the corruption of our country's administration and addiction among our own citizens. In an aid-dependent country, like Bolivia, real economic development depends heavily upon maintaining good relations with the international community, especially the USA that swops cash and trade privileges for coca eradication. No, we have to modernise and "coca zero" is our policy.

"What a bag of nails!"…but I think something could be done. If you will wait one minute whilst I get my soapbox out, its somewhere at the bottom of my rucksack, I'll tell you how I would cut the knot. I don't believe the executive committee that runs our planet should consist of do-gooders, neoliberals, old ladies, media specialists, sociologists, politicians, religious leaders, or if you like, Osama Bin Laden. Unfortunately, up to now, these and others of similar ilk have been in charge. No sirree! It's not working. It isn't working for one very simple reason…

Dammit! I've really buggered my soapbox. But let me just finish, please. I'm not saying that this hodgepodge of mostly reasonable beings shouldn't participate. What I'm saying is: May I ask you a question first? Who do you call when your plumbing packs up? I mean, do you call a politician? A media specialist? The Priest? Of course all three might be useful if we're talking about your colon, and we could be, but I actually meant the pipes in your house. Yes, you call the plumber, "It was on a Monday morning when the plumber came to call". Okay, you call for a plumber who is a real professional. And who are they? The really professional plumbers are the people who really know, and if they don't know they are the people who know how to find out, the Scientists! That poor forgotten crowd. So, yes, I believe the coca problem, and most of the world's problems to boot, should rest on cold scientific analysis. I believe the cure should be scientifically planned. And I believe the treatment should be in the hands of experienced professional: media experts, sociologists, anthropologists and whatever other professional is required. But not politicians. The world is full of politics and not enough science.

Back to the problem at hand: coca leaf farming. With the participation of the Native community an impartial Medical Research Centre should establish clinical trials to investigate the long term effects of chewing coca leaves; after all, if the habit proves favourable Bolivia would have a strong case for its inclusion in GATT (the General Agreement on Trades and Tariffs), with all the subsequent benefits to the poor campesinos the increased demand would provide. If it proves deleterious, coca eradication will benefit everybody. In the meantime, the coca grower's associations, like the Federación de Empleos de Coca (FEC) should be made responsible for the activities of their own members. Only those holding FEC identity cards would be allowed to grow or sell coca; anyone suspected of illicit trading would immediately lose their membership. The Government will continue its policy of eradicating coca forcefully, the number of fields eradicated will be based on a sliding scale, the less evidence of cocaine processing the fewer the hectares eradicated, and vice versa. The international community should help finance and monitor this scheme; and more to the point, both they and the Coca Leaf Growers Association should accept the results of the clinical trial and be willing to act on it.

****

After crossing the Rio Cheyo and one or two other minor streams, I arrived at the Rio Agua Blanca. Unlike most of the rivers in the reserve, which are crystal clear, this river is milky in colour, not rough as the literal translation might suggest. I was relieved, because; I soon found out that I was to follow this river for the next hour or so, during which I had to cross it fourteen times.

Agua Blanca was a small community of twenty Camba families, most of whom had been living there for more than thirty years. There were said to be a number of traditional communities in the reserve but, tradition in this part of the world being of very short duration these communities were often founded by people who had every reason to remove themselves from the jurisdiction of the police or army; or like someone I know, escaping the threat of a vendetta.

I found the people in Agua Blanca very friendly, inviting me to drink coffee, eat fruit, and very willingly showing me their pets. One, a juvenile Howler Monkey, characterised by its deep chestnut colour and short cranial tufts had just been captured. While we were talking a small boy rushed up to tell his father there was a "Ciervo" ("Deer") down by the river. His father promptly grabbed his, already loaded rifle and ran after the boy. I heard the gun being fired and crossed my fingers that he had missed.

I was to be disappointed, for several men and the boy soon returned with a fine looking Red Brocket Deer weighing about forty pounds, its dark red blood oozing from its eyes and mouth. For a moment I was unable to speak, so I covered my sorrow by taking a picture of the hunters with their prey. They all struck poses of the brave hunter with such unconscious parody, I could not help but forget my disgust; very inappropriate to the occasion; and only likely to evoke ridicule of myself, I joined in the general laughter.

I asked Don Julio, as the father was called, if he ever missed a shot. He explained to me he never did because; he never took snap shots and bullets were too precious to waste.

Fortunately, I knew what he said to be true, if it were not, very little wildlife would survive, so keen is the killing instinct among young and old alike. No self-respecting little Camba boy would be seen dead without his catapult. In a country where toys are only available to the rich, and where there is no tradition of making them in the home, the catapult remains almost the only plaything a child will own. God help the wildlife if the air rifle should ever come into common use.

During the, by now, customary conversation about gold hunting I invited them to accompany me, to see for themselves what I really liked doing. They were fascinated by my field glasses that I had waved around as if proof of my intentions; and I let all of them, including a passel of kids try them out. They all "oohed" and "ahed"; but I don't think any of them were able to adjust the focus, and conscious of the greasy-fingered abuse the lenses were receiving I didn't feel in the mood to explain. Though I offered to pay, I didn't secure a guide. The only man who admitted he had the time to spare explained to me how it would be sheer folly to go now, since the moon would be full in two days. They had a rough rule of thumb which invariably saved them much misery: Never travel during the week of the full moon, nor the week of the new moon, since it was almost certain to rain more than usual. After this trip I always used the rule myself.

At the time the sun was shining, there was not a cloud in the sky, and old wives' tales were for old wives. I said my goodbyes, "Hasta luego, Hasta pronto" and set off. "Adios" is rarely used here for other than more permanent partings. I might say it to a friend when, if ever, I leave Bolivia for good.

As I approached the serrania, which now looked close, the river became littered with slippery round stones. Though only knee-deep I couldn't see where I was putting my feet, and more than once I was nearly pitched into the water. Most of the time I was able to walk along the sandy beach, but at each bend the river invariably formed a deep pool, or was deep with mud, or the beach ran out into a small cliff. For one reason or another I was invariably forced to cross to the other side. At one point, where the river split into two separate channels, I took the right hand one, that I had been told was the River Chonta.

Wildlife was still scarce. I saw a solitary cormorant drying its wings in the afternoon sun. I had always thought of them as seabirds, but this Neotropical species is common in any suitable freshwater habitat from Mexico to Cape Horn. A White-necked Heron flew upriver and pairs of Yellow-browed Sparrows foraged in piles of driftwood on the river's bank. On one bend I disturbed a metre-long Tupinambi lizard (or "Peni" as they are called here) digging a hole in the sand.

After an hour, during which I seemed to have rounded countless bends, I stopped for a late lunch. I heard a grunting noise from close by that at first I took for sign of wild pig but, on investigation found to be a sleeping drunkard. There was no doubt about his condition, for he lay sprawled as he had apparently fallen, his face in the sand, his limbs spread out in careless abandon. He smelled strongly of "Cachasa", a crude cane alcohol for which the campesinos have a great predilection. In his book, Fawcett often alludes to the evils of this brew and the consequent tragedies it causes. I can vouch for the mean condition it provokes, and as much as I would have liked to ask him for directions, I let sleeping pigs lie.

I knew that somewhere near here I had to leave the River Chonta and take a path overland to the gateway of the River Saguayo, the one that drained the cliffs I was intending to visit. Once again, fate took a hand. I saw a woman leave the jungle fringe, descend to the opposite bank, and make her way in my direction. Her hair was caught up in a black pigtail, which peeped out from under a brown bowler hat reminiscent of those worn by Edwardian gentlemen. These hats, which are moulded from llama hair, are traditional gear for Colla women, as was the rest of her costume: a cotton blouse, a white overskirt, and seven or eight underskirts, each one a little shorter than the next. The number of skirts varies with the woman's status. The effect is to make them appear large in the rear and thin in the leg and upper body. A bit like ducklings. This woman was very upset, she gabbled at me in her own language, I understood hardly a word. Many Collas, especially women, don't speak Spanish, but I gathered she was looking for her husband, who she said had left the day before to visit a friend and had not come back. I showed her where he lay. The sight of him appeared too much for her, for on the instant she gave a high-pitched wail and belaboured her chest with her fists. At first I thought she had taken him for dead but, then she tried to wake him, without success. After further futile attempts, she indicated that I should help her carry him back to their house. When she saw I was reluctant to oblige her, she produced a grapefruit and a knife from a poncho on her back and with the dexterity of an accomplished magician she cut off the top and handed me the fruit to suck. Thirsty, as I was, it was delicious. Having compromised myself, I acquiesced to her inveigling and agreed to help her. "This was neither going to be pleasant nor easy", I thought. The problem of my pack was resolved when she called to a lad of six or seven years who had been hiding in the long grass. This mighty midget heaved my pack (that weighed close to thirty pounds) on to his head and hopped across the river to disappear with it into the jungle. I momentarily wondered if I would ever see it again, but immediately rebuked myself for being suspicious. Now, faced by the temporary loss of my pack, whatever reluctance still lingered in me evaporated. Between us we man-handled the brute into a more or less upright position. During this manoeuvre I noticed that she was very pregnant; her outline matched by her obese spouse, who remained stubbornly comatose.

At my insistence, we dragged him the score of yards to the river and unceremoniously dropped him into the water. This produced the desired result. Lying there in the shallows he looked like a water-logged frog, a pantomime toad, a "Toad of Toad Hall". His thick black hair was trimmed to a severe crew cut. His yellow skin was stretched tightly across the cheekbones, but his cheeks and jowls remained very fleshy. From deep within this fleshy mass two little black eyes and a small round mouth began to open, releasing a quantity of saliva that dribbled down his chin. This was sprayed in our direction by a spasm of deep, chesty coughs. From somewhere an overpowering smell issued forth. The man, whose name was Felipe, slowly recovered, at least he was able to get to his feet and with his arms around our shoulders we dragged him across the river and then to his hut 200 yards further on.

By now I was beginning to fret at all the delays. It was getting late in the afternoon, and I had still not reached the foot of the mountains. On the positive side, as it turned out, Felipe's hut straddled the path I had been looking for, the one to the River Saguayo, that Felipe's older son told me was twenty minutes walk away. Moreover, the older boy was delegated to carry my pack. So, I said farewell to the youngest lad and his mother, ignored Felipe who sat red-eyed and brooding on a log.

No sooner had we got out of sight than we heard a fierce quarrel break out from the direction of the house. Felipe's hoarse shouting was suddenly punctuated by a piercing scream. The boy dropped my pack ignominiously to the floor and, without so much as a glance in my direction fled back down the path. The altercation picked up again until something heavy fell to the ground with a thud. This was followed by utter silence. I decided to press on and, shouldering my pack hastened away towards the River Saguayo.

Drinks, drugs, violence, hate. I couldn't help brooding over the humanity of it all. Why governments were so concerned to stamp out the cultivation of the coca leaf, with its soothing effects, but encouraged the production of raw alcohol that did nothing but harm, both physical and mental. I knew this was a naive viewpoint and that the true situation was a good deal more complicated. Santa Cruz was at the centre of an international storm because of the trade in base, the raw paste exported to Colombia for the production of pure cocaine. The huge sums of money earned by this illicit business in Bolivia gave financial and political power to the "Pichicateros" (the "Drug Dealers"), largely greedy men in a desperately corrupt country.

Occupied, as I was, by my brooding, and in my haste to press forward, I nearly stepped on a large "Yoperobobo" (the South American "Fer-de-Lance") that lay coiled in the shadow at the edge of the path. The snake was apparently dozing and I didn't attempt to disturb it. Its ugly flat head was about three inches in length, its body about as thick as my forearm. From dead ones I had examined of equal dimensions I knew its fangs would be more than an inch long.

****

A case I investigated some years later involved the death of a young man who had been bitten by one of these snakes, coincidentally not more than a mile from the River Saguayo. He and his brother were harvesting groundnuts, raking the uprooted plants into piles, when one of them saw this sort of snake just behind his brother, who, responding to his brother's cry of warning jumped back on to the snake that bit him deeply in the foot. The young men immediately returned to their village five miles away and reported to the medical post. There the victim received some basic treatment, but no serum was administered. By the following day the patient had sufficiently recovered to dispel any anxiety for his safety.

Four days later, while sitting on the doorstep of his hut, he saw a pregnant young woman standing before him. The local people believe if a snakebite victim sees a pregnant woman his demise is certain, and true to tradition the young man died within the hour. The parents of the young man had no doubts, they were adamant in their belief that his death had been caused by an old woman, a Witch, who bearing the family malice had entranced the young woman to appear before their son.

Following up the rumours of witchcraft, I questioned the rather reticent young woman about the circumstances leading her to take her deadly stroll. Since the young man's house was several hundred yards away, along a path she rarely used, I was interested to know why she had gone there? She said she had been preparing the yuca and rice for the family's midday meal and went to fetch water from the stream behind her house. But instead, she found herself standing before the young man's house. She didn't know how she got there. More than that I couldn't get from her.

So, if not witchcraft and the power of superstition, what did kill him? Relapses in such cases are not uncommon but it is hard to believe that four days after being bitten and by all accounts suffering from little more than a swollen foot, the young man succumbed to the poison. Moreover, in the Buena Vista area very few deaths occur where bites from this species are quite common. Was it sheer coincidence then, that local superstition was so tragically verified? Many people believe that in these cases, when superstition and the fear of witchcraft go hand in hand, fear steals across the mind to rob the hapless victim of some undiscovered vital mental spark that keeps us all alive.

****

So there the mystery must lie, for I had arrived at the River Saguayo feeling tired and not a little depressed. The forest through which I had passed had seemed dark, oppressive and preternaturally quiet. The sky had become progressively cloudy and was now grey and overcast. The threat of rain, the hateful events with Felipe and my own depressing thoughts (intensified by the snake's symbolic evil) combined to drain me of all pleasure.

Few people, however, could resist the appeal of the Saguayo River and, its tumbling, buoyant, playfulness soon lifted my spirits once more. The river had cut its way through the rock to leave the water ten feet below where I stood. The water, tumbling and gurgling in this channel of pale coloured limestone, had carved itself a labyrinth of secret hollows and graceful ledges. Peeping out from this aquatic sculpture were many small fish hanging in the crystal-clear water. It looked like a mural dedicated to Poseidon.

A short distance upstream was the place called "El Porton" ("The Door") because, here the river runs between massive rocky cliffs several hundred feet high. As it reaches this narrow ravine the river has scoured a deep pool, beyond which it bends tantalisingly out of view. And truly, it was like the arrival at some forbidden gateway hiding "The Garden of the Hesperides" from mortal sight.

The overcast sky and the lateness of the hour interrupted my reverie and prompted me to look immediately for a place to make camp. Just below El Porton the turbulence of the river had thrown up a sandy beach, together with a good supply of firewood. I was too tired to consider the perils of falling rocks or flash floods and quickly got myself organised for my evening meal and a good night's sleep. My preparations never took long since I didn't have a tent and my simple meals could be cooked in one saucepan. In case of rain, I carried a tube of plastic sheet that was long enough for my equipment and me to shelter in. It wasn't ideal since my body heat created a good deal of condensation, and it could get too hot for comfort.

I had just finished my supper (boiled rice, a few chillies and a tin of soya-bean-sausages) when it began to drizzle. I was lucky to get the coffee made, and finished, before it came down in earnest. I had no alternative but to crawl inside the tube and lie there like a corpse wrapped for the morgue. The storm that followed was quite spectacular, deep rolling thunder, sudden ear-splitting crashes and crackling blue streaks of lightning. This continued for some time, on and off throughout the night. I remained relatively dry, and I knew little harm would come to my gear in its pack at my feet. Sleep, however, did not come easily and those fitful moments were punctuated by wild dreams in which giant snakes threatened me, rivers washed me helplessly away and angry campesinos chased me, yelling: "Beware the ides of the moon. Beware the storm. Beware."

At one point in the night a good-sized tree gave up the ghost and crashed to the rocks not twenty yards from where I lay. Even though I knew full well "The Door" would block any warning rumble of a flash flood, for much of the time I lay awake, my ears strained to gauge the river's mood.

By morning I was still alive and the rain had reverted to a gentle drizzle. As I'm allergic to rubbing sticks together, especially in the morning, I got the fire going with the help of some newspaper, palm fibre and twigs that I had shredded for the purpose. Now I was ready to consider my position over a welcome cigarette and cup of hot coffee.

The prospect of passing the deep pool guarding further ascent of the river almost convinced me to press on no more but: then came a small break in the clouds accompanied by a touch of 'Fawcett's Folly'. What, after all, lay beyond that tantalising bend? Would I find a "Garden of Eden" full of tame birds and animals? What strange bird now called? "Pee pee-yo, Pee pee-yo, Pee pee-yo" its fast, piercing whistles repeated over and over again. What would Henry Bates or Charles Waterton have said were I to turn back now?

It was still drizzling as I shouldered my pack and faced the dark, uninviting pool. But, now too, the small blue patches in the sky were becoming more frequent, even as the sky ahead looked, if anything, a good deal worse than before. Wispy fragments of torn cloud floated into the ravine and a few heavier drops of rain fell intermittently. I decided to try and climb around the gateway by scrambling up the steep wooded slope to the left. This ill-considered approach was doomed to failure, leaving me tired and hot on the brink of a sheer drop one hundred feet above the river. By the time I descended to the beach again I had lost a precious hour, and my hand was throbbing with pain, for on my way down I had slipped and grabbed the trunk of a Chonta palm. These palms guard themselves well with a dense mat of wicked black spines six or more inches long. The poisonous tips of many now lay embedded in the palm of my hand.

I resolved to wade the pool if I could, swim if I had to. Here again the plastic tube serves me well, for I can put all my clothes and gear inside and by leaving it full of air float it across. Earlier explorers never had the blessings of plastic sheets and plastic jars and had to contend with heavy oilcloths and rubber-backed tarpaulins. I was about to enter the pool when I noticed a narrow ledge running out of sight up the right-hand shoulder of rock. I managed to scramble on to it and after a few yards found it widened considerably, that was just as well for it was steep and slippery with half rotten leaves and a greenish-yellow algae. Providentially the ledge led to a patch of more level ground, and from there I squeezed through the bushes back to the river. I had passed El Porton and now stood on a broad raised beach of boulders.

By now the sun had come out and its warmth imbibed me with a powerful sense of well-being. I spread myself and my things out to dry and sucked the scenery in. I simply lay there, repeating over and over to myself, "God, this is beautiful. God, this is beautiful". My whole body shivered with delight. I was filled with a sense of total happiness, as though I belonged here among the animals and trees; it was as if I could reach out with my mind and enter into every living thing, feel the rhythm of its breathing, share its hopes and fears, touch its spirit. I think it was at this moment that a decision was made for me. I didn't know it then and it was still years ahead but, someone decided I was to be the first Director of Amboró National Park. (but this is to leap ahead of the story.)

Here I was presented with another gift, for the "pee-pee-yo" bird was perched on some creepers above me where I had plenty of time to study it. It was not a distinctive bird, being a bit larger than a thrush and of a uniform grey colour. I now knew it to be the "Screaming Piha", a member of the Cotingidae (the Fruiteaters). Whatever it lacked in colour, it more than made up with its astonishing call, and the energy it expended to produce it. Nearer now, I realised that before its whistle, it made a throaty noise like it was winding itself up with a rusty spring. Then the first whistled note, "pee", uttered with the neck stretched rigidly forward (a microsecond pause during which the head was retracted), then the "pee-yo", both syllables uttered with the head pulled back into the shoulders and the bright yellow gape stretched widely open, it looked exactly like somebody screaming. For a zoologist it is the little experiences like this that make a day worthwhile. But, the day had only just started and many more things were in store.

Above this point the river became decidedly rockier. Where it narrowed and the cliffs closed in huge, house-sized boulders blocked the way. Sandy beaches were few and far between and the forest pressed in on either side. Although now reduced to a babbling brook some thirty feet wide, it was clear from the line of the flood refuse and the huge tree trunks dumped in jumbled piles around the largest rocks, that the river, in full flood would be a raging impassable torrent several metres deep. One day, I promised myself, I would attempt the trip during the rains, when the sight of it would be awe inspiring. Some of the tree trunks were more than sixty-feet long, yet here they were thrown higgledy-piggledy. One monster that had one end wedged between two boulders, had been lifted skyward again in a parody of life after death.

These once mighty giants of the forest had outgrown their shallow system of roots and down they had come, smashing and maiming a dozen others in their fall. I once heard the death throes of such a giant; the roots began to snap, one by one, as the tree began to tilt, each snap (sharp as the crack of small-arms-fire), sent little tremors through the soil. After a few hours the remaining roots fractured with the stutter of a machine-gun as the tree fell to the forest floor.

I had just rounded a bend when I glimpsed a slight movement at the river's edge about sixty-yards upstream. This was definitely a large animal; and there were plenty of tapir, pig, and cat prints to see. I lowered myself slowly behind a rock and, there, through my binoculars I saw a pair of handsome tan-coloured Pumas drinking at the water's edge. As far as I knew they hadn't spotted me, and though I remained frozen in the hope of seeing cubs, none appeared. The South American Puma is a large cat and compares favourably with a leopard in size. I estimated them to be a little over seven feet including the tail, obviously adults. I watched them for a minute or two, before, with a few agile bounds, they surmounted some bigger boulders to cross the river and disappeared into the dense vegetation.

A fair description of the terrain that I was passing would be to compare it to the Grand Canyon in Arizona, a series of steep slopes every now and again giving way to sheer rock faces but, here all covered by dense forest. Impossible as it would seem, given the topography, it was clear that some of the animals travelled in almost straight lines. As the river looped its way around the rocky spurs I kept meeting the tracks of a large party of White-lipped Peccaries. At one place the tracks still following the course of the river disappeared straight into the forest as the river began to bend, reappeared again on the next stretch, crossed the water, and once more disappeared over the ridge on the other side. This had been a fairly large family, with about a score of young and six adults. Wherever they passed they left a wide trail of trampled soil with the marks of their trotters sharply imprinted. The local hunters are very wary of attacking these animals, for they have been known to turn about, and inflict severe wounds on their tormentors. I was hoping I wouldn't run into them myself, for the tracks were clearly very fresh.

In one section of the river I followed the prints of a Tapir for about an hour before, they disappeared into the forest. Tapirs are shy animals and usually only come to the rivers to drink at night, retreating to the tops of ridges during the day. To see them one must be prepared for some pretty energetic climbing. Following in the footsteps of this one saved me a good deal of time, for the animal would know the river well. When left to my own devices I often had to double back because of impassable boulder heaps or deep looking pools that barred my way. I must have crossed the river dozens of times to avoid such obstacles and, at one particularly difficult section, just after the tapir took to the forest, I made a long and successful diversion up a very steep slope.

In another place I came across an Otter actively fishing a deep pool, and thanks to the noise of the river it was not aware of my presence. I ducked down behind a tree trunk from where I was able to watch it with delight as this graceful animal turned and dived, surfaced momentarily and submerged again. Then it must have caught my scent, for it remained with its head out of the water busily working its nostrils in my direction. With a swift, but unhurried dive, it disappeared under some large rocks from where it didn't reappear. I thought if left in peace the future of these harmless animals must be assured; the waters were full of fish and few, if any people seemed to pass this way. Only on my return was I to be sadly disabused of this notion.

Coming around another bend in the river the rocks abruptly gave way to a hard red clay; it was just like a red carpet laid out to welcome me. For several hundred yards the river ran silently over it; the clay, smooth as a table, was seamed with a greenish-coloured substance that gave it a very appetising appearance. Actually, it looked like "Robinson's Liquorice Allsorts." At the head of this feast, another one waited, for a "Red-velvet-cake" cliff rose abruptly from the river. Here, I had to make another diversion by taking to the forest.

On descending to the river I found myself under a massive fig tree, its height amplified by the huge boulder upon which it stood. It's ring of roots, each one thicker than a man's thigh, dropped twenty feet to the ground. The boulder as securely imprisoned as some hapless beast in a Brueghel painting. Unable to take my eyes from this sinister scene, I shambled into a cleft between two rocks and fell through the roof of an underground cavity. Fortunately I suffered no more than a well barked shin, for my pack, having wedged itself in the cleft, prevented me from slipping further, that was just as well for I heard a deep hissing from somewhere below my dangling legs. I felt the hair rise on the back of my neck as if some large serpent had licked me there. Desperate with fear I struggled to free myself. My flailing legs touched something that swayed with the contact and the hissing deepened to an even more disconcerting "shushing". The rocks to either side of the cleft were within my reach but their smooth sloping sides were slick with moisture, and try as I might, I couldn't get a grip on them. "Think! for God's sake think!" I dove my hands into the thick layer of humus and immediately found what I needed, one of the fig tree's roots. Grabbing hold of it and summoning up all my strength I was able to heave myself clear. By the time I did, I was trembling so violently I could not stand and had a massive serpent emerged, as I imagined it might, I would have been at its mercy.

The way up the river had become decidedly more difficult, steeper, the rocks larger and consequently my progress much slower. Now I had to clamber over boulders the size of houses. Once I had to slosh across a fresh mud slide, the grey ooze sucking at my shoes and once, gripping my whole leg. At this spot I picked up a stake to help me safely across. It was an evil-looking place, full of dying and dead trees brought down by the wall of mud.

Later, I was to discover that many such slides occur in the area. Some, much more extensive, when whole hillsides start to creep. The Andes are young geologically and have not been subjected to the intense erosion that has shaped the World's older mountain chains. I like to refer to them as the "World's Biggest Sand Castle", for that is what they appear to be. These landslips start off when the cap of sandy soil (that may be a yard or more in thickness) becomes saturated with rainwater and slips on the impermeable clay beds below. The forest cover is of little help in preventing this since the roots of even the largest trees cannot deeply penetrate this hard clay. The forest slows the process down, transpiring much of the excess rain and binding the sand but, then, down come a few trees, the equilibrium is upset, the relentless process gathers pace, and all is lost.

My close call, with whatever it was dwelling in that hole, had really shaken me. I was in two minds again, whether or not to turn back. But I was now about to make a discovery that would attract many people from faraway places to Amboró and play a vital part in securing the future of the proposed Park. It was clear very few people, even the ubiquitous hunter, ever came this far upriver. There were no human footprints or paths and where the going was difficult, where it was necessary to leave the river and cut a way through the dense vegetation it was me who had to cut it.

Animal prints were more common and large exotic birds less afraid. Through the tops of the trees I caught an occasional glimpse of my cliffs in the distance. Channel-billed Toucans were making their frog-like croaks from all sides. A dozen Blue-and-Yellow Macaws flew over with familiar strident screams and moments later, with their little red cap bands, a platoon of Military Macaws rose from the forest ahead. Two Amazon parrots peered at me from the safety of a hole in a tall dead tree, their raspy "quoinks" mixing with those of other strange birds I did not recognise.

Surrounded by this teeming wildlife I was surprised to hear a soft sneeze, then another and yet once more, repeatedly like those of a hay-fevered child. The sneezing was accompanied by other sounds, humming noises and coughs. "Maybe", I thought 'I really am about to encounter old man Fawcett surrounded by his vassal subjects' for I was quite sure that here in the middle of nowhere I had come upon some hidden settlement. I freed myself of my backpack and picked my way cautiously up the steep overgrown slope. Not daring to use my machete, I squeezed my body through the thicket. Barbed vines tore at my clothes and I became hooked-up in a tangle of thorny "Tacuara". The Shamans stopped humming but: the consumptive had irritated the child, who was now sneezing more forcefully. In fact there were now several sneezing together.

I couldn't see more than a few yards in front of me. A Black-faced Antthrush started up from somewhere close, scolding me with its cry, "bio-bio, bio-bio, bio-bio".

Then, suddenly, in a more open patch on the jungle floor, I saw a group of birds of legendary status. I had found the Southern Horned Curassow (Pauxi unicornis) a large, yard-high Turkey-like bird first discovered fifty years ago when two dying ones were taken to Cochabamba by a pair of Indians.

The Curassows were glossy blue-black with white bellies and white-tipped tails. Each one had an extraordinary bright red bill, for on the base of the maxilla, pointing upwards towards the sky sat a thumb-like casque more than an inch and a half long and sky-blue in colour. Unfortunately I didn't have much time to watch them, for the Antthrush's alarm drew their attention to my presence. The humming and coughing came to an immediate halt and the birds, probably seven or eight, I still could not see them well, flew up with a heavy laboured flight. All but one disappeared into the top of some low scrub nearby. The last one, now sneezing, "k-sup, k-sup", sat watching me suspiciously from a tangle of vines, before it too disappeared. I tried to follow them, but the tangle held me back as surely as a cordon of police, and I saw them no more.

By now the sky had cleared; but it was too late to explore further, and I made camp in a small grove of trees. The deep pulsing of a Band-bellied Owl floated across the late evening air. A family of Blue-throated Piping-Guans made curious rattling noises with their wings as they glided away, probably disconcerted by the smoke from my smouldering fire. Before I knew it I fell fast asleep, until I was woken by heavy drops of rain. Too tired to worry about them I crept into my plastic tube where I slept the sleep of oblivion until the early morning.

It was raining again, and the sky looked unrelievedly depressing. Low cloud partly obscured the cliff face at my back and in the distance I could hear the steady roll of thunder. I decided to forego coffee, a very rare event, leave my pack where it was and make a quick reconnoitre of the upper Saguayo. Following what looked like a Tapir trail it took me less than ten minutes to get there. The river tumbled through a deep narrow gorge in a series of twenty foot-high waterfalls. A strong wind coming down the gorge blew spray and heavy drops of rain into my face. I was soon wet through. The sheer sides of the gorge were colonised by scattered trees that pushed their stunted heads far out into the void. On a stormy a day like this one the place had a terrible beauty, disconcerting in its austere remoteness, deceptive in its beckoning charm. A bit like Bolivia really. Deep pools guarded the ascent, dangerous looking black, rocky sides invited a precarious hold. I edged my way up the side of the first waterfall, my body almost in the water cascading from above. The rock was treacherously slick and the wind tugged at my body. At one point I very nearly fell into a deep pool. I persevered a little further until I was satisfied that a patch of forest that clung to the brow of the ravine could be reached by way of a ledge that shelved into the river. One day, I said to myself I would have to pass, but for now, defeated by the weather and my solitary weariness I decided to head for home.

Two "White-headed Dippers" shot past me as I scrambled back to the mouth of the gorge. Then as I turned around to take a last look, there they were, round feathery balls, chinking at me from the top of the falls. "Farewell, farewell."

Going down river was much easier than the ascent had been. Minutes lost climbing up and around difficult obstacles were over in a moment, for it was safe to jump to the sandbanks below. I lost no time in useless reconnoitres for, unknown to me, my mind had imprinted my memory with the way I had come. It rained hard throughout the morning, the river rising a foot or so made many crossings that much more disagreeable. The mud slide had become active, a moving morass into which my feet sunk with each step. As I was already soaked to the skin there was no need for me to avoid the deeper pools, and I had long given up worrying about my precious belongings stowed in their waterproof bags. After two hours I was well on my way down the upper river. My mind, feeding on all it had seen kept me in a happy mood.

Then came disaster, like a blow to the body. I fell to my knees, the tears welled from my eyes, for there caught on a snag in the current the smashed body of my otter oscillated to and fro. "The bastards. The bastards. "The bloody mindless, murdering bastards" I shouted for all the World to hear. The otter's body was torn asunder, its guts trailed in the water, one leg had almost been severed from its hip. I knew immediately what must have happened, somebody had followed in my footsteps and, as I had done, chanced upon the happy animal as it played under the water. But unable to leave a living thing in peace, and just to amuse himself, had tossed a stick of dynamite into the pool. I buried the otter as best I could and over its pathetic grave, I made a vow to help all the animals here find peace; to give them a home, a sanctuary where no man with gun or trap or explosive would ever pass.

The rest of my journey down the upper river was made with ever-increasing difficulty. The rain came down harder and the river got deeper and muddier. There was no time to enjoy the symphony of sound, the forest and the river having come alive with the rhythm of the rushing water. A thousand little streams had been reborn and each one played a separate note. Slow trickles were now tumbling torrents, mere drips small waterfalls. A large rock now unsteadied by the stronger tug of the river beat out a deep and throbbing bass. The rain a feathered kettle drum, rising in pitch one moment, falling off the next; the plink and click of piano and castanet, rendered by moving pebbles, and from all sides the separate notes were picked up and orchestrated into the deep tone of the swelling river.

No longer was it possible to see through the murky water, and countless times I stubbed my toes or barked my shins on submerged logs and rocks. And hungry for a victim, the current tugging, always tugging in the hope of dragging me down.

"El Porton", my little paradise was now impassable. The water lapped high up the rocky cliffs. The pool had grown, its centre taken over by a swirling, powerful whirlpool. Little waves, bouncing off the rocks, threw themselves into the maelstrom to be absorbed without trace. Logs thrashing about in the turbulent current were greedily sucked in, to disappear without trace under the cocoa-coloured water. The raised beach was being rapidly eaten away and, with the waters still rising I could see that soon it would cease to exist. The ledge by which I had ascended now fell to a swift broad channel that carried the flood and dashed it on to a group of expectant looking rocks downstream. I knew I was never going to pass this way now. Yes, awe-inspiring. I had already fulfilled my earlier promise; I was witnessing the river in flood.

I sat smoking a cigarette absorbed by the problem facing me. If I could not pass, how long could I wait? How much food did I have? Where would I sleep? It would have to be in the jungle, dripping and dark and, later, as the rain abated sure to release a host of unpleasant hungry insects. Only my equipment and my cigarettes snug inside a plastic jar were dry; everything else, my spare clothes, sleeping bag, and the last of my rice were soaking wet. My pack, heavy before, now weighed oppressively. I think I was close to tears.

But why is it, that in the middle of absolutely nowhere, far from the madding crowd, and far away again, just when despair has taken hold someone pops up out of nowhere. Salvation! A man appeared, almost certainly a hunter, probably the same sod who had killed the otter. But by God! I was really pleased to see him, and he knew it.

He was short, probably less than five feet and carried the 'Law of the Land' in the shape of a useful-looking Winchester rifle. He wore a scruffy pair of faded blue shorts and a tattered T-shirt, nothing more. The shirt should have announced in plain English "I Love You" but, the 'u' having folded inside where the shirt was ripped left the message somewhat in doubt, since in Spanish "yo" means 'I'.

I couldn't help but grin and he started to grin with me. What a bizarre looking chap he was to be sure. One side of his face had been broken and had slightly shrivelled to leave a permanent scowl but, as if to explain this lack of civility the other half, with its raised eyebrow seemed to ask, "What did you expect?"

An increase in the strength of the rain cut off any intention we might have had of speaking. The little man simply beckoned for me to follow and set off at a good pace up a swollen stream that entered on the left. After about two minutes we climbed over a steep bank and, following an overgrown path through some secondary growth we returned to the Saguayo where I had come to it on my way in. Obviously the path had been cut to avoid the difficult passage through "El Porton". I realised that had I known about it before I would have saved a lot of time. Then I would have deprived myself of the 'mystical' experience I had there, and that was something I was never going to forget!

"Peligro" ("Dangerous") he muttered. I simply stared, because I fully shared the sentiment. I knew I had to cross the Saguayo but, looking at it now, with the spate crashing over the rocky sides, I felt my confidence ebb away and sheer funk take its place. Had this man not been there, I might have tried it, and that might have been the last thing I ever tried.

Looking at him again, I realised only that single word, "peligro", had passed between us, even though we had been together for the best part of an hour. I also realised he had brought me here to show me, if any doubt remained, that the crossing I knew had become impossible. As if recognising my thought, and shy of further communication he promptly started back up the same path, and again beckoned for me to follow.

After a spell of abrupt climbs and steep descents we arrived back at the River Saguayo. Here it was a lot wider and there were far fewer rocks, but there was still a nasty looking current in the middle. (A few years later we nearly lost a young Englishman here under very similar conditions.)

My companion gestured for me to rest and that I was more than happy to do. I still carried my pack and the ups and downs of the last fifteen minutes as I hurried after him had exhausted me. He entered the river cautiously, edging his way towards the strong current at the centre. The water rose to his ribcage. I could see he was on the point of being swept away, my heart was in my mouth. "Please God let it be all right".

Apparently satisfied, he was grinning that lopsided grin, he returned, took up my pack and turned to face the flood again. At the centre he placed his back to the river and edged his way, gingerly feeling for hidden rocks, across the swiftest part. If he fell, I knew there would be little I could do.

Suddenly, he was across. Now he turned and laughed as he strode towards the far bank, the water no longer covering his waist. Very soon but, with the same concentration and the same caution at the centre of the river, he was back. "Magnifico!" ("Well done!") was all I could say to him. At last I had said something.

Together, arms around shoulders, we stepped into the opaque water. Long before the deepest part I felt my legs being swept from under me. I was afraid. Recognising this he turned down his broken mouth and raised his eyes heavenwards, a bit of pantomime to comfort me. And then, sooner than I expected, we were laughing and splashing through the shallows of the far side. "We did it. We did it". "Pee pee-yo! Pee pee-yo! Gracias, gracias amigo." Everything I had, my camera with its precious film, my sleeping bag, machete, a book of birds, and my irreplaceable binoculars had made it.

I gave him all I could spare, cigarettes, a tin of fish, some torch batteries, the saucepan, all of which he accepted as his due. And, without further delay, picked himself up and crossed the dangerous waters once again, at one point even jumping into the air and waving to me for the last time. Then I remembered I didn't know his name.

"Me llamo Robin," my scream crossing the river. Pointing at himself, "Tadeo," he replied. Good Lord! Saint Thaddeus, patron saint of hopeless cases. After he had retrieved his gun and walked into the jungle, I never saw him again. Sodding hunter he may have been, saviour he certainly was. God! Why can nothing ever be cut and dried?

The long way back was punctuated by river crossings that before would have made me flinch, now I scorned them. Away from the mountains the rivers were deep but not rocky and without the powerful current tugging at my legs I strode across them without hesitation. Only the River Surutú gave me a moment of despondence. But seasoned explorers, one of who I now considered myself to be, do not fuss over such things. What was that you said about escaped lunatics? And, yes, permit me to be demure, to say, "The Surutú was crossed without much difficulty." There, done in true Fawcett fashion.

I arrived back in Toy Town, late, wet and very "hangry". Everything was shut, and only after severely testing my fragile patience did I manage to procure some more cigarettes, a beer, the same African bread, and my old quarters back. The return by the long muddy road had deprived me of my more reasonable half; poor Melon Breasts, I never did become her friend.

2. Preacher Bird

Sitting in my office, shy of sharing my adventure with my colleagues at work, a mood they may have translated as latent hostility. Now I realised that the very basis of the future I had mapped out for myself had changed. A different voice within me could be heard, nudging me towards a path as uncertain as those I had trodden while walking in the Park.

As if to remind me my left hand began to throb with the pain of the Chonta Palm spines, the memory of that took me straight back to El Porton and my mystical experience there. Then that silent promise to the dead otter, such portentous words. Was that really me? That person who had wandered into Narnia? And where, now, was the green cupboard door?

Who was this new voice sitting in the driving seat of my mind? "Hey man, what are you scared of? You know you'd like it. Have you forgotten your youthful ideals? Isn't it the very thing you've always wanted to do?"

"Yes, but I'm forty and youthful ideals belong to youth. What about Cheri, my wife, eh? Who'd pay us? Where would we live? Anyway, don't be daft, I've got a good job and responsibilities."

"What good job? What responsibilities? Your wife would like it. Do the Park. Do it!.

Uh…? If you don't, I will!"

Is this the beginning of schizophrenia? Is this my mid-life crisis? Is it a crisis? Or is it an opportunity? The voice is no longer nudging me, it has become insistent. If in the past I could have been accused of wearing the strait jacket of security, now I permit this thing to undermine me with the sincerity of my own thoughts, bind me by my oath to the dead otter and leave me where?

Is an inner promise as binding as a public one? Is this the voice of another alter ego, or do we only have one as easily influenced by a passing experience as our inconsistent selves? My good job. Crop Protection Officer to the "British Tropical Agricultural Mission" in

Bolivia (that I will refer to as the BTAM or the 'British Mission' or simply the 'Mission' from now on). My bloody good job! Oh yes, no doubt about it, it was the sort of good job my father thought was a good job. It used to be a good job. What was going wrong this time?

****

It was a question I was to ask myself a dozen times a week for the next three years. I had worked for UK's Ministry of Overseas Development (ODM or ODA) in Ethiopia for five years and in Swaziland for three. I was one of the more experienced members of the British Mission in Santa Cruz. I was well qualified for the job. I was making modest progress towards an understanding of the direction in which my research should go. But, as soon as I had anything to do with my Bolivian counterparts, like the proverbial Mummy, everything turned to dust.

I used to get things done. I was allowed to operate with a good deal of freedom. As a Scientific Advisor people thought I knew more than them, and left me free to do my work, Sinatra-like, "My way". But, here in Bolivia all that had changed. I used to work alone, now I was part of a team, and what a team! Mainly career scientists jockeying for position, playing (as I always said) the "Indispensability Game".

"Look. With so much land around who needs to worry about plant pests? The farmers can work a little harder and plant a little more; better than wasting time or money on Pest Control. No?"

Yes. They were right, at least to a certain extent, for low value field crops like rice, maize and beans. Even many campesinos (peasant farmers) held the same point of view. But, this is what I am here for. To find out if such statements were valid. Very little was known about South America's plant pests, even less those in Bolivia. We Brits had centuries of experience to offer in Africa and Asia, but here in the Neotropics we were like fish out of water, and my colleague's asphyxiating generalisations had to be taken with a good draught of seawater, or so I thought. And, there was serious pest damage in some high value crops (like Pineapples) that could not be compensated for by growing more. And in others? Well, I could only answer, we don't know what we don't know. Anyway, the world is full of bankers, bus drivers and businessmen, why shouldn't it employ one more entomologist?

In retrospect, I believe many of the troubles within the British Mission (and there were many) were caused by their failure to state the Mission's objectives in clear and unequivocal terms. Following a request for British Development Aid, a project's terms of reference are drawn up by the Agricultural Advisors at Eland House in London and their counterparts in the receiving country. At my job interview I was categorically told the Mission was there to reduce the loss of tropical forest in the Bolivian lowlands by developing a viable agricultural system to replace the wasteful "slash-and-burn" practice universally adopted by the poor campesinos. And very good sense too, especially to me, a keen conservationist. Had I been aware of the true situation I would have opted to go to Borneo instead. Fate, fate. I should have been less trusting.

So, what did I find when I got to Santa Cruz? I found that the British Mission's main projects were: cotton (with emphasis on expanding the crop to include campesino colonists in the humid forests of the north), sugar cane, ugh!, pasture and cattle breeding. Service sections included Weed control, Crop protection, Soils and Economics. Later we were expected to get into soya, wheat, rice, and maize.

One could hardly imagine a better recipe for ecological disaster. Every one of the Mission's main research programmes was geared to agricultural systems whose prerequisite was the uprooting of trees on a grand scale. All of the crops earmarked for research required large, shade-free, stump-free fields for their profitable management. It was sickening. Where was the Agroforestry component? The Silviculturist? Tree Crop Agronomists? Spice Specialists? Up a gum tree with me. "Look this is not a perfect World. Okay? The Bolivians have a lot to say in things, you know. And what they are saying is they are not interested in your ideas."

"My ideas! Ecology? Conservation?" Why do I never take the hint?

Once again, they were quite right, and I was quite wrong. It did not take me long to find out that the 'powers-that-be' did not want the Mission wasting its time on small-farmer agroforestry. Our counterpart organisation, the "Centro de Investigación en Agricultura Tropical" (that I will refer to as CIAT from now on) was made up of persons with deeply vested interests in the promotion of crops most suitable for large-scale mechanised farming since many of them were already involved in the family's large farm, even the individual research workers.

One week after my arrival in Santa Cruz the die was cast. Even the most inept fortune-teller with one hand tied behind her back could have predicted the inevitable outcome. I was going to have to rock the boat. They were going to feed me to the sharks.

I'd been given an important part to play. Unfortunately for me the script had been rewritten on the Opening Night:

Crop Protection Officer: Enters stage left. Waits for cue; then realises the Rest of the Cast is speaking another language. Titters, like parties of foraging Vireos, run around the audience.

Rest of Cast: Enjoy performing to their utmost; after all it is a "Comedy of Errors".

Crop Protection Officer: Exit stage right.

I don't want to resort to petty niggling; and even were I peevish by nature, my publisher would not permit me to exploit the reader's attention simply for the sake of a good old grouse. But, in order for you to understand the fabric of my story I have to give you the warp and the weft of it. The details of the social jungle in which I trudged are as important to the reader as a full understanding of subatomic particles would be to a nuclear physicist. My dogmatic insistence that the work of the British Mission (and by association that of CIAT), if it were to achieve anything, be founded on a cold examination of the situation both scientific and social, did not make me many friends. New men with new ideas rarely do.

If I considered cotton growing a rock-rolling occupation as useless as the "Labours of Sisyphus", then, as a recognised expert in the matter, I had the right to expect the whole thing to be put through the grinder of economic analysis. Not only was my criticism of the cotton project not popular with my employers in ODA (since the British sold a megabuck cotton mill to the Bolivians) but, it was also a threat to the very existence in the Mission of some of the indispensables who had built their reputations on its shaky foundation.

I was convinced we had to learn to cooperate with Nature, not fight Her. After all, we were not dealing with a high budget approach along the lines of an "Israeli Desert Reclamation Scheme", and never would be; that the agricultural development of the area depended upon a sound understanding of its ecology, its soils and their drainage, and the fickle climatic conditions (sometimes tropical, sometimes temperate) that are the mantissa of transition zones like Santa Cruz. The optimistically myopic opinions of short-contract experts who thought that tomorrow the sun would come out was unexceptionable.

I considered the level of education, and the attitudes of most of our counterpart staff, inadequate. My insistence on calling a spade a spade and my indisposition to believe we could make a silk purse from a sow's ear, was my attempt to shake the Mission out of the hopeless resignation that prevented us from requesting CIAT to take a firm hand and weed out the incompetent members of its staff.

There were good people who needed jobs, but CIAT, like every government institution, was staffed according to political affiliation and with a good dose of nepotism. For my stance I attracted the shrapnel of spite as a magnet does iron filings because, like magnetism itself, the forces at work were as immutable as any Law of Physics. That, instead of weeding out the counterpart staff, the Mission retired some of its brightest team members (even after ODA did an about-face on cutting the size of the team), and that I went soon after will not surprise you.

****

When Cheri first saw a real toucan, she set her heart on having one for a pet. As we knew of an American who was said to supply everything from guns to parrots, we decided to contact him about a toucan. This was a fateful decision, for we got a lot more than a pet bird from him.

"Yep," he said, "as it so heppens I've got a peir of young birds, the sort we cawl Tocos. Mebbe your wife would like to try at rising one." To my English ear the accent over the phone sounded decidedly hillbilly.

He would bring one of them over some time in the evening. "Yo'all sure you wouldn' like a parrot insteid? Towcans, now, are what we cawl soft shitters," he added; the voice almost singsongy, friendly.

"No. The toucan sounds perfect." I thought he had said soft critters, meaning lovable.

He arrived with his wife sometime after nine o'clock, quite a lot later than we had expected. He was a bit tubby, of average height, and wore a permanent slightly mocking grin. I guessed he was in his middle forties. Chachi, his wife, was obviously Bolivian, younger, slim, and beautiful, with finely moulded features and hooded green eyes that gave her an attractive, supercilious expression. Under one arm he was carrying a cardboard box.

They were both dressed as if for a formal dinner, that it was, and they were already late for their engagement.

The toucan, he explained, had been taken from the nest four days ago and was about a month old. It already had feathers; only its fat rounded belly was still naked. It cocked its tail over its back like the handle of a food mixer. It had a naked yellow face, the skin soft and velvety to the touch. The bill was small, not the Guinness trademark we were so familiar with, not yet anyway, more Puffin-like. The bird was about the size of an overgrown thrush, and its feet, which were far too big, were awkwardly folded beneath it like parts of a broken umbrella.

He had brought a little food with him, chopped papaya and little squares of apple. "They minely eat fruit, but every now and agine they like a little mait. We honest to God dun'no much about theym, and mebbe yo'all could record what they reilly do like." Now he looked like a benevolent preacher, his confiding smile running up from the corner of his lips joined another that made its way down from the outer edge of his eyes; he smiled with his whole face.

When I mentioned that one of the justifications for having such an unusual pet would be to find out more about them, his face lit up even more. They didn't have time now, but he would come round tomorrow, probably in the afternoon. This was the beginning of our brief friendship.

But to return to more serious matters, away from the Preacher Bird to the smiling enigmatic preacher himself, Gene and my business with him. Although we didn't see each other often, we gradually became friends. We were both aware that Gene's animal and bird business created difficulties between us; gnawed by our separate consciences, the bone of contention threatened to unearth our buried differences and thwart real friendship. Gene knew very well that I uncompromisingly opposed the commercial exploitation of wildlife, be they dead insects, or live animals. He thought me an extremist, impractical idealist, whose philosophy followed to its logical conclusion would send mankind back to his cave.

Then one day Gene called by to ask me if I would like to see the premises where he kept his animals. This was a most unexpected invitation as those in the trade discouraged visitors. A week or two earlier we had been discussing this very point.

Together with a few friends, Cheri and I had launched the "Santa Cruz Ecological Society". I told Gene that one of its prime activities would be to investigate the traffic of wildlife. I had not expected him to be enthusiastic about the idea, but neither was I prepared for his grim reaction. Gene was always very placid and wore that permanent smile but, when the subject came up (that bone unearthed) he involuntarily gave a start, a tic, and said to me, "Yo'all better be careful about what you're getting yourselves into", then got up, and left.

At the time I thought he was angry because we might upset his business interests. Looking back on it, I saw that this was the least important part, he was afraid for us! When he left in such a hurry, I think he had a premonition of what was coming; our plan convinced him how very naive we were. He wanted to warn me, as his friend, but he didn't want me to despise him; without being completely open, filling me in on the whole dubious business, he could not dissuade me, or thereby, protect me.

Naive I certainly was. I imagined the trade was at the level of, say a "High Street Pet Shop". When Gene invited me to visit his premises it was supposed to be an eye-opener. It was!

The day he came to collect me was a Sunday. It had rained all morning but, by the time he arrived it was a bright, sunny afternoon. His normal bonhomie was replaced by a visible tension. Even the words he used, "Weil, I giss we'd bitter git this thing over with," suggested the visit would be a painful experience for us both.

We took the road to the northern suburbs, but soon jolted off the tarmac (where there was a step which Gene didn't slow down for) and followed a narrow track running between the high walls of private dwellings. After negotiating our way around a great many deep muddy puddles (even here within the city limits it was safer to engage four-wheel drive) we came to a stop in front of a pair of tall gates. These gave access to a spacious high-walled compound. Since we had stopped in a sea of mud, and there had been no reaction to Gene's insistent honking, he motioned to me to stay put while he got out and slithered over to the entrance. His thumping of the sheet metal gates made enough noise to wake up the whole barrio.

Above the noise I could hear the shrill cries of small parrots, the raucous screams of larger ones, the excited ululation of monkeys, and the strange calls of other animals and birds I didn't know. Then the familiar harsh croaking of a toucan caught my attention.

The gates were at last thrown open by a swarthy looking sereno ("guard"), who only partly dressed, must have been, like the rest of

the barrio having his afternoon siesta. He was closely followed by
two elegant Red Brocket Deer, their long erect ears nervously flicking
one way then the other, their large liquid eyes staring out past the
gates as if in hope of freedom.

While Gene got back into the car and drove it inside. I had time to
absorb the scene in front of me. The compound was well protected by
four high walls topped by the regulation variety of broken bottles; the
sun winking off their many colours dazzled my eyes. The floor of the
compound was wet but, the mud had been covered with a thick layer
of sawdust. What caught the eye was row after row of cages made
from heavy-duty cyclone fencing. Each cage, standing on tall legs,
was raised a yard or so from the ground. To protect them from the
rain corrugated tin roofs ran from one side of the yard to the other. A
number of windowless brick buildings occupied the whole of one
side. I estimated the area to be about an acre and with the exception
of one or two small open corrals, that contained tapirs and more deer,
the whole compound was filled with cages. This was no "High Street
Pet Shop!"

Gene told me later that he had invested more than $20,000 in cages
alone. "You see Rabin, I don't like to see theise here animals asuffrin.
Not like others, yo'all be surprised how baid some of theym trait their
animals." (Later I was to learn the truth of his words.) One of the
Brocket Deer was nuzzling the small of his back and he
absentmindedly reached behind him to caress its ears. "Look around,
yo'all see that they'r teken good cair of." He went off to discuss
something with the sereno.

I started walking down the long lines of cages, my mind racing
with the implications of what I was seeing. Gene had told me there
were others who had well-established businesses in the animal trade.
I tried to make an estimate of what Gene's stock of animals might be
worth. I could hear Gene, inside a building on my left, talking sharply
to the Sereno. From what little I inadvertently overheard, by now he
was making no secret of it, there should have been some other
workers around to prepare crates for the next day's shipment.

In common species alone: parrotlets, conures, some brightly
coloured jays, tanagers, tortoises, squirrels, and monkeys of one sort
or another I reckoned the stock to be worth at least three or four
hundred dollars a cage. Maybe the large one full of marmosets, Gene
told me they were "Saddleback Tamarinds" was worth fifty times as
much. Since there were more than one hundred cages, I already had

a figure of $30,000. This did not include any of the more valuable stuff; some of the larger parrots (Amazons and Macaws) were worth thousands of dollars each, and there were dozens of them. And what anyone would give for that splendid Puma, or those two fine looking Ocelots, or one of the Tapirs, or a cage full of brightly coloured Trogons, or over there, that lithe Tamandua, or…or…I had no idea, my head was full of six-figure numbers.

My mind had been in a whirl. I retain only a few fleeting memories of what I saw that day. I remember, of course, a pair of magnificent Toco Toucans, fluffing themselves up with pride and eyeing me in the way I came to know so well. A friendly Coatimundi that snuffled my fingers as it stood on its hind legs, its front ones pushing at the wire of the cage. The soft nasal boom of a Curassow with a thin red blade on its bill. Then, incongruously, amongst all the native species, a "Sulphur-crested Cockatoo", where had he come from? The somnolence of a snake of some sort or other, a Viper, very similar to the one I had nearly trodden on near the Saguayo River. And the amazed look of a Spectacled Owl, seeming to see over my shoulder visions of the past.

Maybe I was not supposed to, but I opened a door to one of the small side buildings and switched on the light. A full-grown Jaguar came bounding towards me, its teeth bared in a snarl. I just managed to get the door shut in time. I leant against it to quieten my heart that was pounding with absurd force.

Not to be put off but, much more cautiously, I opened the door of the adjacent room. My nerves, already on edge, were once again subjected to a jolt, if anything worse than before, for no sooner had I put my head round the door, than the most terrific screaming, like an army of Harpies falling on their victim, broke out. The cacophony was excruciating, for inside the room, that measured about five square yards, there must have been more than one hundred "Hyacinthine Macaws", each one leaning towards me, screaming with rage, or fright, I knew not. The noise was so painful I covered my ears with my hands and collapsed to a sitting position on a box by the door. I knew these large parrots, like others I had already seen were protected birds, protected by National and International laws. What was Gene doing with a roomful of them? How did he expect to get rid of them? I knew a little at the time, admittedly very little about the rules governing the sale of protected species. Gene came in and stood staring at me for a moment, and then, in search of relief from the noise we left the room together. "Yep. Worth more'n a million bucks back in the good U.S.of A", was all he said as we regained the relative peace outside. Relative for the Macaws were still screaming and the noise carried through the walls as if they didn't exist.

"A million dollars!" Like a well-trained parrot I kept repeating it to myself, a million-dollars-worth of feathers. Who were these damn pet lovers willing to buy the world into extinction? What made them tick? Surely this could all be stopped. Why can't these majestic creatures be allowed a life of their own? I was astounded and angry. With Gene. With everybody.

We hardly spoke on the way back to my house, and when we arrived Gene declined to come in. He knew I wanted to be alone, alone to think things out, put some perspective into the whole lot. "Yo'all betta take a luk at theise," passing me a folder of documents, and with a throw away gesture of his hand, Gene was gone.

I put the folder aside and lit another cigarette. In no mood to study papers, I went to the fridge and began to drink one beer after the other. I really wanted to get drunk; forget what I had seen and drown the disturbing implications, that like Jackals trailing a sickly beast, were creeping into my thoughts.

I could think of nothing but Gene's part in the business. How could he do it? He told me he loved the very animals he dealt in. How then did he rationalise all this to himself? Did he, or was I quite wrong? Why had he decided to show me it all? Was I meant to see it all? "A million dollars! A million dollars!"

Then I began to think of the wider aspects. Of course! That is why Gene had taken me there. Hadn't I been boasting to him? Hadn't I said we were going to investigate the whole rotten thing? "Good afternoon, Mr So-and-so, we have come to ask you a few simple questions. Don't mind us, pretend we are not here. How many Macaws do you, Mr So-and-So, export every week? How much money do you make every month, Mr So-and-So? When do you ship your animals out? How big are the bribes you pay? Who do you pay them to? Who do you send your animals to? Will you show us around your premises, Mr So-and-So? You see we are from the Ecological Society, Mr So-and-So, and we want to investigate what you are up to. Do you have any endangered species here? Do you have anything to say in your defence, Mr So-and-so?"

"Si, Señor". Blam! Blam! Blam!

My eyes shot open. Good God, yes! You bloody fool. People get killed for a lot less, and up to now you have been a complete idiot!

The next morning was office day, Monday, and what's worse the last Monday of the month, good old "Team-meeting Day". All through the morning I could think of little else except the happenings of the day before. At lunchtime, the meeting over, I rushed directly back home and piled straight into an examination of the documents Gene had given me. I didn't show up for work that afternoon.

The first document was a list of animal dealers, most of them in Santa Cruz; a footnote said "Main ones anyway." Beside each firm was the name of the owner. There were twelve firms on the list that looked like this:

REGISTERED ANIMAL AND BIRD DEALERS

1.	MERCK SHARP Y DHOME	Jiro Onishi
2.	ZOOLOGICAL GARDEN SUPPLIERS S.R.L.	Jiro Onishi/José Roig
3.	BOLIVIAN EXPORT FISH + BIRD	Luis Claure
4.	FAUNA ITENEZ	Juan López Zuñiga
5.	FAUNA DE BOLIVIA	Alberto Carillo Salinas
6.	AVIS IMPORT EXPORT	Dr. Rolando Romero
7.	FAUNA SANTA CRUZ	Marco Antonio Antelo
8.	MINFAUNA	Joseph Minten
9.	RESEARCH ANIMAL BREEDING CENTRE	[illegible]
10.	FAUNA SUDAMERICANA IMPORT EXPORT	Alejandro Aguilera
11.	LABORATORY SUPPLY	Enrique Jordan/G.Harris
12.	SAN MARTIN (Cochabamba)	José Humerez/T. Sabia

Twelve. Supposing Gene was Mr. Big, somehow, I couldn't imagine him being anything else, and that his stock of animals was worth more than a million dollars. What of the others? How much turnover of stock was there?

I quickly glanced at the other papers. One was a very poorly photocopied contract, or rather, I saw it now, a permit made out in the name of the eleventh firm on the list, Gene's "Laboratory Supply". It was basically a long list of birds and animals, together with the quantity of each species, the value of each species, and the total value of the permit. The regulating authority was the Centro Desarollo Forestal, Departamento de Vida Silvestre y Parques Nacionales, La Paz. Yes, I knew the CDF, as I had had dealings with them over Amboró. I noticed all the prices were in local currency and valued the animals at Santa Cruz rates. The first paragraph, which was poorly copied, took me some time to decipher it. It seemed to deal with the duty payable on the animals listed. I had it now, this was a quota of animals and birds that Laboratory Supply could collect for the year 1980, the sum beside each species was not its value, but the duty to be paid on each one. The grand total was 480,359.00 Bolivian Pesos, worth the equivalent of $24,000. I quickly scanned the list to see what duty was payable on a Hyacinthine Macaw. It wasn't there; even the CDF couldn't put this species on an official quota. Unfortunately, it did not mention how the duty was calculated, or what percentage of the total value it represented. So, I was no closer to understanding the value of the business, its turnover and profits for one year but, what I did notice about the other forty-one species listed (4,423 birds) was that only sixteen of them (377 birds) were not protected by law.

I looked down the list again. The fourth item caught my eye: 25 "Red-fronted Macaws", precio unitario 287 pesos, precio total 7,175 pesos. The Bolivian Government would sell you twenty-five Ara rubrogenys, one of its rare endemic parrots for a little under $400. And twenty-five of them would represent about 5% of the known population!

Furthermore, this species is specifically protected by Bolivian law, Decreto Supremo 16646 (17 May 1979) and by the "Convention for International Trade in Endangered Species". (The CITES Convention). It could not be killed or captured, and certainly not exported.

Something was very wrong. The system was being manipulated. Clearly, this legal document, drawn up, signed and sealed by the very people supposed to protect Bolivia's wildlife was illegal. There it was, clearly signed by Dr. Herman Montero Zankyz, Asesor Legal, CDF, La Paz.

My thoughts went racing on. Was this lawyer just incompetent? Or was he bent? But, good God, this man was an important government official, if he was taking bribes and, a little voice told me it could be no other way, how far up the ladder did this thing go? How much was Gene going to pay for his million-dollars-worth of Macaws?

I went back to the rest of the papers. The next was a list of documents required by law before a shipment could be made. The list looked like this:

1. CERTIFICATE OF ORIGIN (CITES PERMIT).
2. OFFICIAL HEALTH CERTIFICATE.
3. COMMERCIAL INVOICE.
4. AIRWAYS BILL.
5. EXPORT CERTIFICATE.

The next set of documents was official forms, examples of each of the forms on this list. They had been made out in the name of "Zoological Garden Suppliers" of Santa Cruz. Quick reference to Gene's list of firms showed the owner to be Señor Jiro Onishi. One of the most important documents, the CITES permit to export had been completed with relevant licence numbers, consignee's name, name of Wildlife Chief (in this case Carlos Monasterio) and signed by him at the bottom. It also bore the official CDF stamp over his name. But the thing that was missing, the part left blank, could you believe it, was the list of animals and birds to be exported!

I couldn't concentrate any longer. I kept getting up and walking around the room, swearing profanely to myself, returning every few seconds to look at the documents again. I wanted to ring Gene, the CDF, Noel Kempff at the zoo, the newspapers, send off letters to CITES, the USA, make a bloody great fuss. Expose the lot of them.

Then I thought about why Gene had given me the documents. As a warning? No, not that ... or, ... Yes, that; though not to warn me off; but to stop me getting myself into trouble through my stupid naivety.

I rang Gene. No, he couldn't come out right now. No, he didn't think it a good idea to talk over the phone. Yes, he would come over as soon as possible, but it would have to be in a couple of days; he would let me know.

By now I was feeling bloody-minded and returned to the fridge for its replenished supply of cold beer. "The second day," I thought to myself. "How many more revelations will I have to contend with if I try to follow this thing through?" By now I knew I had no option, my resolve had firmed.

I went back to look at the documents, for there were still more. First I examined the official invoices: Zoological Garden Suppliers, invoice made out to Pacific Bird Co., Van Nuys, USA for five boxes containing 100 Hyacinthine Macaws, dated 27 March 1982. Just a week ago! ZGS (again) to Golden Gate Birds Inc., Foster City, California, for 240 Toco Toucans, dated 19 December 1981; and another (ZGS to Golden Gate) for eight crates of Chilean Flamingos, 150 birds, dated 3 December 1981. International laws protected all of these birds.

The last two documents were more examples of the most important forms, those issued by CITES based in Switzerland, and no higher authority in the business. The first, a permit made out to Bolivian Export Fish and Bird, Claure's setup, for 1,065 parrots to be sent to Pet Farm Inc/Bird Haven, Miami (some bloody haven!). The list contained 265 Macaws protected by Bolivian Law, and, the number was not legible, could be, yes! 32 Red-fronted Macaws protected by CITES. But this was a CITES form! This was utterly impossible!

The second form proved to be even more impossible. I just could not believe my eyes. CITES was giving permission to Fauna Santa Cruz to illegally export 134 Macaws, again to that wretched bird haven, and furthermore, permission to export 10 Hyacinthine Macaws! Both permits were dated 5 January 1982.

Something was obviously very wrong indeed. CITES would not permit the exportation of a single Hyacinthine Macaw, yet alone ten. "Somehow", I thought "I had to get to the bottom of this." Further examination of the second form showed permission was given to export a total of 1,194 birds (I presumed all birds), but only 534 of them were specifically named. What, then, were the other 660 unnamed items, Pterodactyls? Impossible. Ludicrous. NOT possible!

I was off, walking around the room again, off for yet another beer. Empty the ashtray. Back to the phone. Who this time? Of course! Noel, he ought to know about this. He would be prepared to make a fuss. "But wait".

Back to the forms again. Looking at them was as compulsive as worrying the Chonta thorns still in my hand. Permit valid to...shipper's licence No...Ministerio de Asuntos Campesinos y Agropecuarios, CDF seal, Number of boxes in shipment...

Description or Marks for their Identification...CITES Appendix No....Common Name...Scientific Name...Airways Bill No.... Approval of Exportation... Port of Exportation... Cargo Stamp...and then, there it was, NO, NOT POSSIBLE, PLEASE, it could not be! Another look...there was no mistaking it, clear as bloody daylight on both CITES documents, black as the ace of spades, guaranteed by his own sweet hand. Yes! the signature of my friend Noel Kempff Mercado, approved by Santa Cruz 5 January 1982.

I lit my umpteenth cigarette and got the whisky out. I very rarely touch spirits of any kind, but as I said, I got more than a pet bird from Gene. When Cheri came home it was still early, but me long gone.

The next morning I got up before six. The rising sun was already painting the house and garden with warm golden light. I could hear the Purple Jays screaming at each other in the trees. The dog was scratching itself (flea powder needed today), under the bedroom window; the sound of his claws striking the collar as he worked around his neck imitated the "chink-chink-chink" of a Bush Cricket nearby. A flat banana-spider ran across the wall as I got dressed, and I squashed a cockroach that had unwisely taken refuge under my slippers. My head seemed to follow me around the room, but always a few inches behind. I went to brush my teeth and wash my face. There was no water, the tap just let out a disrespectful "ggghh", then let go a dribble, about enough to wet the brush and no more. I made myself a cup of coffee from the emergency water supply we kept in a four-gallon container. Somehow or other a bloody cockroach had drowned in it and now lay at the bottom half eaten away by grey mould. I made my coffee anyway and had another one, a large one!

Tuesday. I decided this was going to be another bad day at work. Did I have any appointments? No. This was April, the quiet time of the year for crops, the beginning of the dry winter, or so it should have been. But this was a year of "El Niño", the Pacific Ocean current that drives the weather crazy, and it had not really stopped raining yet. Today the sun was out, or at least, out for the moment. Who could tell what it would be doing in another hour? Probably raining again. I lurched into the side of the coffee table, spilling most of my drink. Perfidious Scots, was it Scots or was it Scotch? My nouns and adjectives were mixed up, but why did they have to manufacture whisky?

I went outside and playfully kicked the dog, which in turn left a long glob of slobber down my flies. Disgusting animal. We were never sure what sort it was meant to be. Assured it was a German Shepherd, it looked more like one of the German shepherd's sheep.

The ants were busy undermining the patio, making crater-shaped heaps of sand on top of the polished brown tiles. A Hummingbird kissed a Frangipani flower not two feet from my aching head. A Spider-hunting wasp, about the same size as the bird, finding me attractive (glad something did) drove me back inside the house. I made myself another cockroach-flavoured coffee; it had an oily slick on top, "ugh".

Item 13. This permit is approved by the following authority: Noel Kempff Mercado, Director, Zoo Fauna Sudamericana, Santa Cruz, 5 January 1982. I rubbed my eyes in the hope it would go away. It stayed. It had not been a sleeping nightmare, it was a waking one. Right! I felt bolshie, no bloody office for me today. I had to think. I had to admit I wasn't in the best shape for doing that. Probably better suited to the office. In the end I got into the Land Rover and went to see how my Pineapple Trials were getting along. They were pretty well-scoured by the rain. Everything this year was a wash out.

By lunchtime I was back. Poring over the papers again. And this time, I had a few more painful thorns in my fingers from the pineapple plants. I sat there brooding over the permits, beseeching them to give me a break, erase their sinister implications: man's betrayal of the innocent world around him. I got a beer from the fridge, "hair of the dog", and felt a bit better.

The phone rang. It was Marcelo. Would I come to the airport tonight? Didn't I remember we were going to start a watch at the airport to see what went into the baggage hold? "Yes. Okay. What time?" I put the phone down.

As a start to our investigations we had agreed to begin at the airport. Of course, I had not forgotten. I hesitated because I wanted to put him off, say no; protect him from what I now realised was going to become a shadowy threat to which he, as a Bolivian, was more vulnerable than I, or so I thought. I met him there at eight-thirty, feeling silly and self-conscious carrying a pair of binoculars in the dark. We made our way up to the observation deck that was crowded with the excited relatives and friends that the occasion demands. We hung around until their flight departed, and the crowds with it. Only then did anything happen near the belly of the aircraft we had come to spy on: Lloyd Aereo Boliviano's Miami flight, ETD 2200 hours.

The loaders were working in a dull pool of light some eighty yards from us. It was too far. It was too dark. I didn't want to talk to Marcelo. I didn't want him dragged in. He was too nice, too straight. Under different circumstances maybe, but there were too many "toos." "Look, Marcelo, they could be shipping out Blue Whales for all we know. Can you tell me what those crates contain?" No, he couldn't.

Dispirited, defeated, we left. If only Marcelo would sit down and have a beer. Maybe we could get around to talking about it. But no. He didn't drink. He said he had too many responsibilities. More "toos." I looked at him: short black hair stuck to his narrow skull with grease, delicate dark features and large front teeth. Slim and fragile-looking in his respectable navy blue suit. He looked Mediterranean; he looked the part he was, a Vet. I really couldn't imagine him climbing over walls in the dark and running away from trigger-happy watchmen or vicious guard dogs. I had to say something before we left. But I couldn't tell him that our plan had now to be mine, and mine alone. So, I asked him to see if he could get us into the loading warehouse officially. I didn't think this would get him into trouble. If he was successful, we could at least check the shipments in peace, if he wasn't I would find a way by hook or by crook.

I resolved to see Noel in the morning. Now all I wanted was some sleep. For some reason I started to think about John MacDonald. What was his hero's name? What had he said? "The early bird which gets the worm works for the man who owns the worm farm." Yes, that was it. As I drove home, I tossed this around in my mind, occasionally repeating it out loud to myself. Why had the quote come to me? Was it relevant to my situation? I batted the words around my brain like I batted a ball around the squash court; frontal lobe, left lobe, right lobe, frontal lobe, I still couldn't see the relevance. Maybe I was just feeling like Travis. Yes, Travis McGee, all this skulduggery. I drove a bit faster, brushing the slower traffic aside like so many phantoms.

When I got to the house it was close to ten. The phone was ringing. It was Gene. Could I meet him for dinner at Don Miguel's in half an hour? He sounded brittle, skittish and even nervous. That was not like him.

I got there at 10:40. I had thought I knew exactly where it was. But, not being a big fan of Bolivian cooking, I had only been there once before. I wasted fifteen minutes finding the place. When I finally did, Gene and Chachi were waiting for me, sipping at tall glasses of fresh orange juice. I ordered a beer on my way to their table ("man of action, rude man") and went through the ritual kissing and handshaking before taking my seat. If anything, Gene looked more tired than I felt.

As soon as we had ordered our meals, I tried to get an omelette but had to settle for spare ribs. Gene started with, "Weil, I guiss yo'all have seen the docs and understan' the containts."

Yes, I had. And I had brought them back. Could I take copies? "No neid to, the'r'all for you to keip." Did I have a pen, paper? No, I didn't. "Weil, then you could use theise," passing me some blank sheets of paper and a blue biro (that I see, on referring back to the notes I made that evening, must have run out and been replaced by another one and, finally ending, with a black felt-tip).

"We'all don't have much time, Rabin, so I'm gonna teil you a few thaings that you prob'ly don't know."

I didn't. By the time we left it was past 1am. The waiters had cleaned every square inch of the restaurant except for the square metre where we were hunched over the table like something out of the "Gunpowder Plot." I had no time to eat my spare ribs (actually I didn't eat meat) so I took them back home with me for the "sheep."

At the time I was not able to think about the things Gene told me. Most of it was too outlandish not to take seriously: tales of murder, bribery and corruption at the very highest levels. The night swept by like a ride in the underground. Incidents quickly scribbled down flew past like brightly lit station's names of people and businesses like the names of those train stations; half gathered, half spelt.

****

The following pages are an exact copy of them:

Chilean Flamingos: stolen from lake in Argentina/export permit from Bolivia/boxes (empty) dispatched from Santa Cruz/birds secretly waiting in hanger in B.Aires/sent out. Also illegal in Bolivia.

Oscar Llado: gets paid/box animals/no diff. what. He's Director CDF Santa Cruz.

Jiro Onishi: Jap/works for Mafia in US/boss David Molof in Los Angeles/has agents: Pujold in B.Aires, Twinstra in Paraguay, Montoya in Lima/sends across borders to Bolivia/from here exported legal-illegal.

José (Pepe) Roig: frontman for Onishi/ex-politico Falangista/pays off local newspapers—no stories/also S.Cruz thru CDF lawyer (name?) paid $500 mo. for services.

Freedom of Information Bureau/Washington/will supply all copies documents/visit??

Reynaldo Rodriguez: Pando (?), news journalist on Onishi payroll.

Gaston Bejarano, Jorge Borja, Carlos Villaroel: all in league with [illegible] CDF La Paz.

Rolando Romero: sold Gene's birds [the Hyancinthine Macaws I had seen] to these three guys—split money between them.

Vladamir Rivera: lawyer, started Animal Protection Soc. with journalist Victor Callau BUT Rivera gets kick back from illegal exporters.

Father Gallagher: says WWF [World Wildlife Fund] went to investigate exports at Riberalta but WWF never left hotel to make survey as Gall. would not let them use his boat.

CDF clerk at El Trompillo [airport] paid $25 for signing airways bill per shipment.

Onishi: is Molof's major S.Amer. supplier NOT frontman.

Twinstra: Dutch/own plane/works out of Asuncion/[illegible] to Paraguay sending all to Europe. Bertha: ex-mistress Hans Meyer/worked for Gene/got G. all fouled up.

Wayne Green: has sawmill near Comarapa/wants [illegible] bird business.

Carlos Carillo: with Onishi.

Noel Kempff: why did he sign illegal CITES forms/he knows Bolivian Law/why he sign?

Juan López: very dangerous man/ex-campesino/started off as hunter/sly (not smart) fox/managed to get Parra Bros. into partnership/he's still [illegible] to make it alone.

Theodor Blakener: screwed everybody/now out of business.

Charles Courdier: German Swiss/very nice/knows ALL, real animal lover/zoologist-collector/in Cbba/ex Katanga/first to send gorillas to zoos/sent ten.

Dr Rolando Rivero: ex Cbba [Cochabamba]/no lawyer/exporting?

Dr Rolando Romero: cardiologist (don't mix with above) was guarantor for Gene when G. was in prison. He has stopped many shipments.

Marco (Poncho) Antelo/Claure/Harris [Gene]: going into partnership with Romero/G. 10% also involved illegal export gemstones from Gaiba.

Antelo: fired gun at Onishi when he (A) was drunk. Playboy/talks lot about killing people but really coward + not much action.

Juada: spell? Kuada? doesn't take money.

Bruno Emma: wanted all over world for everything/very dangerous—involved with Italian terrorism as professional killer/now is Romero's right-hand man in Bolivia since two years.

Spies put into other people's compounds to KILL their stock. Put them out of business.

Phillip Pan American Export Co.: co-owned by 2 French guys/both killed (reported in El Mundo '79-'80?) in fight with Milton and Chichi Parra/both wounded also.

Marcos Haupfmann: (Halfmann?) 15% partner with Onishi/worked with Hans Stellfeld who died in S.Cruz from bullet in head in '81/from Paraguay/animal exporter in '70's/in prison 2 yrs./Onishi then offered Haupfmann partnership if he would bring his trained Paraguayan hunters with him to show Bolivians how to collect.

Onishi: owes Bolivian Govt. $20 million in taxes over last 12 yrs/can prove/Gov. offers 15% reward if anybody proves.

Flamingos were sent on a phoney airways bill + shipped out on private plane/ask Hugo Paz friendly manager El Mundo to make denuncia [expose]:

1. When did he get Alcalde's [the Mayor's] approval?
2. When did N.K.M. [Noel Kempff] sign papers?
3. Nota de remision—where is it?
4. Poliza de exportacion—where?
5. Produce signature of inspector at airport.
6. Produce airways bill from legitimate carrier.

Papa Camacho: as part of Comite Pro-Santa Cruz should have done something/Gene gave him all necc. documents to close up illegal exports/nothing happened.

Bejarano: signed flamingo papers for Onishi and is NOT to be trusted/careful/has relation called Tania/she is girlfriend of Hernan Siles Zuazo [Bolivia's President to be].

McAlpin + Richard Barnley: pretend represent Texas A&M University—rented Brownsville Quarantine Station (owned by Gregory?)/landed two shipments from Bolivia before Dec. (1981?).

Luis Añez: Minister in charge of Agriculture/wants Bolivia open [to exports] and wants to stop all adverse publicity.

Senator Barrientos: President Banzer's partyman wanted Bolivia closed.

Avilla: silent partner with Carillo [Fauna de Bolivia] another illegal exporter.

Villarroel: was Minister second time [illegible]/got Gene out [of prison] last time/all the guys in the regulatory field are related.

El MUNDO: [one of Bolivia's main newspapers] printed illegal CITES forms for Bejarano and Onishi (who has Bolivian citizenship, goes to the STATES a lot?).

Readings (Ridings?): from WWF interviewed Romero and L— pez (1982) but Onishi wouldn't talk to him.

Why has Alcaldia [the Mayor's Office] not closed down the company? [Onishi's, over the flamingo affair].

Onishi: forges Cert. of Origin + Vet certificates/Alberto Vasquez CDF vet [and later, as MACA-CDF Director, to become my boss] is another Onishi man.

Carlos Monta–a [Montoya?]: applying for export licence now/will offer Gene 5%/intends to do it legally.

Onishi: 1970 in Brazil/then Paraguay/then Bolivia.

Montoya: exports monkeys out of Peru/Onishi merged with him.

Natusch Busch: 1974 Minister of Agriculture [later President of Bolivia]/brother-in-law Pepe Roig who partnered Onishi/together they made the Laws!

Gaston Bejarano: was paid $3 per monkey and is Onishi man/that is for each one exported from Bolivia. [Not bad considering more than 10.000 monkeys were exported annually.]

Minister of Agriculture changed and two groups got a resolution to export/then blackmailed Onishi to be allowed to export???

Gene: got permit to breed Blue-fronted Amazons/Chachi could not get licence from Bejarano.

Aneus (Aeez spell?): another new Minister got Gene his licence.

Onishi: made Law to establish licence for exporters.

Jordan: (CDF Santa Cruz Wildlife Chief) was paid by Onishi. Now an agronomist/everybody had to pay Jordan $100/he spread the money around/now same permits cost $5,000 per shipment/in old days $5!/even have to pay the clerk $50 [this is the CDF typist].

Kempff: said Onishi gave zoo big gift (news article in El Mundo), this was some of the stolen flamingos.

Oscar Mendez: fired from Ministry of Ag. after fiasco with Gene/he 'lost' $30,000.

Pepe Roig: probably man who will ruin your Ecological Soc. his address 688 24 Septiembre, Santa Cruz. (See?).

Onishi: WILL HAVE U ELIMINATED!

****

Golly! The runaway train had finally crashed into the buffers. Fortunately, I suffered nothing worse than some very stiff fingers. Gene, with occasional interruptions from Chachi had talked non-stop for two hours.

But that was last night, and feeling back on form after a deep sleep, I was now sitting in my office smoking one cigarette after the other: my legs propped up on an insect cabinet, my back to the window. Every now and again I turned around in my seat to watch the rain trying to get in. I was trying to make head or tail of my Bolivian student's scribbled field notes. They were like my own, taken down so hurriedly at Don Miguel's. And with this reminder I checked out of my office an hour early and went home. I needed more time to think. It was all so serious now.

That evening an American friend of ours, Peter, dropped in. As usual we got around to talking about Amboró, and finally ... I needed to confide in someone: bird and animal exports. I showed him some of the documents Gene had given me.

Since we were due for leave, he suggested we pass by New York and talk with Mr William Conway, Director of the New York Zoological Society. They were friends: something to do with protecting the flamingos of Laguna Colorada. And, then and there, Peter insisted on sitting down to draft a letter of introduction for us. I let him go ahead even though I was far from sure what I was going to do about the whole thing. My inner voice appealed to me to proceed with extreme caution.

I still had to sort out in my own mind the Gene connection. Could I believe all he had told me? Onishi was obviously not one of his favourites. Was Gene simply using me to get at him? Gene himself was not going to come out of this smelling like roses. How would the whole business, say I took it further, affect my carefully nurtured friendship with Noel Kempff? I kept gnawing the bone of apprehension without getting to the marrow of the problem. Noel,

Gene, Onishi, Bejarano, walked like silent ghosts through all my thoughts.

****

Noel. Profesor Noel Kempff Mercado. (Note: the title signifies a 'teacher' in Spanish, not professor.) Age 58. Origin: German, born in Santa Cruz, Bolivia. Positions: Director of the Botanical Gardens and Director of the local zoo, "Fauna Sudamericana", sounding too much like an export company for my liking.

Appearance: short, slightly running to fat but, not unfit; grey thinning hair, always a little untidy; complexion reddish, veined; nose wedge shaped; eyes pale blue, spectacles. Other characteristics: often needed a shave (like me). Personality: intelligent, morose, trust-worthy? Probably.

I sat reviewing what more I knew about him. At first when I went to see him, to help him with his book on Bolivia's birds, he used to hide behind his typewriter after a few minutes of conversation. Reticent? Maybe, he also made it quite plain that he didn't suffer fools gladly and he was capable of telling them so. Phlegmatic? Yes, but later on I saw the better side of his character, how, when you had touched on something that amused him, his face would light up with an unexpectedly beatific smile. I liked him.

In a place like Santa Cruz it was no small achievement to success-fully run a zoo, where any one of the thirty blockheads who worked there could intentionally, or through pig ignorance, cause the death of one of the animals. And Bolivians are not renowned for their punctiliousness, or their reliability, factors that mitigate against a healthy happy menagerie, as would leprosy against becoming Miss Universe.

Noel's manifest shortcoming was his lack of English. He received many foreign visitors prepared to help the zoo and conservation, but was unable to really impress them, or sort out the sheep from the goats. Neither could he read English, this was where I was able to help him. But helping Noel was a bit like helping General de Gaulle administer France; he resented interference, or dependence on others. He was the only Cruceñean who really knew what it was all about. I respected him. His chief asset was the admiration he engendered among the local population. If Noel attended a meeting he was heard with deference and respect. And nobody could effectively participate in the fields of conservation and natural history without Don Noel's blessing. At the time I was very willing to play third base; I thought of myself as catalyst not choreographer, ward not father, and yet, if ever a child were father to the man, I was to Noel. As a trained zoologist and taxonomist I was able to help him with the kind of expertise unavailable in Santa Cruz.

Now and again, he came over to our house, usually to drop off some reference I required for the Bolivian bird list and once to see an audio-visual presentation Cheri had prepared about Amboró. He was always very busy and never stayed longer than necessary. In addition to his duties at the zoo and the Botanical Gardens, and his penchant for writing booklets, he had many civic duties to perform I knew nothing about. He also had innumerable hobbies: bee keeping, flower growing and orchid collecting to fill his spare time. He was a very active man.

Occasionally I would go round to his cramped apartment in the centre of town but, I preferred to call on him at the zoo. There I could stroll through the gardens, past cages of pacing animals or noisy colourful birds. And spend a few quiet moments in the huge walk-through aviary, one of the biggest in the World, before reaching his office. Here Noel would sit over interminable cups of coffee, his desk littered with papers and books, tapping away on an ancient typewriter. Like me, he smoked endless cigarettes and never seemed to empty the ashtrays. We were fellow spirits.

****

But I was still hesitant to confront Noel with the information at my disposal. While reviewing my relationship with him (as set out above) and how I should approach him on the matter, I kept asking myself the same basic questions. Could I ignore his part in it all? No, not if I wished to continue working with him, and I couldn't work in my chosen field without him. Was he my friend? Yes, and later, maybe more than I realised. Yes, because we needed each other, and I think he respected me as I did him. Could I, therefore, afford to be frank with him? I believed so; the problem was how to be tactful, put it in the right words without a really good working knowledge of Spanish, a language rooted in Middle Ages' Cchivalry.

Apart from my own wish to see the export of birds and animals permanently closed down, that I am sure he was aware of, were there any other factors that might affect his response to my oblique accusations? Yes, I could think of several. Whilst a gang of mafia-backed hoodlums was running around Santa Cruz suborning every echelon of power, anybody who didn't agree with them, or obstructed their patently lucrative business was likely to end up with size twenty lead boots.

Presumably Noel was well aware of this. Was he too scared to do anything about it? His signature on those forms suggested he was in league with the exporters. Was he being blackmailed? "Um". Was his interest in promoting conservation being adversely affected by what was going on? Yes! How could Bolivia expect to obtain International funding for its National Parks and Forest Reserves while it officially encouraged, nay, conspired with the illegal exploitation of its wildlife and the uncontrolled exploitation of its timber? This was a MAJOR factor in my plans to establish the Park.

With so many conflicts I knew nothing about, I decided to put the matter aside until we came back from leave. I would go to New York and talk with Conway, see what he had to say. In retrospect and I think incorrectly, I decided not to confront Noel yet. I needed some corroboration of the details supplied to me by Gene.

I also had to consider my official obligations. If I allowed the business to take me over I would end up having no time for anything else. My initial contract with the British Government had been extended for a further eighteen months. My leave had been approved but, I had to prepare my summer work programme for review before we could go. I was due to get another research student from England. I already had one. And I needed to pay more attention to my Bolivian counterpart (the dreadful Mario who was proving more difficult then usual.)

Also, at the request of CIAT, the British Mission had agreed to move our offices to the Saavedra Research Station. The problem was that Saavedra was fifty miles north of Santa Cruz. And apart from the time and money wasted travelling each day, an extremely contentious issue that set everyone's teeth on edge.

Notwithstanding a tight schedule I didn't entirely drop the business to which Gene had opened my eyes. With Cheri, I spent a weekend driving around the city limits locating the dealer's premises. As it turned out there were no high walls to climb over; most of the ones we traced were, as Gene had implied, rather shabby affairs. Only those which belonged to Onishi and Antelo even approached the orderliness of Gene's facility. It was not my intention to do a Travis McGee, climb over walls or crawl under wires at night. The session with Gene at Don Miguel's had convinced me that would be foolhardy. Besides, had I wanted to, I am sure I could have posed as a potential customer, or through the CDF (where I was persona grata through my work with Amboró) I could have accompanied one of their inspectors. In fact, as you will see I did both later on.

Above all, it seemed, the wisest course of action was for me to keep a very low profile, to be, in Burgess parlance, a Mole. Our clandestine visits were not simply to locate each dealer's compound but, to obtain a rough idea of the magnitude of their businesses. Later, using a few safe friends from the Ecological Society we could visit them if we chose to.

I say safe, but Gene had told me the dealer's employees and their friends had, like other societies before ours, infiltrated and destroyed them, and would almost certainly destroy our's. He was to be proved right yet again.

It was not difficult to find the dealers' premises. With Marcelo's help I had made a rough sketch map of where the main ones were situated. By asking the locals and then following the noise: the loud screams of Macaws with their, "we're here, we're here", were a dead giveaway. I soon had Onishi, Minten, Carillo, Romero, Claure, Lopez and Antelo accurately marked on the map, and in some cases I took photographs to remind us which was which.

I made brief notes on each one as I went along: Onishi carretera Cbba. [Cochabamba Road] km 9. Large compound (4 acres) beside main road, hidden by high wall and large wooden entrance gates. Converted chicken farm with long wide cages for bulk handling with v. many Macaws. At back, large wooden shack containing c. 200 Squirrel Monkeys. Security good but easy access at back. Dogs. Guard. Lights.

Minten: carr. Cbba km 10. Open compound off road on right (opp. water tower). Barbed wire fence. Number of small and large cages containing mainly Macaws and some Capuchin Monkeys. One cramped cage with a sad looking Spider Monkey. Security poor, access on all sides. No sign of a guard. No(?) lights. Dogs, and note Seriemas (South America cranelike birds used as guard dogs.)

Carillo: carr. Cbba km 6 (then turn right and follow poor road for 2 km). Many cages scattered throughout citrus orchard at back and side of house. Mainly Macaws (lots!). No security(?). Lighting poor. Easy access on all sides. Dogs (small ones).

Romero: visible from road to Palmasola km 6 (off road on right). Small compound with indifferent wire fence. Two ramshackle clapboard buildings (one containing Macaws in a lean-to) and some open cages (Monkeys). Armed security guard but easy access. Few lights. Alsatian dog.

The rest were similar to the setup owned by Carillo. Antelo had a large open-sided barn with many small cages in rows that contained almost entirely Parrots. Here and there were a few Coatis, Peccaries, Cats, a cage full of Brown Capuchin Monkeys, and some other Birds.

As far as I could see all these men were dealing in illegal animals and birds. I had to leave to imagination what was hidden in the buildings and cages screened from view but, it would be reasonable to suppose they contained the more sensitive stuff. At all the places I visited there were dogs, mainly pie dogs, poor specimens but good barkers. Minton, as I had noted, had some "Red-legged Seriemas" that, if disturbed, would make a lot more noise than the dogs. At least his weren't destined to become feather dusters, the fate of so many of these fascinating and useful birds when they fell into the hands of Campesinos, who would sell them to the people in Santa Cruz.

During the few days it took us to gather the information we were only seen once, at Romero's. There I decided to test the feasibility of getting into the compound by posing as a prospective customer. I stopped the car right outside the gates. The Sereno, a consumptive looking individual, was carrying lumps of meat into one of the clapboard buildings. I got out of the car and called him over, ostensibly to ask him for directions to a fictitious person. While he answered me, "No conozco, Señor" ("I don't know him. Mister") I gave him the impression I had just noticed all the animals and birds. "Caramba!" ("Gosh!"). "Es un zoologico publico, Señor?". No, he shook his head, it wasn't. It was private, not a zoo. Then I pretended I had just cottoned on and was now interested in buying a Parrot as if the thought had just occurred to me. After a short fit of coughing, he patiently explained that if I wanted to buy a Parrot I would have to talk to the owner. "Quien es?" I ventured. "Dr Romero," he said. Not very discrete I thought. I considered bribing him to show me around; but decided against it since he was by now getting a bit nervous, a condition that set him off on a long spasm of coughing. As I went back to the car I noticed a handy looking rifle propped up outside one of the buildings. No need for the rifle I thought, his cough was deterrent enough.

Then we drove off. Back home, back to the Busted Flush. Travis needed a cold beer; my head was full of moles, worm farms and birds. And that rifle.

At least I now had an idea of the scale of the wildlife trade. Based on what I had seen, both at Gene's and through our surreptitious visits, I was able to 'guesstimate' what the trade was worth in terms of stock in trade. I used Gene as my standard, who by his own admission held more than a million-dollars-worth of stock at any one time. Onishi's business, and Gene had confirmed this, was probably worth twice as much. Maybe López, too, made as much as Gene. As for most of the rest, they held about half as much. Romero's facilities appeared somewhat amateurish. Maybe, as Gene had intimated, it was more of a hobby than a business. I had now visited eight of the twelve exporters on Gene's list. Based on the price they would get once their animals were delivered to the importing country, they had a total stock in trade of six or seven million dollars!

This gave me a round figure of ten million dollars for all the companies combined. What I still did not know was how often the stock was turned over. I guessed, and later on I found out this was nearly right, about every four months. This gave me a yearly gross turnover of thirty million American dollars. My figures were very rough indeed but, anything like half this amount, shared between twelve individuals was a colossal amount, and to me a worrying revelation.

"Onishi will have you eliminated, Rabin", the last thing Gene ever said to me, disturbed my thoughts.

I now needed to calculate the probable net profit. Thirty million dollars left a large leeway for overheads, but I was sure these couldn't be in proportion to the money to be made. After all, according to Gene, some of these people were directly linked to the Mafia, and they didn't waste time on enterprises with small profit margins. I worked out a theoretical balance sheet for a shipment of 100 Squirrel Monkeys.

It looked something like this scribbled down version.

EXPENDITURE	US$
a. Hunters fee. $2 per monkey x100	2
b. Hire of aeroplane, out and return Santa Cruz	1,000
c. Holding costs $8/monkey/month x4 months x100 (this includes overheads for facility, food, wages, etc.)	3,200
d. Export permits—illegal (ref. Gene's notes)	5,000
e. Boxes for monkeys	1,000
f. Government duties (Here I had a good idea from Gene's CDF quota permit)	1,000
g. Shipping costs (Again I had some idea from airway bills Gene had given me)	4,000
h. 4% USD f.o.b. tax, veterinary certificate, etc.	200
i. Hidden costs (Bribes paid to Govt. officials,—like Bejarano's $3 per monkey)	2,000
j. Lawyer's fees	400
k. Staff salaries (guards, fixers, drivers, etc.)	1,000
TOTAL COSTS	20,000
GROSS PROFIT: 80 monkeys (assuming 20% mortality) at $500 each.	40,000
NETT PROFIT; Gross profit less costs ($20,000):	20,000

Rough as it was, I thought I now had a better idea of what I was up against. A thirty-million-dollar business with net profits of $15 million, shared between twelve individuals and their Mafia bosses: who would make even more money when they retailed the merchandise by way of those good old family High Street Pet Shops for pet lovers!

****

Marcelo called me the following Sunday morning. The phone rang just as I was setting off to locate one last exporter's premises. He wanted me to meet him at the airport again. He had got a tip-off that a large consignment of birds was going out on the night flight to Miami. The shipment was expected to reach the customs warehouse sometime in the morning and there would be no difficulty getting in. He thought we should meet at about eleven to allow time for the shipment to reach the airport.

That didn't give me time to go out, so I decided to stay at home and attend to the garden that was suffering from neglect and to write long overdue letters to my family. Since it was still early, I thought I would go to the filling station, as I had intended doing, before the queues got too long. A Sunday morning like this one: warming quickly under the rising sun was not the best time to do this because; the many people leaving town for a day by the river kept the petrol pumps busy. When I arrived, there were only six or seven vehicles, but only one attendant at the pumps.

To pass the time I lit a cigarette and picked up an old newspaper, one I hadn't got around to reading. It was an El Mundo for Wednesday 31st March. Being out of date, I skipped the international section and turned to the centre pages devoted to the local news, that I thought might offer some entertaining scandal. There wasn't much on the first page, but turning to the next, an article headed "Increase of illegal Animal Exports" immediately attracted my attention. It was very rare to see anything about the animal trade in the newspapers. I had to put the paper down to move the Land Rover forwards. On picking it up again my eye fell upon the heading of a short article directly below: "Bird dealer arrested." I caught the words, "Laboratory Supply," and, "Jim Harris". I knew they must mean Gene, but before I could read it the noise of car horns behind interrupted me.

Begrudgingly, I put the paper aside, slid the vehicle forward and got out to unlock the petrol cap. As the nozzle went in, I checked that the counter was at zero, it wasn't; it rarely was. Just the next racket, the next rip-off, if you were careless. I wasn't. I casually indicated to the attendant he had forgotten to zero it. He tried to tell me some had already gone in, twenty litres! The horns were starting up once more. Here we go again. The point of no return. Make a scene, or take it? Had my mind not been preoccupied with what I'd just read, I wouldn't have put myself in this position. I would have made damn sure the pump was zeroed before the nozzle went anywhere near the tank. I wanted to fill the twenty-litre jerrycan as well, so I let him go ahead. When I paid him, I gave him the total indicated less the cost of fifteen litres; I hadn't tipped him for some time anyway. Honour had been satisfied. Blood had been drawn, but not much. I got the jerry filled, he accepted without demurring.

****

War and Peace. Thrust and parry. Life here was one long fencing match, like Fiddler Crabs in a mangrove swamp. The timid get pushed under, the bold held the beach. Whenever I got annoyed, I reminded myself that in all their wisdom Her Majesty's Government recognised the realities and officially made it up to us by paying us a tax-free bonus, what they called "The difficult post allowance" for accepting work in one of the countries officially listed as being difficult. In Bolivia it was worth £1,500 a year.

My mulling over the niceties of the rules was brought to a sudden stop when I recalled the news article. Its cold facts drove me home as a chill wind drives a Leaf-cutter Ant to its nest: dropping everything on the way.

For five days that El Mundo had been kicking around the back of the Land Rover; meanwhile my friend had gone to gaol, paid his way out and quietly left the country.

The article about the increase of illegal wildlife exports didn't mention the other animal exporters. And the wording implied that Gene was almost the only one making illegal shipments: Harris ... habría logrado centralizar toda, o la mayor parte, de la exportación illegal de animales silvestres con destino a EE.UU. y Europa. ("Harris ... managed to corner the largest share of the illegal market exporting wild animals to the USA and Europe").

I found this disturbing for it suggested a cover up. The article had less to do with calling the public's attention to the illegal trade in general, and more to do with a personal attack on Gene. Somebody wanted Gene out of the way without creating a scandal over the whole dirty business and, I viewed the article as being yet another example of the Animal Mafia's power over the press.

My suspicions were reinforced when I got hold of the following six issues of El Mundo. Only once, on April Fool's Day, was the matter officially referred to in a bland interview with the CDF's Santa Cruz Director, Oscar Llado. He, at least, but incorrectly, stated that there were five firms exporting wildlife from Santa Cruz.

On the 7th of April El Mundo carried an article headed "North American declares that he was not implicated in the illegal traffic in animals." The declaration, it said, had been made in Miami where Gene (they got his name right this time) Harris, was now living.

I picked up the phone and called Gene's number. Chachi answered; she too, was about to leave for a while. For "How long?" I asked. "I don't know, Robin. But we'll be back. Adios."

****

I looked at my watch. It was five-to-eleven, time to meet Marcelo. No time, yet again, to write to my poor family.

When I got to the air terminal Marcelo was waiting for me. He had been to the warehouse and seen the boxes of birds. He was disappointed because he didn't think I would be able to recognise the species being shipped out.

We went over to take a look. He was right; the solid plywood boxes were perforated by finger-sized holes but the birds, packed like sardines, were too crowded to be identifiable. And the tops were nailed firmly down. Another failure.

****

The week before we went on leave we received a letter from the New York Zoological Society inviting us to talk to the Director, William Conway, on the 9th September. Peter had come up trumps. On our way back from London we stopped off in New York and took the subway out to the Bronx. We had a short discussion with Conway and Dr Donald Bruning, Curator of Birds, about our hopes for Amboró and a long one about the animal export trade in Bolivia.

We gave them copies of all the documents in our possession and a resumé of the most important bits of our information. They were impressed, especially by the authenticity of our statements, the details of which, they said, corresponded with what they already knew, and others that filled important gaps in their knowledge. The clincher was probably our information on Robert Molof and his connection with the Bolivian Mafia. Apparently, this dubious American gentleman had been federally indicted many times but for lack of sufficient evidence had evaded further prosecution.

In Bolivia one becomes inured to scandal; the newspapers print so many stories of outrageous incidents, involving well-known people and government officials, that the population has become conditioned by their sheer frequency. Bribery, corruption, complicity in murder, kidnapping, and drug trafficking, were almost daily news. Enough to embarrass a city the size of Hong Kong, Santa Cruz shrugged these revelations off its collective back as casually as a duck does water.

We agreed to send William Conway a written account of our investigations as soon as we returned to Santa Cruz. "Be very careful. Don't take any risks," he warned us.

I never worried them with my problems or told them about this dirty little brush war I was involved in, at least, not until it was long over. For my part, I was gratified to learn the information given to us by Gene and a few other contacts, had proven to be accurate and of great value. It was refreshing to discuss the whole rotten business with two people who genuinely shared our concerns and who were influential enough to ensure our disclosures would not be wasted.

3. Things Start Getting Hot

Even though we were heading for home, the uneasy sensation of becoming lost, like following, contrary to one's own timidity, a little used path in an unfamiliar wood. The feeling had grown persistent in me as the plane slid ever southwards towards our destination. Lowry says it's a special feeling reserved for those travelling to the Southern Hemisphere: that we were approaching the very rim of civilisation, getting too close to the edge.

****

We got back from leave in October 1982; just two short years had gone by since we first set foot in this alien land.

My baggage contained little more than books. Santa Cruz had few bookstores and nothing available in English. Reading appeared to be an unpopular pastime in Bolivia, maybe because books were very expensive, but even many well-to-do families had no more than the odd novel or encyclopaedia in their homes. Noel... perhaps he was an exception.

It was just as well we had brought our own entertainment. When, with profound relief, we got back to the house, we were surrounded by the three English sisters who had been baby-sitting it during our absence. Only the day before, thieves had cut a hole in the perimeter fence and while everybody had been asleep (including the sheepdog), they had broken in and stolen the TV and video. My first reaction was, hooray! for quite honestly, I was glad to see the back of them. During my forty years, much of it spent in developing countries, I had never been burgled before. This was the third time in Santa Cruz.

I was happily contemplating the word processor we could buy with the insurance claim, when I remembered we were no longer covered. Just before we left, I'd tried to renew it, but the spirit of Mañana ("Tomorrow"), which pervades even the best organised businesses foiled my attempts to do so.

A loud familiar squawking was coming from the house, at least Cyrana that damned toucan, had survived. I was glad for Cheri's sake.

As soon as we got inside, it repaid my felicity by shaking my shoelaces like a terrier with a rat. I love pets, the dog had already christened the front of my trousers with an extra-large dose of his slobber.

Pets were to be the dominant theme of the day. In the afternoon our friend Oscar called in and told us there was a kid down the road trying to peddle what looked like a Fox, but whatever it was it needed a home. So off we went in his jeep and retrieved it for a few pesos. It was indeed a Fox, a Crab-eating Fox. As we all know, foxes eat just about anything, and this species is no exception: the misnomer originates from the time of its discovery when the first ones were seen eating crabs on the banks of the Paraná River.

The boy told us he had had it for some time and, truly, it was very tame. We named him "Don Juan" in honour of Castaneda's shaman teacher. Don Juan's only disconcerting trait was his habit of 'spray scenting' people he didn't know...and by the time we got back to the house, where a party was already in progress, my trousers already fouled with the dog's saliva were now soaked in fox's urine. I really stank. I released the fox amongst the revellers and hurried inside for a shower and change of clothes. The last thing I saw was a trolley of beer-filled glasses fall over as one of our friends sprang clear of Don Juan's greeting.

Don Juan eventually settled down and became a very endearing addition to our menagerie. The only really naughty thing he ever did was to steal Cyrana's tail. Up until then, Cyrana, unlike the dog, had always got the best of their confrontations. From that day he was banned from the house.

But all good things come to an end, and one day Don Juan never came back. We made enquiries; we searched everywhere, calling his name to which he had quickly learned to respond. Our neighbours thought he had been stolen by one of a gang working on the road outside. Maybe, but there were many pitfalls waiting for naive little foxes. Yes, and naive little people too.

****

After the social flurry following our return from leave had died down, and the loose threads at work picked up and sorted, I got down to prepare the report for the New York Zoological Society. After that Travis McGee was to be laid to rest. I was going to retract into my shell and wait for the tide of events to take their course. I would broach the subject with Noel at a suitable moment, but would not seek the moment out.

We sent the report a month before Christmas. It was entitled "Provisional Notes on the Export of Live Animals and Birds from Bolivia." We decided to keep to the main issues and make it brief. If they wanted more, they could ask for it, and, depending upon the situation we would do what we could. The report consisted of four foolscap pages divided into three sections: a list of exporters and their assistants, a list of corrupt government officials and a list of documents required by the Bolivian licensing system. Beside each of the people mentioned was added a summary of their illegal activities. With this done we sat back to await events.

****

If 1982 had been a wet year, 1983 was even worse. Normally, the heavy rains start with the advent of summer in December and stop sometime in March. During autumn it should get progressively drier, and winter (from the 1st of June to the end of August) was the driest time of all.

The 1982-83 rainy season started just after we got back from leave in October. As the rains had only stopped two months before, the soil was still unusually moist. I had a nagging feeling some sort of natural disaster was on its way. After a reconnoitre of the river Piraí and its flood plain, I wrote a report for the British Mission. I reasoned that another wet summer would cause unprecedented flooding in the Santa Cruz area... a danger exacerbated by the unabated felling of riparian forest in the river valley. Although my report was discussed and endorsed at the December team meeting, it was decided there was nothing more we could do with it.

By January news started to come in of minor disasters in the outlying districts. Places that had never been subject to flooding before were now inundated. In the Amboró area the Yapacaní River caused the death of nineteen people and a further twenty-seven were missing. By February the western suburbs of the City were swamped and many of the main streets were under a foot of water. People in the worst affected areas were appealing for help, many of whom had been forced to abandon their homes. Those living in the Beni, several hundred miles downriver, were marooned on the roofs of their houses; some were rescued after a few days, others died there.

By March, when the rain should have been intermittent, a new rhythm developed, and the rain came down without respite. The sky took on the colour of a ripe plum and the dim light that filtered through cast a deathly pallor over everything. For days on end the rain fell with unremitting force, as it only does in the tropics. People jokingly said, "It isn't just the fault of El Niño, the child, it's the whole bloody family."

Everything in our house was damp, and everything went mouldy. Even in the cupboards, supplied with light bulbs to keep things dry, our clothes became dusted with grey fungus and our leather shoes looked like specimens of "Mossy Foot". The walls glistened with moisture and took on a green hue. In one room, where a crack had developed in the ceiling, two groups of mushrooms sprouted from the crevice. They looked like miniature white umbrellas. Nature's little joke.

Outside it was chaos. To get to our vehicles we were forced to wade through foot-deep water. Everywhere we looked it was the same. Some days we were entirely marooned. Shops were beginning to run out of supplies and the marketplaces were almost deserted. Nobody knew what was happening in the surrounding districts as only vague rumours reached the city. No planes could land or take off. And reports confirmed that all the main roads had been cut.

And still it went on raining. One particular week it rained for the twenty-four hours of each day. We sheltered in our houses. The constant roll of thunder was hardly audible above the noise of the deluge. All the wildlife had disappeared. No birds were calling in the garden and only once did I hear the plaintive cries of monkeys. Large trees began to list and when the waterlogged soil could hold them no longer, they fell with a disquieting splash. Branches and all manners of rubbish went floating down the streets.

Three blocks away the flooding was even worse. Oscar rang us up in the middle of the night to come and rescue him, as his wooden house was afloat. Since the poor man had recently broken both his legs while landing his plane in the middle of a forest, he sounded a bit panicky. It took me four hours in my Land Rover to bring him back to our house. It created a tale I would like to relate sometime.

The newspapers were frantically trying to keep their presses rolling to report on all the natural disasters and personal tragedies. A mother and child trying to cross the street, swept away. A woman alighting from a bus near the city centre suddenly disappeared when she stepped into a manhole that had lost its cover; her body was found hours later floating down one of the drainage canals. Terrible, ghastly things were happening all around us.

The radio was jammed with jabbered accounts of the cataclysm. As far as we could gather the River Piraí had wiped out a fair chunk of the city. The clouds having lifted momentarily, the civic authorities had made a hurried aerial survey of the situation. The film taken from the plane was scheduled to be broadcast at ten in the morning. The city had been declared a 'Disaster Area' and appeals for International Help had been sent out to cope with the emergency. Nobody would know the details until the return of the plane, but it was believed many people had perished. They were calling for volunteers, boats, vehicles, chain saws, and tents, anything that formed the usual paraphernalia of relief and rescue work.

I called the head of the British Mission but, he had been down to the most affected part of the city and they didn't have any use for Land Rovers. They were short of tractors and boats alone. There was nothing we could do, but he would keep us informed. He advised us to stay home. Some of the Mission were in trouble and our help might be needed later.

The video showed the predictable situation: the raging river and the flooded city with square miles of urban housing surrounded by water. The plane went lower, and the pictures became more dramatic. Every road was a swirling torrent. Some poor wretches sat on the roofs of their sagging grass-roofed huts, the children too forlorn to wave to the passing aircraft. One bridge, more than a hundred yards long, had washed away on the highway out of main city. Another, at La Belgica, even longer, had disappeared without trace, not even one of its massive concrete pillars remained. The River Piraí, normally a sand-banked corseted sluice 400 yards wide, had grown into a ravening monster ten miles across. One small town had ceased to exist. Noel's precious Botanical Gardens, so painstakingly assembled over the years, together with the population that ran the many restaurants to cater for lovers and botanists alike, had in a single night been obliterated from the map. The wealthy barrio of Equipetrol had been buried in sand; the roofs of what were once luxury villas now protruding here and there like pieces of flotsam.

We sat there stunned. It had all come to pass. The river, deprived of its protective mantle of trees, had given us all a message as free of ambiguity as a punch in the face: "Leave me alone, or else!" I sat there thinking about my report. The 'dreamtime' of the Aborigines was it possible? Could a man think a thing and make it happen? "Crede quod habes, et habes".

Later the reason for the river's belligerence was made known to us. The source of the tragedy was sixty miles upstream, but its cause was rooted in prehistory. Over the millennia the Piraí had carved itself a deep ravine, the opposing cliffs and rocky outcrops standing like the pieces of a fossilised chess set. The tangled woodlands on the upper slopes secured for the valley a brooding peace: a stalemate between the forces of gravity and the grip of the forest's roots. The Giant Sloths and Glyptodonts, the original inhabitants of the land, gave way to island-hopping Lemurs and rodents and later, isthmus-creeping land animals among which Homo sapiens came last. Homo commandeered a process of extermination which is still going on today, a desperate game of hide and seek in which, maybe, the only living members of the ancient past to survive into the future will be the modern Sloths.

Having taken all the easy meat, it was not long before these men turned to agriculture. They started exterminating the forests and sooner or later they were bound to stumble across the petrified chess set. The game commenced: man moves his first pawn, he starts to denude the cliffs of their trees. Queen's bishop to King's knight five, check, nature responds with the first landslides. Man clears the whole watershed. Nature answers: King's knight to take white Queen, Checkmate, a whole hillside of tomato and maize fields slides into the gorge of the River Piraí.

Destabilized by the continuous rain, several million cubic feet of rock and soil tumbled to the riverbed 200 yards below. There it came to a rest, a wall thirty yards high. A temporarily effective dam behind which the river gathered up its destructive force, all its pent-up fury. Then, unable to hold out, the precarious dam gave way to release upon the unsuspecting little town of Taruma a wall of savage hate, an unstoppable, all-enveloping torrent of rocks, mud, and water.

Taruma was swept away as casually as a man brushes flies from his face. Even the revered Virgin, protectress of the village, high up in her niche in the rock face vanished without trace. More than twenty families went with her.

The next in line was the vital road bridge spanning the river. This too was swept away in its entirety. With its loss Santa Cruz was deprived of its only substantial road link with the outside world.

Thirty miles downstream the Piraí overflowed its banks to return
to its ancient bed. The original settlers of Santa Cruz, those wise
descendants of Nuflo de Chavez, had built the city on the river's
archaic eastern bank. Less wisely, its present incumbents had
extended the city and much of it (including the wealthy district of
Equipetrol, the primary ring of the city's spider-web plan) now lay
below the bank, what in truth was an old ox-bow lake.

Even today, little has been done to control the deforestation of the
Piraí's watershed. The river's message of "Leave me alone!" went
unheeded. True, further downstream many of the "Damnificados"
(the "Victims") have been relocated to new barrios, and with the help
of the European Community, many millions of pounds have been
sunk into a scheme to protect the city. This project, Servicio de
Encausamiento del Rio Piraí (or SEARPI for short) has placed great
emphasis on the construction of many concrete canals in the
expectation of keeping the river between its banks.

Many Cruceñans, having demonstrated their faith in SEARPI, or
maybe, working on the theory lightning never strikes the same place
twice, have moved back to occupy homes devastated in the floods.
Some of these people may have had no option but, I find it difficult
to understand the confidence, Fatalism? shown by those who have
returned to the worst affected areas, for once again many posh new
villas are rising 'gram by gram' (my little jibe at the cocaine trade)
from the ruins of the old ones. I wish them the best of luck. (Après
moi, le deluge).

****

By March it was becoming clear to us that the "Sociedad
Ecologica Cruceña", that we had founded with the help of our friends
in December 1981 was beginning to collapse for lack of direction.
During its first twelve months the society had been very successful.
It had attracted more than forty regular members, and at the
bimonthly meetings that were held in the "Casa de Cultura",
interested members of the public and students from the University
augmented our numbers. These meetings consisted of the usual
minutes and notices, and a guest speaker to give an illustrated talk on
a relevant topic.

In January 1983 the society held its first annual election.
Unfortunately, the core of the group, those who had made the society
such a success, was unable to stand for office, because they were due

to leave Bolivia. When we arrived at the venue chosen for the elections the number of people present immediately struck me; normally, as I have said, about forty people would attend the meetings, here was about one hundred. Moreover, someone had planned for the increase by providing a much larger room than usual, and with a commensurate number of chairs. I couldn't believe my eyes. Who were all these people? I expected the outgoing committee to say something about what appeared to be the presence of so many non-members, but in the event, nothing was said, and the ballot went ahead without delay.

Roberto Unterladstatter, the Dean of Agriculture at the University, won an easy victory, and Victor Callau, a well-known radio broadcaster, was elected Secretary. What surprised me was the nomination for Vice-president, a man I knew well from the office, and one, as far as I was aware, who had never been to a society meeting before. As head of investigations at CIAT, this man, Oscar Paniagua, had he really been interested in ecology and conservation, could have championed my fight to introduce a more appropriate dogma into the institution, but he was instead one of my chief critics.

You may, then imagine my discomfort when Oscar was elected, apparently swept to office by a strong showing among the strangers present. Something was wrong. Someone appeared to have manipulated the situation. The riddle was, who would invest the time and effort to rig the elections of what was, after all, essentially an amateur society? Who was it that felt so strongly that our little society had to be infiltrated and taken over by people of his own choosing? And what was to be achieved by doing so?

We rushed back home with a feeling of bewilderment. Gene, what had he said to me? Something about the society being torpedoed. I got out the original notes and found it right at the end, "Pepe Roig probably the man who will ruin your Ecological Soc, and back near the beginning: Pepe Roig frontman for Onishi/ex politico Falangista/ pays off local newspapers, no stories/ also S.Cruz thru CDF lawyer (name?) paid $500 mo. for his services.

There it was. Yet another bad night. I went to the fridge for a beer and broke out a fresh packet of cigarettes. Ecology and conservation, protection of animals and birds, bird and animal exporters, Onishi powerful interests rigged elections. Victor, news journalist, Pepe Roig, unexplained hostilities at work, Oscar Paniagua of CIAT, hostile dogma, ecology and conservation. The perfect circle?

Unterladstatter, did he know? Was he in it, or was he, too, a dupe? Victor? No. I couldn't believe he was Roig's man, but dupe, yes. Gene had told me he was suckered by a man called Vladimir Ribera when Victor started up a society for the protection of animals.

By March, it looked like our fears had been justified. The new committee had not held a single meeting, certainly not a public one, and from what I gathered from Victor, not amongst themselves either. In fact, the society never came together again. We had been routed. Gene's first prediction had come true.

Today, amongst the founder members only I am left. Noel, who was reluctant to become a member, but who cast his benevolence upon the Society, died under mysterious circumstances in his beloved Huanchaca National Park; a story I have yet to come to. My wife, Cheri, is gone. Gavan and Garvin the two English lads, student advisors of the British Mission left soon after those fateful elections. Marilla, my Spanish teacher, and her husband, Florencio, struggled on for a time, but unable to make a dignified living as teachers, they left for Buenos Aires; may they found peace and happiness there.

Victor Callau disappeared without trace. Neither his family, nor any of his friends knew where he had gone. A well-known local personality and candidate for the central committee that administers the city of Santa Cruz had vanished off the face of the earth. Notwithstanding a massive public appeal for information, nobody knew if he was alive or dead. Some believed he might have been aboard a light aeroplane that disappeared without trace. The plane, on its way to Charagua (several hundred miles south of Santa Cruz) was assumed to have come down in the vast forests of the Chaco. In spite of the untiring efforts of both the Bolivian and the Brazilian air forces it was never found. Rumour has it that many of these missing aircraft lie in the remote fastness of Amboró. I was really worried he had been liquidated.

Who's next?
And the little one said, roll over, roll over,
And they all rolled over and one fell out,
Till there was nine in the bed,
And the little one said,
Roll over, roll over.

****

To coax my memory back to the events of early 1983, a period I have already stated was one of the most depressing, presents me with unforeseen difficulty. It is said that we forget what we want to forget, and maybe it is true, for here I sit smoking far into the night, my table lit by candlelight, my pencil poised and sharpened, trying to cross over the void which separates me from the events of that period as surely as time separates me from memories of last Sunday's lunch. However, if time is the great destroyer of recollection, writing about it is the overgrown path back.

And yet memory is such a strange beast, for I can remember from my childhood the charm and the toil of many a country ramble better than a field trip to Peru only two years ago. I clearly recall the details and the excitement of finding my first Lapwing's nest, the beauty of a Ground Beetle, its violet sheen contrasting with the chalk-white path. I happily recall the wearisome slog up the hill to my childhood home, a hedged-in lane to the summit of the moorland three miles away, and on my paper round a tedious track across a large field of stubble. I remember: "Spud Guns", "Conkers", Gas masks and "Sherbet fizz."

But what did I do in Peru? Part of the reason, doubtless, resides with our mood at the time; however grandiose the mountain peak, however esoteric the ancient ruins, or awe inspiring the raging torrent, a happy heart remembers well, a fearful one forgets.

In South America the humdrum of the English countryside, its safe intimacy, is replaced by the slight tension of, say, a séance. One becomes aware of something a little threatening, strangely impersonal, mystical, as unexplored as the spirit world itself.

Here, there is no mossy hummock to recline on in the sun's warming rays. Here, if you sit on the grass you get bitten by ants, or pierced by thorns, or you pick up ticks, or whatever but, it's rarely restful. A walk in a tropical forest can be monotonous, as it rarely offers the relief of splendid views, even when the land is hilly the foliage hems you in. Raising one's spirits depends upon the inner peace to be wrung from the allure of an umbral glade, or the cathedral quiet of huge trees, or even the gay flippancy of courting butterflies, in other words, the catharsis of one's fears.

And so to return down memory lane, that begins way back in the innocence of a leafy glade of an English past and ends in the uncertain menace of Bolivia's social jungle.

****

By May 1983 my position at work had become untenable. The
Falkland Islands War, or more particularly the sinking of the
Belgrano, sent shrapnel to lodge in the body of tolerance that existed
between myself and my fellow countrymen, and between all of us
Brits and some of our Bolivian colleagues. The majority of Bolivians
kindly hid their personal sympathies over the war behind the good
humour of a shared joke. One I thought particularly apt saved me
from any serious discourse on the subject: "To gain a honourable
peace, and in exchange for the Falklands, Mrs Thatcher has offered
the Argentinean people Ulster."

With truly difficult numbskulls I would talk seriously about
turning the Falklands, the "Islas Malvinas" as the South Americans
call them, into a wildlife sanctuary for penguins. This specious
argument, coming from a recognised conservationist, usually
deflected any tendency for the hostile to turn openly belligerent. But
you had to watch out for those stealthy rock hoppers fishing for a
fight.

The British Embassy took the situation seriously enough to warn
us all to keep our heads down. Apparently a few mindless patriots had
made threats against some of the British residents. If so, there must
have been rather few of them, for the Bolivians are usually slow to
take sides in any argument. The Embassy told us not to go out, or if
we had to, to avoid public places. Then, after it was known that threats
to bomb houses occupied by Britons had been made, the Embassy
told us not to stay in.

This was followed by a so-called 'incident' in which patriotic
Bolivians went to the house of one of the British geologists and
knocked him around a bit. Then it transpired that this had rather less
to do with the Falklands and rather more to do with someone's sister
he'd been abusing.

****

More seriously. Yes, let's have something military, I did notice that
our house was being watched. Not round the clock, just now and
again. We saw one or two shifty characters hanging about in the
shadows under the trees, and once or twice they took snaps of our
house and our visitors. On one occasion, we took one of them by
surprise and nearly caught him. In retrospect we should have made a
real effort to find out what it was they wanted, but in the event our

will to do so was no stronger than our half-hearted pursuit of the one who got away.

Some of our friends thought the snoopers were casing the house. Burglary had become a booming business, and nobody could know that inside our expensive British-paid-for villa there was nothing left to steal; even poor Don Juan had disappeared by this time. The newspapers reported a number of serious incidents in which a gang of escaped convicts had combined robbery with rape and murder. It was alleged that the Governor of the local prison regularly furloughed a number of these on the condition they return with something for him.

Personally, in my case, I didn't go for the robber theory. These snoopers were too persistent and seemed much more interested in our friends and us than they were in the contents of the house. In fact, there were a number of people who might have had reason to spy on us. Apart from the Animal Mafia, it was quite likely we had aroused the interests of the Narcotic's Police because; we had a strange crowd of friends, and Cheri brought home her daily quota of waifs and strays as a cat will bring home dead mice.

So, at the time we thought these "Watchers at the Gate" were more likely to be connected with the narcos (whom we had no reason to fear) than the animal dealers (whom we did). After all, it had been about eight months since I'd done any active snooping myself and, if I had been rumbled, we would have been spied upon then, not now when lying low. How wrong we were was to be proved later. We had forgotten the report to New York.

****

Inflation at this time, and up to the end of 1985 not only made it impossible to sell things, it also made it difficult to buy things. When we first arrived in Bolivia the exchange rate was twenty pesos to the dollar and the largest note was the one-hundred one. In the days before inflation two people could go out on the town with fifty pesos in their pocket, eat at a good restaurant, drink in expensive bars, catch a taxi home, and still have change for a tip. The town was a boisterous, rollicking playground for anybody with a little money, and the population as a whole, at the time seeming to sense the presence of dark clouds on the horizon, threw themselves into the holiday atmosphere with the abandon of Cook's crew back from a voyage. Bright eyed and bushy-tailed, old and young alike, they filled the restaurants, bars and nightclubs with their noisy chatter. Pretty girls in bright new dresses vied with each other for attention. Life was one long carnival, relationships casual, goodwill and bonhomie expressed in every face. Shoeshine boys, street vendors, taxi drivers and whores thronged the streets to cater for the all-night revellers. Every morning was the one after the night before. Cures for hangovers became a universal topic of conversation. An article in a Colombian newspaper, "El Siglo", described Santa Cruz as "El Nuevo Sodom y Gomorra".

By 1983 the storm clouds that had hovered over the horizon lay directly overhead. The happy-go-lucky swings and roundabouts of fortune turned to misfortune, the only free ride the roller coaster of despair. The economic tide of disaster scoured the joy from men's hearts: the jolly, vibrant social structure swept away as suddenly as the adobe houses in Taruma. And, as if in parody of the climate of depression, the real weather unleashed a torrent of rain that left the festive atmosphere in ruins.

By the end of 1985 the dollar was worth nearly two million pesos. Any attempt to calculate the astronomical rate of inflation was as meaningless as counting the stars themselves. The population turned surly, suspicious and mean: and many didn't know where their next meal was coming from. Salaries were so low (twenty dollars a month for a teacher) that many normally honest citizens turned to petty crime. Most transactions were made in American dollars, and dollars could be freely exchanged in the shops and streets, a legacy remaining today. A new profession sprang up, the "Cambista", men or women who would openly change money on the street at black market rates. The situation had gotten so bad, a shopping trip required two people: one to carry the purchases, the other to carry the money. As the largest note would barely buy a box of matches, everybody walked around with plastic bags, or for more expensive items, holdalls filled with the useless peso bills, the queues at the pay desks were enormous because of the time required to count them. The streets were littered with discarded money nobody bothered to pick up.

The value of the peso, changed with dollars in the morning, could be halved by the afternoon. The owners of stores couldn't keep up with the zeros and many assistants gave you more change than the value of the money they received. A lot of shopkeepers gave up trying to keep track and simply put down their shutters for good. Others stayed open but were reluctant to sell anything. Lots of people went rushing off to settle old debts; those who owed small fortunes now repaid them at the drop of a hat. Public utilities gave up sending out accounts. The Post Office had trouble with stamps: a letter sent to my family required thirty-two, and even by overlapping rows of stamps the envelope was too small.

Desperate to keep up, the Government went on issuing more and more inflated bills. After the 50's and 100's issued in 1962, came the 500 in 1981, the 1,000 in 1982 and then in quick succession: the 5,000, 10,000, 50,000, 100,000, 500,000, and finally the million-peso note. Bolivia had more millionaires than water closets. Every day, it was reported, the "Banco del Estado" received a plane load of new notes from the printers in Europe. By the time they were unloaded they were almost worthless. Bank notes gave way to "Cheques de Gerencia": poor quality replacements.

Bolivia, the archetypal country of the coup d'état, suddenly found a new form of political peace. After 160 years of independence and 180 changes of Government, nobody wanted to be President. Siles Zuazo was obliged to continue as leader of the ruling coalition. Labour unrest, a few isolated bomb strikes and food riots, and roads blocked by protesting Campesinoa became the order of the day. A Government minister officially declared that a state of anarchy existed. An article in The Wall Street Journal (2nd March 1984) said: "The coalition was run by three obstinate old men who have hated each other for thirty years." And, at 69, Siles was the youngest of the trio!

Smuggling goods into Bolivia had always been a traditional activity. Trains laden down with electrical goods, whisky, food and clothing arrived daily from Brazil and Argentina. During Siles' reign, severe price controls led to a reverse of the flow: gasoline, tyres, cigarettes, food and drink were smuggled out. Some employees of the state-owned petrol company, the YPFB, drove fully laden tankers across the border and retired for life on the profit. Even before the worst of the inflation, The Wall Street Journal (ibid) reported: "a gallon of petrol that cost seventy-five American cents could be sold in Brazil for more than four dollars."

These, then, were the difficult conditions under which we struggled. Sudden social upheaval aggravated by an impotent governing coalition. The Americans, jumpy as ever, were even talking about Civil War and, even more unlikely a Moscow-backed communist takeover (The Wall Street Journal, ibid). Sitting in my vantage point of Santa Cruz I could not see how these statements bore any relation to the character of the people.

In recent times (1954) the "Ucureños" incident, that was essentially made up by a Government-backed peasant militia, is the only event remotely approaching a Civil War. In order to stifle an incipient independence movement, this rabble from the altiplano was loosed upon the poorly prepared citizens of Santa Cruz: a shameful incident rarely referred to in the history books.

The stolid stoicism of the bulk of the population does not predispose them to armed conflict. Unlike Peru, Colombia, and most of Central America, Bolivians are generally a peace-loving people and guerrilla organisations of any real account are unknown. John Hewko in The Wall Street Journal (28th April 1989) had this to say: "Fortunately, Bolivia has been spared the horrors of rural guerrilla movements, in part because of its 1952 agrarian reform. The reason for this is very simple and should be a lesson to land reformers everywhere, peasants who participated in the Bolivian land reform were given individual titles to redistributed land; titles were not retained by the state, as was the case in other countries in the region."

This contributed directly to Guevara's failure (and demise) in the 1960's, when he tried to convince the Bolivian farmer of the benefits of state-owned agriculture, and it has also prevented the penetration of Maoist terrorists from neighbouring Peru.

The Agrarian Reform Hewko refers to was hurried through Congress to quell an uprising by twelve-thousand Quechua Indians who seized a seven-thousand-acre estate in the town of Ucureña, not far from the city of Cochabamba. That these brave Ucureños were subsequently identified with the rabble that pillaged Santa Cruz two years later is as unjust as it is ironical.

The duality of the Bolivian national character: the Latino's fierce independence inherited from his Spanish forebears and the docile acceptance of the peasants...their legacy from the Incas is too unlikely to prove compatible with communist philosophy. As long as those who might stand to gain from it (the highland Indians) remain stubborn and backward. Imported ideologies will have little impact for some time to come. Certainly, the future holds surprises in store, but everything I have seen suggests that the words of South America's great liberator, Simón Bolivar are nearer to the truth: "America is ungovernable. He who sows a revolution ploughs the sea". And, as it is said, "Today is fiesta, Mañana siesta". And, in truth, Bolivia approximates a Gilbert and Sullivan opera more than it does a revolutionary Republic. This is why it still retains a certain charm.

****

On the 8th January 1983 the opening skirmish against the animal exporters was made by PRODENA ("Pro Defensa de la Naturaleza", a La Paz based conservation society) in a letter to Presencia (generally considered to be Bolivia's most influential newspaper). Corresponding to documents I had given to Conway, the article denounced the illegality of many of the exports and referred to four particular shipments made by Onishi's firm ("Zoological Garden Suppliers") that contained a total of 6,450 birds, all protected species. It also referred to three Los Angeles based firms that had imported protected species of birds: Pacific Bird, Polly Imports and MPM-Bird Company.

This article was followed by a three-month hiatus. It was not until the 25th March that a follow-up article appeared (based on an interview with the CDF Santa Cruz) to explain how the CITES registration worked. It identified Noel Kempff Mercado as one of the authorised signatories of the official export permits. The article included denials by the CDF that they were allowing the export of protected species.

On several occasions, I got together with Noel to prepare the groundwork for the establishment of Amboró National Park. The area had been legally protected since 1973, when the government of General Hugo Banzer Suarez named it the "Lieutenant-Colonel German-Busch Nature Reserve" with Decreto Supremo 11254 20th December 1973. (Typing out this long name appears to have exhausted his Government's energy, for after the decree was signed the Reserve was abandoned to Campesinos, loggers and hunters of both animals and gold.) As a rider, the decree stated that on further consideration, the Reserve might justify the creation of a National Park in the future. That further consideration was what I had been doing.

The opportune moment to discuss the trade in animals with Noel did not immediately materialise. Only once did I raise the issue when I suggested to him that it would be better if he called for the whole shabby business to be closed down. I went on to ask him how the licensing system worked and if he was always consulted.

"Yes," he said, "I have to sign the export documents to indicate that none of the animals and birds listed was wanted by the zoo." In other words, he had the right to expropriate anything the zoo required for its own collection.

Then Noel followed this remark with one I never forgot: "the zoo depends upon confiscated animals for its exhibitions."

What he did not refer to was the power invested in him to stop a shipment being made. The CITES permits were clearly signed by him in the space left for Stamp of Wildlife Inspector. I saw he over-stamped these words with "Director, Zoo Fauna Sudamericana." To me it looked like Noel was unfamiliar with the CITES system and maybe in good faith was signing the documents in the belief that he only represented the Zoo's interests in the matter. Why, otherwise, would he over stamp his signature in the way he did? Surely it should have been franked with the words "CITES Inspectorate" or something like that. Was it possible the exporters knew of his misconception and used the prestige his signature gave any document to continue their illegitimate activities unmolested?

Curious and more curious.

As so often in my early relationship with Noel, I had come away with something of value, but his diffidence discouraged me from questioning him further. I needed to dot the i's and cross the t's; without doing so, there could be many misunderstandings that might inadvertently make enemies of my friends and friends of my enemies.

****

By this time an affair that came to be known as the "Texas A&M Scandal" was about to come to the attention of the public. I had written about it in my report to Conway, but at the time I didn't remember this.

Gene had accurately forewarned me of the impending scandal during that long night at Don Miguel's: "McAlpin + Richard Barnley: pretend represent Texas A&M University — rented Brownsville Quarantine Station (owned by Gregory?)/landed two shipments from Bolivia before Dec. (1981?)."

I said I had retired Travis McGee, and I did on an active footing, but when a casual opportunity presented itself, I took advantage of it. On the evening of the Barnley debacle, Cheri and I had gone to the New Orleans bar for a pre-supper drink. The bar, ENGLISH SPOKEN, GOOD DRINKS, BEST IN POP & COUNTRY MUSIC, was one of our favourite venues. When we got there, Mr Barnley, at heart a good man that needed a job, was talking to John Dinn, the handsome, tall Texan landlord. John knew we were compiling information relating to the trade in animals and birds, and being a bit of a nature lover himself, approved of our efforts. We confided in him because John knew just about everybody in town (and their business) through the loose chatter and gaiety his place encouraged, and from time to time he passed us valuable titbits. Something about John, perhaps the friendly shine in his blue eyes, or maybe the inviting curve of his droopy moustache, compelled people to divulge their sins to him more readily than they would to a priest in a confessional.

So, imagine our surprise when we realised we were talking to Richard Barnley. As soon as we sat down John brought him over and introduced us. As Richard turned towards Cheri, John gave me a covert wink behind his back, and said to him, "Why don't you talk to these nice folks, you'll find some interests in common." To Richard, his words were meant to imply that John had things to attend to (the bar was crowded) and wished to take his leave. To us they meant Richard was something to do with the bird trade, so tread carefully.

After a moment chatting about nothing in particular, we got around to asking him what he was doing in Bolivia, a loaded question in a country where too many people have something to hide. He said he was a zoologist (maybe this was not the word he used) and that he was working for a Texas A&M research programme. As I said, at the time I did not recall Barnley by name, but after a bit more wheedling I gathered he was here to obtain the birds needed for the programme. It was at that moment: Cheri, who was making signals to me, had gotten there first but, I suddenly remembered there was a Barnley working with McAlpin. While I digested this, I switched the conversation to our efforts in Amboró, a ploy meant to give him the impression we were not especially interested in his work, and gave me the opportunity to consider the best way to handle him. He expressed a genuine interest in our project and after half an hour or so, the bar having become too noisy for serious discourse, we invited him home for a bite to eat.

I clearly recall our discussion when we got back to the house, for it ended for him in high drama, and for me in unexpected black comedy.

Back at our house we took up the conversation from where we had left it. From what had been said so far, it was evident that he remained unaware of our conflicting interests. During the long wait for supper, we became more relaxed, our tongues loosened by the lagers we had drunk. Even so, Richard's mood swung around like a weathervane in a force four gale. I gradually steered the conversation back to his work in Bolivia. McAlpin's name was now mentioned for the first time, and Richard categorically stated that he was here to hunt down the macaws for Exotic Birds Research Associates. From the way he talked ("I'm supposed to... they want me to... I'm expected to ...") he appeared to have serious misgivings about his part in it all and was unable to hide the symptoms of his apparent neurosis: fidgeting with his shirt buttons, playing with the table lighter, swirling the beer in his glass. It suddenly dawned on me the man was scared. What of? I was not sure.

I thought he would be more forthcoming after supper and tried to change the subject again. Patting at air, I signalled to Cheri to calm down, for she had begun to bristle with anger, and looked as if she might bury the ice pick in Richard's throat at any moment. Refusing to acknowledge my signals, which had become anything but covert, she now demanded to know how many Macaws were needed for the project. "Several thousands," he croaked, to which Cheri jumped up from her chair and started screaming at him, "You crook, You crook! You bloody bastard."

I managed to restore a moment's fragile peace and forestall Richard's immediate departure by humouring him with what I said was "Cheri's unsophisticated attitude". I thought I had saved the situation, and to my relief, our maid entered at that moment and put the supper on the table. But Cheri, stung by my duplicitous comment, rushed to the office and brought back a copy of our report to the New York Zoological Society that she thrust into Richard's face, stabbing her finger into the relevant paragraph, which he was forced to read:

"Gordon McAlpin posing as a representative of Texas A&M University and pretending to be a rich benefactor of research into diseases of Tropical birds and animals. Employed Richard Barnley (English?) to be his agent here."

By now his complexion had turned grey and without any warning he jumped up from the table, where he had only just sat down, and ran out of the house. I never saw him again, but wherever he is I wish him no ill.

I was angry with Cheri, for I was sure Richard would have told us a great deal more had she not frightened him off. The evening turned sour, and we launched into one of those violent quarrels that had become tiresomely frequent of late. She stomped off to the bedroom. I sat down to a supper that had gone as cold as our marriage. Whilst I picked at my plate, I went through everything Richard had said. I decided the time had now come to discuss the whole export issue with Noel.

****

I went to see him first thing in the morning before the zoo opened to the public , and when I knew he would be alone in his office. I strolled through the deserted gardens and lingered in the giant aviary to admire Onishi's silent group of pink flamingos. I used these blissful moments to compose my thoughts and to remind myself that all these enchanting creatures were worth fighting for. I could not imagine a life devoid of bird song or the quiet rustle of small mammals in dry leaves. I had to find out where Noel stood and somehow persuade him to publicly denounce the animal dealers and to promote Amboró.

Noel was sitting behind his desk squinting over the typewriter at a manuscript he was revising. He waved me forward with cheery impatience. As I sat down, I was pleased to note that he seemed to be in a good mood. It was some time since I had seen him and given that I liked to start our meetings with something to please him, I handed over the finished bird list I had been preparing for his book. He congratulated me on the speed with which I had completed the labour. "Estupendo, hombre," he said.

Over a cup of thick black coffee, I related the incident with Richard Barnley and what I knew of the McAlpin's Texas A&M project. By repeatedly nodding his head he indicated he already knew about it. Then I took out some of the documents Gene had given me, those bearing Noel's signature. While he quickly scanned these (peering over the top of his spectacles in a way that clearly indicated he had seen them before) I told him I believed the business was ruining his reputation, and I thought he should explain to the public his part in the licensing procedures. I mentioned that quite a lot of people, including visiting gringo biologists, assumed he was behind the racket. I said this in a way that clearly indicated my personal support for him. I told him, as he let me run on, the best way for him to clear his name was to publicly denounce the trade and call for it to be closed down. I gave him a brief summary of the legal infractions each one of the exporters was guilty of. As I referred to Onishi, I passed Noel a book he had lent me: Forshaw's "Parrots of the World". Inside the cover Onishi had written something like, "To my good friend for his valuable studies", I did not remark upon this, and there I stopped.

Noel looked angry, or maybe not angry, but wary. His good humour, or so I thought, had evaporated. He accepted a cigarette. As I passed it to him, I could almost see the thoughts racing through his mind. His baleful glare lost its hold on me and was momentarily transferred to the page he had been revising. He asked me where I had got hold of this information. I answered him truthfully. He went on to speak badly of Gene and emphasised that in his opinion Harris was the worst of a bad bunch; he repeated what I already knew about his deportation. He mumbled something about gringos being behind a lot of the deals. In a rising voice he then laid into Gaston Bejarano, the Chief of Wildlife in La Paz. Gene had already told me Bejarano was deeply involved in the illegal side of the animal export trade. He went on to explain that Bejarano was determined to cause him as much trouble as possible because Noel had denounced him over his part in the export business. As far as the documents were concerned, he assured me, his signature had been forged. No mention was made of his official seal. He asked me if he could keep the documents. I said he could, I had, I said, already photocopied them.

He told me to leave the whole matter in his hands and affably came around his desk to pat me lightly on the back. He thanked me for the obvious trouble I had been to and once more told me to leave the matter with him.

Just before I left, after shaking his hand, I remembered another piece of good news; a well-known American bird photographer was going to pay us a visit. He hardly seemed to register the fact and shuffled back behind his desk in apparently revived spirits. Mine were too. I was happy to have gotten it over with, and though I still harboured some doubts, I felt Noel had been telling the truth. Whether or not he would take any action against the trade or do anything about the Texas T&M project, I did not know. Only time would tell.

The 'Texas A&M Scandal' had its origins in early 1981 when a wealthy sixty-nine-year-old Texan known as Gordon McAlpin, of Exotic Birds Research Associates Incorporated, based in Cuero, Texas, conceived an idea for a moneymaking scam. This jovial ex-oilman, rancher of Brahman cattle, approached the Gabriel Rene Moreno University in Santa Cruz with a plan as badly thought out as it was illegal. His proposal was that the prestigious Texas A&M University, through its College of veterinary medicine, should get together with the University in Santa Cruz and himself to study the possibility of establishing a commercial Macaw-breeding Centre.

Here I copy his undated proposal under the heading, "Participation of Exotic Birds Research Associates, Inc.":

1. To construct and pay for a building on the campus of Texas A&M University suitable for housing and maintaining numbers of parrots and parrot-like birds being used in the research to be carried out by Texas A&M University.

2. To construct a facility near the campus of Texas A&M University suitable for experimental work related to reproduction of parrots in captivity. This facility to be of a size to adequately qualify as "commercial" in scope as one phase of this study is to determine whether parrots can be made to reproduce in a large operation. This facility is to be supervised as an experimental and teaching aid by Texas A&M University.

3. Exotic Birds Research Associates, Inc. is committed to acquire free of any cost to Texas A&M University permits, where available, for capture or purchase, maintenance in quarantine, condition and export from host countries the birds required to fill Texas A&M University's research program.

4. Exotic Birds Research Associates, Inc. has agreed to furnish funds as required for travel of Texas A&M University personnel working on the parrot research program.

And, curiously, in spite of the letterhead announcing him to be "G.T. McALPHIN, DIRECTOR", was signed "G.T. McALPIN". Apart from a natural hesitancy to confide in a man who is not sure of his own name, the document would at first sight appear free of any nefarious intentions. Looking more closely at it one cannot help but be struck by Mr McAlpin's apparent altruism, for nowhere does it state in what way he would benefit, on the contrary, his company declares its intention of providing the (not inconsiderable) funds required for the venture.

To anybody conversant with the background, the proposal would look highly suspicious. All the more surprising, then, that the Dean of Veterinary Medicine at Texas A&M should write a letter (8th June 1981) to Mr McAlpin expressing his extreme interest in the project proposal.

How could anybody in a position to know better take McAlpin's proposal seriously? What does he mean by parrots and parrot-like birds? Why does a research programme to study methods of breeding these birds need to be on a commercial scale? How many birds would be required to fill the research programme? How near is near, in point two? And who would control the disposal of the (by inference) large number of birds produced? And in the event that Mr McAlpin's enthusiasm led him to import far more birds than needed for the programme, what would become of these birds? If everything were truly on the level, why does Exotic Birds Research Associates undertake the difficult business of obtaining the export permits, when it could be done so much more simply under the aegis of the two Universities?

On the 24 May, after a newspaper article based on my report to the NYZS, El Mundo published a statement written by the public relations department of the Gabriel Rene Moreno University (UGRM). It filled one quarter of an entire page (El Mundo was big format) and was set out in bold print. The announcement stated that the University had, indeed, been prepared to participate in the project, but had now been made aware of several facts that gave rise to doubts regarding McAlpin's intentions. It categorically stated that the parrots to be exported were exclusively for research purposes; that Exotic Birds Research Associates, Inc. had obtained export permits without the assistance of the UGRM, and, moreover, contrary to the University's instructions to the Ministry of Agriculture (read CDF La Paz) to cancel all transactions. It went on to say that Exotic Birds Research Associates (in Spanish it looks grandiose: "Corporación de Investigadores en Aves Exóticas") had paid the CDF a little over fifteen thousand pesos for the permit (Resolución Ministerial No. 210-11-82) to export the birds. It mentioned that two shipments of birds had already been sent to Texas A&M (as Gene had said) and that on the 22nd March the Dean of the Agricultural Department at the UGRM had written to the Texas A&M warning them that this had been done without their knowledge; a letter to which they had not yet received a reply.

The following day a cartoon appeared in El Mundo by a Señor Rod Bal that summed up the public attitude to the whole business. It showed two aeroplanes leaving Bolivia, one full of parrots destined for Miami, the other with monkeys for Japan. The air above the two planes were covered in question marks, as much as to say: "What is going on?"

These public announcements were very interesting in that they confirmed Mr McAlpin's nefarious intentions. It was obvious he had not made it clear to the UGRM that the project had a marked commercial flavour. The withdrawal by the UGRM was also very significant, at least somebody at an official level had rumbled Mr. McAlpin.

Maybe this had something to do with a letter William Conway wrote to the President of Texas A&M University on the 15th October 1982, just one month after our interview with him:

"Dear Dr. Vandivar",
"Enclosed, please find materials made available to [NYZS]. In brief, it is contended that your university has entered into a cooperative agreement with the University Gabriel Rene Moreno of Bolivia to study macaws in various ways and that in doing so large numbers of endangered species will be captured from nature."

A paragraph follows expressing his scepticism regarding the project's value; and continues:

"We ask that you investigate the propriety of this project from the standpoint of your university and that you make available the name of those [T&M] faculty members who intend to participate.
Sincerely, William Conway."

The story appeared in international journals over the following months. The New Scientist for the 8th September left nobody to doubt the whole project was planned for McAlpin's personal profit: "But the only definite 'research' project so far is McAlpin's plan to breed the birds...some species are very popular as pets in the U.S., where a single Hyacinth Macaw can fetch as much as $10,000." It emphasised that his interest was in valuable macaws, not just plain parrots: "... McAlpin was able to secure export permits from Bolivia for up to 2,770 macaws." As was to be expected, it also referred to a copy it had obtained of an internal A&M memo to the effect that the macaws to be given to the University [to fill the research programme] would only be 12% of the total imported by McAlpin.

Another article, "The Parrots' Predicament" appeared in the September issue of the Texas Monthly and mentioned further, very significant details. It stated that 126 Hyacinth Macaws had been received by Texas A&M in the spring of 1983 (when I visited Gene's premises, I counted 125 in that one room. Were these the same ones? It mentioned that McAlpin's first dealings with A&M had been through Dr Dean Brown, professor of small-animal medicine and surgery. It stated that McAlpin had paid $40,000 to obtain the Bolivian permits signed by Benigno Rodriguez of the CDF in La Paz, whereas the UGRM in Santa Cruz had copies that showed he paid the equivalent of $5,000. Was the balance used for bribes to obtain the (strictly illegal) export permits?

(Dr Brown committed suicide in December 1982, was he driven to it by the impending macaw scandal?)

Throughout the many newspaper articles and letters referring to the "Texas A&M Scandal", nobody mentioned Bolivia's point of view. When in late 1983 it seemed that the project was not entirely dead, I wrote a letter to Texas A&M and sent copies to the various organisations that had become involved in the affair. Apart from supporting the denouncements made by the conservationists, I submitted a counterproposal that I thought would entirely overcome the objections. I did not envisage the participation by Exotic Birds Research Associates, Inc. My proposal was a modest one, suggesting that Texas A&M centre a much smaller project in Amboró to be jointly run by the UGRM, Noel Kempff's zoo, and PRODENA (the Bolivian Wildlife Society that at that time I represented). One of my reasons for this proposal was to remind the hierophants of the avian world that future projects, no matter where, should consider the interests of the country supplying the birds:

"I would also like to add that a locally based research and breeding unit would benefit Bolivia as a whole, providing some jobs and educational facilities as well as bringing in much needed foreign exchange through tourism and the sale of birds officially. It could be said that every bird bred in captivity outside Bolivia is a loss of revenue to Bolivia. Why should Bolivians cooperate with such schemes like the Texas A&M one?"

By reminding them that any profit to be made would have to be shared in this way, I hoped there wouldn't be any future plan. I really did not expect many replies, and only received any reaction at all after I was passed a copy of a letter addressed to PRODENA in January 1984. It was from Texas A&M stating that they had received a communication from Mr Clarke in Santa Cruz, but the university did not have the resources to implement the project in Bolivia. "Maybe," it said, "Mr McAlpin, a private citizen, might support the work."

Looking back on events, I saw that a lot of the potential to corrupt the system for exporting live animals lay in the regulatory procedures. Permission to export wildlife was given in La Paz by the Minister of Agriculture through Ministerial Resolutions. As a department of the Ministry, the CDF's sole responsibility was to ensure the shipments contained those animals listed in these Resolutions, and that the number of animals did not go beyond the specified quota. The CITES documents which Noel signed apparently under the impression he was only acting on behalf of the zoo was the ultimate authorisation, and this was the document of interest to the importing country. By CITES rules any protected species appearing on these forms had to be returned to the country of origin. The fact that this was rarely done was a matter of practicality, who was going to organise, and pay for, their return? And, after their return, what was to become of the, by now, semi-tame animals? In the case of a few score, they would probably end up in the zoo. But what could be done with the tens of thousands of animals and birds illegally exported from Bolivia each year?

The system was probably as good as it could be and worked where the registering authorities, the CDF through the Ministry, the Airlines and the Exporters themselves, played fair. When this was not the case (as was obvious in Bolivia) the CITES system demonstrated its fallibility. Another weakness was that participation in CITES was a voluntary commitment; some countries were not signatories to the agreement, and those that were not made capital out of those which were, especially if they were neighbours. Just to give one example: Paraguay, which was not a CITES member at the time, openly exported wildlife smuggled across the border from Bolivia.

Noel, like the CDF itself, accepted a Ministerial Resolution as the principal guiding authorisation, "a fait accompli". The fact that many of these had been issued contrary to the law, procured through bribery, surely annoyed him; his remarks to me about Bejarano proved this. But in a country where almost everything, and maybe to him more important issues, was tainted in the same way, he probably would have told himself that it was impossible for one man to do anything. Add to this his natural desire (here I am speculating again) to remain aloof from the whole sordid business, and that he was a very busy man with many important responsibilities, one can see why he acted with apparent hesitation.

His error was that he had agreed to participate in the CITES system, albeit due to his miscomprehension of what that entailed, and his apparent failure to recognise that he was not alone, for a good part of the educated citizenry looked to him to correct the situation. But as it was, Noel's failure to call for a ban on all exports was disappointing to all of us.

In any case, on the 19th of May 1983, a picture of him seated behind his desk appeared in El Mundo, and he was quoted as having said, "... these things [the evasion of the rules and the corruption of the whole regulatory system] must stop..." and that "the shipments being made contain protected species."

Four days later, jubilant at this article, the Association of Animal Exporters addressed a statement to Public Opinion that occupied half of an entire page and was printed in huge letters. It declared: "the birds and animals referred to were not protected species, and therefore were not illegally exported; that shipments could only be made if they conformed to the respective Ministerial Resolutions and international agreements and were sanctioned by the CDF through its own wildlife inspectors."

It went on to say: "The honest exporter's point of view did not conflict with that of Prof. Noel Kempff Mercado, guardian of the national ecology." It ended (with reference to another article) "It was untrue to say that Bolivia was the only country making such exports; and in South America alone, Ecuador, Chile, Argentina, Peru, Guyana and French Guiana also exported their wildlife." [But note, no mention of Paraguay, where the Association had close links with the animal exporters.]

The newspaper editor added at the bottom that "the corresponding signatures of their clients were illegible, but belonged to identity card numbers 721623 (from Cochabamba) and 109001 (from La Paz)."

On the 4th of June another article appeared after an extensive interview with the President of the Association of Animal Exporters, Juan Lopez (the same man Gene had warned me was a dangerous, ignorant campesino). Now he was calling himself "Ingeniero", a respected title that was only used by University graduates and sometimes incorrectly by college graduates.

The article was basically a recapitulation of statements concerning the legality of the exports being made, but it also went on to justify them by pointing out: "The foreign exchange earned was to the good of the country, and two thousand peasant families benefited by earning money as animal collectors."

Señor Lopez, I thought, had made a big mistake. By his own words he had shocked the public by alerting them to the scale of the business. I was beginning to like him.

On the same day, none other than the Director of the CDF in La Paz, Ing. Benigno Rodriguez (the man paid by Bolivian society to protect their natural treasures) was quoted as stating: "What the Association of Animal Exporters was saying was correct, and even Noel had said a ban on official exports of live animals and birds would only lead to a contraband trade with a consequent loss of fiscal revenue."

It looked like the Exporters were going to win; Onishi's friends in La Paz (Bejarano et al.) and his pet Falangista, Pepe Roig, were doing a fine job.

Then came unexpected and most welcome support. It was the 5th of June—World Environment Day. The declaration occupied a quarter of a page (again in large print) calling for those charged with protecting the natural resources and the ecology of the country to do their job. It was no less than a "Presidential Address". "Compatriots," it began "This day, dedicated by all the nations of the World in contemplation of the effect that man's social development has had on the environment in which we live, has a special significance for all Bolivians because of the natural disasters [floods] that have caused huge losses in the Departments of the Beni and Santa Cruz, and because of the drought in the Altiplano".

The President, Hernan Siles Zuazo, went on to declare the importance of protecting the land from floods and conserving the wildlife, and asked that those in charge take stronger measures to control the exploitation of wildlife by smugglers, and that the population denounce these people to the authorities.

It ended with a call to preserve the forests from over exploitation and illegal logging and asked that the population help keep the urban areas free of rubbish.

Benigno must have shit his pants.

On the day after the Presidential address, back came the Animal Exporters with yet another public announcement (outdoing the President with their usual half-page format) to say they were totally behind the President; that clandestine animal exporters were giving their trade a bad image, and they would never export animals in danger of extinction. They fully supported the adjudication of Prof. Noel Kempff Mercado, and others responsible for the protection of the environment. This time it was signed "La Directiva CI 1466566."

A week later, clearly in a bid to place the whole matter on the level by making the appropriate innuendoes, the Exporters were at it yet again with another half page full of bold print in El Mundo. It began with an outline of CITES regulations and included the odd statement: "the CITES convention did not aim at stopping the trade in the fauna and flora, but to rather to regulate it to the advantage of the fauna and flora."

It went on to refer to a 1975 law that went a long way to protecting vulnerable species. The rest was a repetition of what had gone before, but included another odd statement, that to paraphrase, read:

The Bolivian wildlife belongs to the people, and if the humble Campesinos (living on the frontiers of the country) are not allowed to exploit it they will have to abandon the countryside and invade the city of Santa Cruz in search of a living.

The author? Guess who? Yes, Juan López (without the Ing. this time) was threatening the Cambas of Santa Cruz with a massive immigration of campesinos into their city unless he was allowed to go on exporting animals. Gene said he was a fox. But he also implied he was an idiot!

It would serve no further purpose to illustrate my account with even more articles that appeared in the local press, and there were quite a few. If those referred to have served to do anything, they should have demonstrated that the issue was developing into a campaign to close down the exports. This was formally proposed by a group of Parliamentary Deputies in an article that appeared in "La Presencia" of the 15th August. That the campaign was not immediately successful cannot be blamed on Noel's hesitation or the failure of those actively fanning the flames.

****

I, at least, had other things on my mind. So, if you will indulge me in a little narcissism, I will show you a letter I received a good deal later than the date on the letterhead:

OVERSEAS DEVELOPMENT ADMINISTRATION
Abercrombie House, Eaglesham Road
East Kilbride, Glasgow G75 8EA
Mr ROS Clarke, c/o British Embassy LA PAZ
Our ref: SA/P 23283   Bolivia 14 July 1983
Sir
    I am directed by the Secretary of State for Foreign and
Commonwealth Affairs to refer to your appointment on loan to the
Government of Bolivia under Technical Cooperation arrangements,
the conditions of which are set out in this Office's letter dated 23
September 1980 and in the Memorandum accompanying that letter
and in this office's further letters dated 23 October 1981, 19 August
1982, 25 August 1982 and 7 February 1983 and to inform you with
regret, that it has been decided to terminate your appointment.
    In accordance with the terms of condition 20(1)a of the Memor-
andum referred to above, you are hereby given 2 months notice of
termination of your appointment, this period of notice to begin on the
day on which you receive this letter. You will also be granted the
privilege leave for which you are eligible under condition 18 of the
Memorandum referred to above.
    I am to request you to acknowledge in writing the receipt of this
letter by signing the acknowledgement below and returning this letter
to this office. A duplicate is enclosed for your retention.
    I am, Sir, your obedient Servant,
    P W Little
    Personnel Services Executive

And, whilst we are talking about narcissism, let me make
reference to just one more article that appeared in El Mundo on the
27th September 1983. It was titled, innocently enough: "CIAT is
carrying out pilot programmes in citrus, coffee and banana
cultivation." Halfway through the three columns of print was a
subheading in bold letters: "About an Entomologist," and here, as this
part of the article was of such importance to me, I will translate it in
full, complete with the original syntax:
    "Referring to this, Ing. Rolando Paz Flores [Director of CIAT]
maintained that due to the irresponsibility demonstrated by Robin
Clark [spelt wrongly], entomologist working with the British Mission
which lends technical assistance to CIAT, it has been requested that
he be dismissed, because he has not complied with the work
programme outlined in this subject, [but has been] dedicating himself
to other activities which have nothing to do with his specific tasks,
for example the study of the habitat and behaviour of parrots and

macaws in Amboró Park's River Ichilo. He [Ing. Paz] said that this study had unknown objectives and was outside the lines of work specified by CIAT and the [British] Mission's advisors."

"Actually, Clarke has diplomatic immunity because he holds an official British Government passport, however [Paz said) his contract with the British Mission will run out during the next few days and that, surely, the British Ambassador will decide that he [Clarke] should be returned to his country, on account of the fact that the specific functions for which he was recruited are already concluded."

"With respect to his real work in CIAT, he [Paz] said, that he [Clarke] had failed to carry out his duties two and a half months ago and at the express wish of the Executive Director of CIAT [it was requested he be dismissed] because his work had not given any [cause for] satisfaction and moreover his divergent interests constituted a problem for the national counterparts [CIAT researchers], particularly with the Bolivian entomologist, assigned to the entomology programme."

My! I didn't realise I was so important.

4. 1984

About the only statement of fact in the El Mundo article about me, what the Spanish call a "Denuncia", was that, indeed, I was about to finish my foreshortened contract extension.

By the middle of September, Cheri and I were getting our house packed up in readiness to leave the country. We planned to make a quick excursion through Bolivia; a pleasure we had postponed because we had spent a lot of our weekends in Amboró and the rest of our free time fighting the animal exporters. By the end of the month, just after the denuncia had appeared in the newspaper, we were all packed-up and ready to say our goodbyes.

John Dunning, the well-known bird photographer, and his wife, Harriet, had been and gone. John, a tall, gangly, seventy-year-old American, had become independently wealthy some forty years ago when he discovered that New York city was keen to buy the sand and gravel his business provided. I had written to him about his book (South American Land Birds) that contained a photographic record of more than a thousand species, an impressive labour which had taken the Dunnings many years to complete. He wrote back to say he had always wanted to do some photographic work in Bolivia. If he and his wife came would I help them organise their field work?

I did not have much time to spend with the Dunnings because I was still at work, and at the time very busy with our pineapple trials. The value to me of John's visit was that he was a board member of the World Wildlife Fund and I thought he might be able to get something done for Amboró. Before he left I did have the opportunity to discuss the Park with him in some detail, and at his suggestion we wrote up a brief résumé of its immediate needs. We also paid a visit to Noel. He told John the most important thing for the Park was for me to stay on in Santa Cruz so that we could complete the studies and prepare the paperwork the Senate would need for its legal establishment.

101

Eight weeks after the departure of the Dunnings we received a letter from John: "Thank you for making our trip possible." He also said he had been trying to talk to some people (presumably he was referring to our need for funds to stay on in Bolivia) and finished rather oddly with "I will write him [Robin] about the Amboró Project."

The day before we were to set out on our trip through Bolivia, we got a second letter from John: "And I can't delay further in telling you that I am prepared to guarantee your basic need for $8,000 for your first year on your Amboró project provided you continue in Bolivia." The letter, dated 20th August 1983, had taken six weeks to arrive from Florida.

We were ecstatic. The first thing we did was to call all our friends for a celebration party. The second was to get over the hangover the next day. The third, unpack some of the boxes awaiting delivery to the shipping agent and get the house put back into some sort of order. We didn't unpack everything because we knew we would have to move to another house. We simply couldn't afford the $500 a month the British Mission had been paying, and we didn't think the landlord would be willing to take less. He wasn't.

As luck would have it, we found another house less than 500 yards away which fulfilled all our requirements. Not only was it as big (just as well as we now needed an extra room for the Park office) but, also had a large galpon (a barn) behind the house. The smaller garden was compensated for by open fields on two sides, and it boasted two magnificent Paquio trees. Those two beautiful trees, similar in general appearance to an English Beech, attracted many birds to shelter in their foilage.

The landlord, a rather silly old buffer, was General something or other, I've forgotten now. His Persian wife (and truly she was a bit of a Tartar) walked around with blue shadow-smudged eyes. We got the house for $200 a month, rather low for the neighbourhood but, we had to pay a rather hefty deposit that we thought would never see again [we didn't]. Even so, we felt the expense was justified, if only for those Paquio trees.

But would you believe it, just before we moved in the old buffer cut one of them down because, he said, it dropped leaves on the galpon's tin roof!

Later we learnt that the rent was moderate because nobody wanted the house; it had been occupied by a twenty-eight-year-old Italian terrorist, Pier Luigi Pagliani (infamous for his part in the Bologna train station bombing) and who was mortally wounded during a gun fight with the police on the streets of Santa Cruz. People believed he might walk about the house in ghostly form.

But in our new house the only ghosts from the past were the watchers. They were back. Sometimes only one, standing half hidden in the hedge, other times two, sitting in a pale blue car. Once, when we were out, one of them tried to question one of our maids but, sensing he meant us no good the girl refused to talk to him. The boldness of the approach had me worried and soon after this we took in a third girl, a typical Camba beauty who wanted to work in exchange for lessons in English. This was fortunate, for the newspapers often gave gruesome accounts of what happened to maids left alone on the premises. With three girls in the house, it was possible to have two of them there at all times.

At about the same time as we moved into our new house, Noel kept his second promise to me. The whole animal export issue had been brought out into the open, and though he did not publicly call for an embargo on the trade, he did call for much tighter controls. And, as I have said, the resulting publicity led to the article of the 15th August in which the Chamber of Deputies declared their support for a total ban of all wildlife exports. With Noel's second promise, to promote the establishment of Amboró National Park, we seemed to be on the verge of make meaningful progress at last.

The first article appeared in El Mundo on the 29th of October 1983, "En procura del Parque Nacional Amboró" ("In support of Amboró National Park") For the first time, as this title proclaimed, the area was publicly referred to by its new name. The article said that parliamentary deputies from Santa Cruz, together with Noel Kempff, had petitioned Government to support the Park's creation because of its almost virgin forests and the great variety of plants and animals found there (among which it did not mention one of the most important, the Spectacled Bear). The news item was illustrated by a map prepared for me by the British Geological Mission but, in spite of its accuracy, the article stated that the Park lay ninety kilometres to the south of Santa Cruz, instead of thirty kilometres to the west.

We were, of course, elated Amboró had made the news but, I couldn't help feeling a bit piqued that no mention was made of my personal contribution. At the same time, I realised the omission was my fault. The article was clearly based on a two-part report Noel and

I had prepared; his, a lengthy account of the geography and wildlife of the area was widely read because it was in Spanish; mine, a proposal for the development of the Park, was in English. Nobody read it. Hardly anybody in Santa Cruz read or spoke English.

I vowed to put the situation right in the future. The problem was commercial translations, which cost as much as five dollars a page, were extremely poor, little better than my own efforts. I reasoned that a badly translated article was worse than nothing at all. For a time, my problems were resolved by Marilla, my lovable Spanish tutor, but as I have already told you, she and her husband eventually returned to Argentina in search of a more dignified wage.

The second article came out five weeks later in the form of an editorial. The first part was a criticism of the Government in La Paz for its failure to institute measures for the protection of the environment, and a call for decentralisation "Imperium in imperio" to include division of responsibility for the environment. You see, they just couldn't leave politics out of anything. The second part was in support of the proposed Park, and this time, among the natural riches cited, mention was made of the Bear.

A third article published on the 5th of February 1984 was headlined, "Amboró National Park will be an Important Centre for Tourism". It added little to the previous articles, but to me it was the best of the three because it was based on my report (at least somebody read English). It reproduced two of my photographs taken in the Park (without credits) and it mentioned me by name as co-worker with Noel on the project.

This was important to me for several practical reasons: it would satisfy the Immigration Department's officials that I, and my wife, were entitled to residents' visas, or any one of the bureaucratic obligations we would have to fulfil now that we had been ousted from under the diplomatic umbrella. Together with a letter of recommendation Noel had prepared for me, it would also help us raise funds for the Park.

More gratuitously, it represented a renewed social status for reasons that you, safely tucked up in your beds at home, might find difficult to appreciate. Let me explain.

Noel had publicly stated that I was working with him. Noel was an extinct species on the level, say, of a nineteenth century English parson-cum-naturalist. He was a trained economist with a degree from the local university and an inveterate dabbler in all things natural. With time he turned himself into an internationally recognised apiculturist, and a horticulturist of national repute, subjects he eventually taught at the university. In 1974 he was made a member of the "Academy of Sciences" (Bolivia's most prestigious scientific institution), and by dint of his many and varied publications, and aware of his work embellishing the city's streets with trees and many public gardens with trees, he was much esteemed by the people of Santa Cruz. In short, to draw a tenuous comparison, Noel was to Bolivia what Gilbert White had once been to England. By identifying me as his co-worker he gave me social status, and here I should add, an honour rarely extended to a gringo in a xenophobic society like that of Santa Cruz.

By the same token, intentionally or not (I believe the former), he may have given me some protection from the Animal Mafia.

When Cheri and I started to talk seriously about Amboró as a project, all our friends said the same thing: the more successful we were, the less credit we would get for our efforts; and in the unlikely event the Park became a going concern, the Bolivians would take over, and we would be thrown out on our ear. We were well aware of this, but Noel's public affirmation went a little way to allaying these suspicions. And finally, so as not to try your patience any more, my new social debut neutralised some of the doubts about me that Rolando Paz's denuncia had planted.

So, the idea of Amboró as a future National Park had at last been launched. It mattered not one whit that Noel, Cheri, myself, and a few of our friends were the only people really interested.

****

By the beginning of 1984 it looked as if we had left our troubles behind; how naive I was at that time. Except on a casual basis, I no longer involved myself with the animal exports, and despite the presence of the watchers, I didn't expect serious trouble from that quarter. I had completed my final report for the British Mission, which was later severely edited, and I was relieved to be free of the intimidation I had been subjected to at the office. Noel was turning out to be a really good friend, and an influential one. When I showed him the denuncia against me, he laughed and told me to ignore it. "At least they are taking notice of you, Robin." I also received many, hitherto unexpected expressions of support and sympathy from my Bolivian ex-colleagues at CIAT. Dunning's financial backing had

demonstrated to one and all that there were people on the outside who
had confidence in us and took the Amboró project seriously. We had
got tired of listening to the general consensus of opinion: "A pie in
the sky." "Have you lost your marbles?" That it had something to do
with my midlife crisis. "Don't you know that idealistic projects like
yours stand no chance in Bolivia?" "Don't you know that others have
tried before you and failed?" Or there were some whose words: "Yes,
I think it's a great idea, best of luck with it" were betrayed by the look
of patient sympathy with which they eyed us.

Cheri and I were certainly not insensitive to these opinions; we
knew we had taken on a monster. Sometimes I felt a little embarrassed
by my decision to give up a well-paid career in exchange for such a
fragile aim. We both very much wanted to achieve something of
importance to us personally, to fulfil ourselves. I devoured books that
would help me harden my resolve, ones that expressed my innermost
desires; my belief that nothing was ever achieved by being timid, that
one had to crawl out on the insecure branches to gather the Golden
Apples. I wrote down some of the passages that were to me the most
poignant. Werther's, "I have come to appreciate how it is that
extraordinary people who have achieved something great, something
apparently impossible, have been decried as drunkards or madmen",
has remained my favourite. I am certainly not extraordinary in the
sense that Werther meant, no Mozart, no Napoleon, not even a Bates
but, I have been called a drunkard and madman. And to some extent,
justifiably too.

I had been forced to spend much more time social-drinking than
was my inclination because it was the only covert means to gather
information about the people behind the animal exports. I had to grab
the opportunity when it presented itself; a discussion in a bar with one
person frequently led me to visit another person in another bar, and
so on. In Santa Cruz to put trust in a civilised social system, one based
on punctuality and appointments, does not work; indeed, no such
system exists. People invited to the house for supper just don't turn
up. An invitation to lunch with friends will lead to embarrassed
admissions of forgetfulness. Few Bolivians think about mañana;
today's good intentions are the morrow's casualties; the spirit of
alcohol-induced goodwill the apathy of the morning's sobriety. No
one here would undertake the self-discipline required to achieve the
selfless task that Cheri and I had embarked upon, nobody in his or her

right mind anyway. So, drunkards and madmen, yes we were. One had to be to live here. Werther would have been proud of us.

But not all our friends thought us mad. A few rallied 'round and gave us both intellectual and practical help. Some dug into already empty pockets and donated a few dollars to the cause. Guy Deuel gave us his Ford-250 pickup; old and abandoned for more than a year, we soon had it going, and it was a blessing to our fledgling project. Tom Hackett donated a mule and a horse for our first park guard, Occidental Oil an electric typewriter, and Tesoro Oil a filing cabinet for the office.

****

In early 1984 Mr Reginald Hardy arrived on the scene. Founder of PRODENA (Bolivia) Reginald had retired to Wales after spending more than twenty years in and out of Bolivia selling machinery to the troubled mining industry. He had elected to spend some of his profits and most of his tremendous energy (that he retained even though he was in his seventies) to resuscitate Bolivia's conservation programme. His appearance can best be summed up by an amusing anecdote, the circumstances of which took place later in the year. Cheri had taken Reg up to Buena Vista where they went to meet two Dutch couples living there. The Dutch ladies were at home when they arrived and one said loudly to the other, "Jesus. Kijk die man, dat moet een tuinkabouter zijn." ("Jesus. Look at that man, he looks like a garden gnome"). Five-feet-four, rounded, white haired and red of face, the ladies could be forgiven their levity so appropriate their observation. What they didn't know was that Reg, having spent many years in South Africa, also spoke Dutch. He gallantly answered them in their own language and took no apparent offence.

He had contacted us through Noel, who in response to a letter from Reg, had mentioned that I was trying to get Amboró going. You may think it odd we did not already know Reg through PRODENA (Pro-Defensa de la Naturaleza), a society with aims very similar to our own. In England I had always belonged to three or four different natural history societies. They were places where active people with shared interests got together to discuss what they were doing, to compare notes, and to organise conferences, exhibitions and field trips together. Nearly everyone who joined did so because it was a channel for his or her enthusiasm. Few regarded the society meeting as a social event; it was more an Exchange and Mart of ideas, where the traditional British amateur could rub shoulders with the one or two knowledgeable professionals who acknowledged the valuable contribution the amateur made, and attended the society's functions in recognition of them.

The Latin Americans do not understand the British concept of an amateur society like PRODENA. This attitude is reflected by the regulations that apply to the establishment of any society. To found a society, one must apply for formal recognition. It has to declare its purpose in a legally worded manifesto; and it has to pay a fee to become officially constituted. Nobody considers it a truly amateur association, on the contrary, any society has the right to advise on, even intervene in, Government affairs. It has political power. In reality, they have taken on the guise of Non-Governmental Organisations, without the legislature that controls these. This system could work well in Europe where admission to a society is granted through a system of references and is only given to those who can demonstrate serious intent; the bungling amateur who obtains membership is likely to be excluded from the solemn business of the society, or at least his influence would be severely curtailed.

In Bolivia these auto-controls are absent. Any Juan, Dick or Harry can become a member. Many do so for the wrong reason (as you have seen) and others, having read one book on the subject, consider themselves experts. The influence these pretentious people have is lamentable. With a dearth of genuine professionals to control them, together with the power given to them by the legal framework of the society, their potential for doing damage can be considerable. I do not mean to say this "Mad Hatter's Tea Party" is the rule, for incompetence is recognised the whole World over, but it is, prevalent in Bolivia.

Another factor is the natural ease with which Bolivians will address an assembly of people. Without self-consciousness, one is struck by the confidence with which children and adults alike will talk on the television or demand their say at public meetings. Lacking the constraints of the true scholar, they are all born orators.

I began all this in explanation of my reticence to join PRODENA, and along the way have found it necessary to set down the weaknesses of a system, which does not differentiate between informed opinion and science. Noel told me this was at the root of his own reluctance to join any of the societies in Santa Cruz, including our own. He said he did not want to give a society, largely made up of bungling amateurs, credence by his participation.

I had a similar attitude myself. Given that PRODENA had no representation in Santa Cruz, and that a Bolivian society tends to be a talk shop not a workshop, we were set up for a disappointing experience with our own venture; and I think you will understand how it was we were not members of PRODENA at the time Reg Hardy arrived on the scene.

All this long-windedness is not set down here for the sake of filling up space, or for pedantry. As you will see later, the inherent quackery of any Bolivian society was to be manipulated by one or two people in a bid to undermine my professional influence and strip me of the executive responsibility the authorities had allowed me.

My initial relationship with PRODENA was excellent. No sooner did I meet Mr Hardy at the airport in January 1984 than I was holding a post-dated cheque for $1.000: "To spend as you see fit. I don't need to know what on." Reg was going to be the sort of person I liked to do business with, or so I thought at the time. He had already stated the prospects for obtaining funds (I thought he meant for Amboró), were good, and that we should join forces. As he put it, "A number of entities are showing a positive interest." He was always talking about entities; it was his way of holding his cards close to his chest. He was also negotiating with the Overseas Development Administration to see if they would donate two of their redundant Land Rovers to the Amboró project. They never did, not even one.

We made plans to go to La Paz and pass by way of Cochabamba. Reg wanted to get me introduced to PRODENA now that he had made me a member. In Cochabamba we were to visit a locally managed project (financed through PRODENA) to protect the Oilbirds of the Chaparé. The oilbird is a highly specialised nightjar that, having evolved a system of echo-location, has adapted to living in caves. The Chaparé is more famous as a centre of coca production than it is for its colony of these very local birds. I was very willing to participate in this project, but Reg wanted me to coordinate it, but I could see the problem of doing so.

The Ford pickup was unreliable and a visit to the Chaparé involved a round trip, over rough roads, of more than 800 miles! After I returned from La Paz, Mr Hardy sent me a letter in which he formally noted the provision of the $1,000 cheque and, as a rider, added, "This money is to be spent at your discretion, but it is understood that in return you will assist with our Oilbird project." I saw why his cheque had been post-dated; bonhomie was all very well on first meeting, Reg believed in first impressions, but nobody gives something for nothing. I also saw that his following paragraph: "I feel that if we work as a team, we will be able to achieve a great deal conservation-wise in Bolivia," could read, "If you work for me...".

I set aside my career to do something on my own account; I was not about to swap one set of bosses for another, especially an amateur like Reg, no matter how good their intentions. This was not something I had thought necessary to explain to him, it was implicit in all I had said and done. When he downplayed this tacit understanding by inundating me with instructions, and referring to me, even in official documents relating to Amboró, not by name, but as PRODENA's representative in Santa Cruz, he sowed the seeds of our angry separation.

They say the unaided human eye can see a star 600,000 light-years away. Had I been blessed with the ability to see beyond the end of my nose, I would have returned the cheque, but that would be unwise. Afterall, with the money I could get the Ford (referred to by one and all as "Clyde") reconditioned and I could at least make it to Buena Vista and back. Imported spare parts, especially American ones, were very expensive and a thousand dollars was just not enough to put Clyde in really good order, let alone expedite the work in the Chaparé.

Conservation can be a thankless business at the best of times. The problem was all sorts of people had ulterior motives for posing as private benefactors and as struggling entrepreneurs, lacking the facts to sort out the sheep from the goats we couldn't afford to say no. We thought Reg might be testing us, prior to becoming serious.

So, as I have said, I should have been a lot more careful in my dealings with this strangely cantankerous old man. At the time his surly nature was not so apparent; he liked to work hard, then play hard, a philosophy I myself believed in. During our trip we got along together, visiting many officials and PRODENA members during the day, talking and drinking too much in the evening. Reg did not believe in stinting himself and very generously accorded me the same indulgences. We stayed at the best hotel in La Paz. Reg, with his tendency to show off at times, told me he kept his room for the whole time he was in Bolivia whether or not he was there to use it; little wonder I got the impression he was well-heeled. Because he still had business interests in mining machinery, I thought these indulgences were paid for by his expense account. They probably were. I never found out, but others obviously thought not. In a letter to the President of PRODENA, Reg wrote a long paragraph refuting an accusation that he was spending PRODENA's funds in needless expense.

I became suspicious when he mentioned, first in a telex, then in a letter the following October, that PRODENA owed him $13,000. Unless he was including all his hotel bills, and his airline tickets, how was this much owed to him? PRODENA, as far as I knew, only had one project on hand, the Oilbirds, that was entirely financed by the International Fund for Animal Welfare. And we had received this cheque. That still left a lot to be accounted for.

No doubt by his reckoning, the remainder was swallowed up by his penchant for producing full-colour, glossy newsletters. These quarterly masterpieces of misapplied enthusiasm were to me as a red rag to a bull. And contained a lot of that bull's shit.

One example will suffice to demonstrate the self-centred publicity and unabashed expenditure with which Reg embellished these productions. The January 1985 issue consisted of just one legal-sized sheet, most of them were two to four sheets. On the centre of the front page was a quality reproduction of a colour photo showing Reg and his wife petting a Cheetah, and down the left-hand margin a coloured strip of Inca artwork. The reverse side was entirely given over to coloured pictures of Bolivia's eleven species of macaw. Below the bold title, COMUNICACIONES DE LA PRODENA, Reg's only lip service to Spanish was the subtitle: "The Year of the Park." The whole newsletter consisted of 430 words: starting with the season's greetings, announcing the charitable status of the Society, asking for donations to the Amboró project which he referred to as our project, meaning PRODENA's pigeon, an explanation of the photo: "Many of you do not know what my wife, Laura and I look like..." (that reminded me of "The Beauty and the Beast" as, maybe, Reg's sense of humour intended), and ending with two addresses to which donations could be sent. Unfortunately, the main part (of which I reproduce a little below) was made incomprehensible by the format, Reg having split the script by inserting the photograph thus:

In point	of fact we
have a	cheetah
problem	in Namibia
on which	we are
working,	but that is
another	story and
Bolivia	definitely
has and	deserves
priority.	So please
help us	make 1985
The Year	of the Park!

Apart from the inappropriate chummy nature of these newsletters, that after all was said and done were the voice of a serious society, the luxury of the format was more likely to convince the would-be-donor that PRODENA had money to burn, rather than persuade them their help was needed to keep it alive. Add Reg's claim that Amboró was the Society's project, spice it with the knowledge that, by this time, we were once more running the Park with my own meagre savings, and you will understand how I felt about the money wasted on these newsletters. The blatant self-propaganda was the sauce which cooked the goose, the expense the straw that broke the camel's back. I regarded them with such contempt I was completely unable to discuss them with Reg. What I did do was to send him an article about conservation philosophy in Bolivia with a request he place it in his next bulletin. He never did.

During our time in La Paz, an incident occurred, that even to this day I find hard to believe. It was my second night in the swish Hotel Plaza. Reg informed me that Gaston Bejarano, advisor to the "Honorable Camera de Diputados" on wildlife and conservation, had left a message to say he would like to visit us. I was intrigued. Doctor Bejarano was, as you may remember, one of the most important government officials named by me in the report to the New York Zoological Society: Gaston Bejarano: lives in La Paz, has been and is again (?) Chief of Wildlife under the Director of the Dept. Forestal CDF. He has to sign all export certificates. It is said that he signs these certificates in advance, receiving a fee for doing this. Also, said to be secretly in partnership with Onishi.

He was also the man so disliked by Noel, and according to Gene, the one who received three dollars for every monkey exported from Bolivia. Our rooms were on the fifth floor and he rung from reception to say he was on his way up. It was 9.30 p.m. "This is going to be some meeting," I thought, as we waited for him to ascend in the lift.

Whilst he and Reg went through the ritual of greeting each other, I had a moment to assess Gaston. He looked like a fifty-year-old Man from the Ministry, any Ministry, a British Minister. At five-foot-five he was of average height for a Bolivian, and stood with a Richard III slant to one side. He had a very pointed nose, sharp as a chisel, and a mouth as if he had been chewing lemons or unexpectedly sharp bones in perfect harmony with his eyes, which being very short of sight, were provided with thick lenses that gave them a grey, dead-fish look. What immediately caught my attention was his pleasant, soft voice, enunciating perfect English. It took me completely by surprise, and said to myself: "here is a man you can trust".

After some time spent in discussing mutual interests, Reg announced that he was too tired to continue and begged leave to retire for the night. For some reason or other, I sensed Gaston had been waiting for this moment and that he had really come to see me alone. I thought, mostly wrongly, that he wanted to talk about Noel. He had won me over by the eminently sensible arguments he had put forward during our discussion with Reg. I welcomed the opportunity to talk with a competent scientist, as Gaston most certainly was. My impression that he was a Wildlife Chief under the Director of the CDF now appeared a grave underestimation of the man. Gaston was a boss, a self-assured and suave man-about-town. He commanded respect. And he immediately got mine.

When Reg drained his glass of whisky, I suggested to Gaston we could finish our discussion in my room, a proposition to which he readily agreed. My room was a luxurious one with a spacious reception area furnished with three comfortable armchairs and a plate-glass coffee table. Through huge windows it looked towards Mount Illimani twenty miles to the east. The view, so clear and crisp in the dark, looked like a hologram of the nativity: the pendant sickle of the moon, the natural stars, the Christmas-tree Mountain with its lustrous snow-covered slopes, and spread out at our feet the whole magic of a city at night, the lighted buildings like so many gifts awaiting the ravages of privileged children.

As we sat down, Gaston produced a large bottle of 'Black Label' from the pocket of his overcoat. Now I knew he had come to see me. I felt relaxed, we were like two friends about to sit down and relate their experiences of times spent apart. We chatted easily for a while, as will two people who have deep interests in common when freed from the social necessity of being polite to others. We talked about birds, people we knew in common, including Noel and the strained relationship he had with Gaston, who wished Noel no ill; and, finally, we started to talk about the bird and animal exports. It was early in the morning; the town had gone quiet, only the sound of the occasional vehicle floated up to us. We were both past being sober and the talk had narrowed down to this one topic as unremittingly as a missile homes on its target. The explosion when it came was just as sudden.

I was listening to Gaston's summary of the export situation, his comments about some of the people involved, and his synopsis of the likely outcome of motions now standing before the Senate. "Let me now tell you something funny," he continued, "In April last year I went to the States and was invited to talk with David Mack at TRAFFIC. You know them. Well, whilst I was there, I was shown the report you sent to Conway and I read what you said about me. They showed me the lot."

Oh my God! Gaston, if he had waited all this time to amaze me, had certainly achieved that. I was thunderstruck. Struck dumb, "Puta de mierda! Hijos de perra!" ... I was trying to pull myself together, but my thoughts continued to blaspheme.

"Robin," he went on, "I want to tell you I don't blame you for writing that report. Most of it is true, including the little about me. I really appreciate what you have been trying to do, and on behalf of Bolivia I thank you. But you have seen what is going on here. Wages aren't worth a damn and we have to find other ways of making a dollar. Apart from all that, you had better consider your situation here. You may think people don't know about your part in it, or maybe you think they won't do anything about it. Be careful, Robin. The next few months are going to be difficult ones for everybody."

With those words he got up to leave. The whisky bottle was dead; my ashtray had long overflowed its capacity. It was four in the morning. I was glad he was leaving. I needed to think about what he had just told me, anything more would have been superfluous. We parted under the best of terms. Reg was staying on in La Paz; I had to get to the airport early to catch my plane home. It was not worthwhile going to bed, as I would have been unable to sleep anyway.

Now he had gone, I thought of a dozen questions I should have asked him. Who, exactly, showed him my report? What had they agreed to do about it? How was it possible that he had seen it? No, Gaston could wring the secrets out of an archbishop. Had he told Mack my report was true? Did Onishi know about the report? Did either of them have anything to do with the watchers at the house? Was Onishi behind the threats I had already received? Did any of this have anything to do with the troubles at the Ecological Society? Or, to come to think of it again, my troubles at work? How did Noel fit into all this? More and more questions filled my troubled mind.

Daylight was creeping into the room; the nativity had given way to the wan light of dawn. I stood at the window for a long time, the mountains in the east a dark wall of sombre eternity as foreboding as my own thoughts. The pastel shades of nearby buildings were melting into the dusky haze of the town beyond. To the south, the ragged outline of the huge erosion gully, flanked by angular pinnacles of more resistant shale, resembled a clawed hand rising from the valley floor. Immediately below, the steep deserted street, the paving, and the scaly-looking cobblestones looked like the skin of a sloughing snake. I could see down into the bare courtyard of a church school, the open space surrounded by ancient cloisters on all sides. The absolute stillness of the foreground was now broken by the meditative passage of a grey-robed priest as he slowly made his way across the yard. The simple beauty was given to the scene by the hundreds of pale tiled roofs, the disorder of their fluted surfaces broken here and there by the emergence of ugly blocks of flats. From around the corner beyond my line of sight, I heard the impatient hoot and beep of early traffic, and across the roofs floated the clang of a church bell calling the devout to early mass.

Life goes on, things never change; the problems of today are the tragedies of tomorrow; one loves and lives, hates and dies; we are all looking for something, but what? Is it good enough to settle for happiness? And what does one do if the foundation of this happiness is being slowly destroyed, eroded away as surely as the distant hills? Is tolerance evil? Is it right to be right? Justification. Education. God is within us. God is us. What place has love in the real circumstances of today's world? Fight the good fight, or wander away seeking relief from disillusionment? What are the grey-robed priest's desires? How does he justify his loneliness shut up behind the cloister's walls? Has he found peace in the solace of his calling? Does he suffer the little children to come unto him? Does he really love his fellow men? How can he, when men are such brutes? How can he, when men revel in their sin? Talk, talk, talk; has he ever truly changed anyone, put them on the path to heaven? And if heaven is on earth, where is it? If heaven is inside us, can it only be obtained, really, truly, by loving mankind? Where do I stand in all this? Why can't I shut my eyes? Why do I fight? What am I trying to save? Myself? Nature? The world? Save the world, Mr. Right, and see heaven. Whose heaven? For whom the bell tolls, does that bell now toll for me?

****

When I got back to Santa Cruz it was early in the afternoon and the sun was shining down from an almost clear sky. I normally preferred to take the direct flight from La Paz. It took only fifty minutes, and if you sat on the right-hand side of the aircraft, you got a good view of the mountains of Amboró looming closer as the plane made its landing approach.

As it turned out, I had to go by way of Cochabamba. Here, sitting at the lounge table next to mine, there was a small group of, late thirtyish, German tourists who had been on the same flight as mine. About every half-hour, a listless feminine voice could be heard through the loudspeakers informing us that there would be further delay but gave no reason for it. Every twenty minutes, one or the other of the Germans would get up from their table to go in search of more information. Every twenty minutes, they returned with the same result: the Santa Cruz flight was not leaving yet, and they had failed to elicit anything more. The two large surly looking men were becoming irritated and kept pointing to our deserted aircraft shimmering in the morning heat. Although my German is very limited, it was obvious from the little I knew that they couldn't understand why, if there was something wrong with the plane, nobody was trying to fix it. A typical South American situation was developing. The Germans, perhaps aggravated by the unruffled lethargy of the rest of the passengers, which they probably took for brute apathy, were getting more annoyed as the long minutes ticked by. What began with conspiratorial whispers between them now turned into louder abuse of the airline, the country, and its people. While they had spoken in German, nobody took much notice, but now the men began to make fools of themselves by attempting to express their insults in Spanish, "Got, es wirklich mierda pais: God, it's really a shitty country".

When I thought the embarrassed good humour of the waiting people was about to turn to hostile reaction and when I noticed two policemen eyeing the frustrated Germans with puzzled distaste, and when the waitress (by then almost in tears), refused to serve them any longer, I leant over my table and told them what everybody else knew, that we were waiting for the arrival of a connecting flight. Just for a second, I thought one of the men was going to hit me, so completely deflating was this perfectly logical explanation. Sniggers ran around the lounge and the, by then, somewhat aroused Bolivians could be heard passing insulting comments about gringo tourists from table to table. The two policemen, encouraged by the turn of events, became bold enough to go up to the Germans and ask to see their air tickets and passports. The humiliation of the circumstances transformed the truculence of the foreigners into impotent quiet. Red with rage before, their faces now became crimson with shame. The lounge, too crowded to permit them to make a dignified exit, became an open court, a silent inquisition; the mortified tourists, the object of one hundred pairs of eyes, stared at the white Formica top of their table in complete silence, until, at last, our flight was called.

The Germans had lost face; once again the locals had proven their own superiority. One more lot of tourists would never return; more empty seats on Lloyd-Aereo Boliviano. More precious foreign exchange would stay at home. And, one can add to AASANA (the airport workers union), a few others (police, immigration) who conspire to destroy the foreign tourist's enjoyment of their fascinating country. As far as the airport workers are concerned, we would go to France for a suitable epigram that they might adopt for their motto: "Hardi comme un coq sur son fumier","Bold as a cock on his own dung hill".

****

As I walked up the short driveway to the house, I could tell from the beery talk emanating from the windows that we had visitors. It turned out to be a Swedish chap who had come over to buy our car. It was to be a sad parting, but short of cash, we saw no justification for keeping it. Taxis were plentiful and cheap, and we had Clyde. I liked the car, a Renault-16; as the only one of its kind in Santa Cruz, maybe in Bolivia, it didn't attract the attention of car thieves. The poor old thing had suffered a series of traumas since we had it shipped over from England. Delayed on a train, surrounded by stubborn floodwater in Brazil, it arrived nearly a year after we did. Then, in order to get it ready for use I delivered it to a local Englishman who was said to be a good mechanic. The night it was ready, he rang me up to say I could collect it in the morning. When I did so, it was ready—ready for the scrap heap. True to form, so I learnt later, the mechanic had first got very drunk, then high, before taking the Renault for a spin. Driving in the dark at eighty miles an hour, he drove it straight into an articulated lorry. He and his female passenger were lucky to live. The police report stated that the skid marks were sixty yards long.

Anywhere else in the world it would have been a write-off, but not here. Although never the same again, we had it back on the road after a further twelve months. One night, Cheri, together with the three English sisters who had babysat our house, were rammed by agents of the infamous Ciento-diez, because it was claimed they had broken the curfew. As everybody else knew, Ciento-diez saw an opportunity to molest four fair-haired gringo girls. The fact that it was ten minutes before curfew (11.50 p.m.) and that the car bore courtesy registration plates did not matter one whit. The four girls spent the night in jail; the Renault went back for some much-needed panel beating.

Then, driving too fast to beat the curfew I drove it up the back of another car, that I just didn't see. And in she went again to have the panel-beaten panel panel-beaten.

A week after our Swedish friend bought her the much-abused Renault found fame in the local newspaper. There it was, centre page, looking like a casualty from the Beirut bombing. "Cuatro heridos, four wounded" announced the caption. I think that was the end of it.

******

The battle against the wildlife dealers received a vital boost on the 9th of November 1983 with the promulgation of Resolución Ministerial No. 538/83, which announced a provisional one-year ban on all animal exports from the 1st of January 1984. In effect, this also implied that the dealers had to dispose of their stock of captive animals and birds by the 31st December 1983. The one-year ban was, of course, very welcome, but still fell short of our hopes of getting a permanent one established. Even so, we celebrated the news in a suitable manner.

However, on the 19th of January 1984 (Resolución Ministerial No 15/84) it was announced that the export ban had been postponed because the period of grace allowed the exporters to clear their stock of animals awaiting shipment had been extended to the 31st of March 1984. The dealers claimed they had not been able to comply with the original deadline because of insufficient cargo space on the aircraft.

Many saw the extra period of grace as an unjustifiable concession that would allow the exporters time to fill their cages once more and this time, go out with a bang.

On the 11th of February, further news appeared in the La Paz daily Presencia after a meeting of government officials with Obdulio Menghi, the South American Scientific Coordinator for CITES. This article openly stated that corrupt officials had made a mockery of the CITES agreement, and specifically mentioned the sale of blank and forged documents for $3,000 each. It also referred to the part played by Benigno Rodriguez who by now had been sacked for the part he had played in the illegal export of Hyacinthine Macaws worth one and a half million dollars in the United States. Lastly it stated that a letter from a presidential secretary [was this Bejarano?] to his boss, President Siles Zuazo, had influenced the regrettable decision to extend the period of grace set for the end of further exports.

During my visit to La Paz in January, Dr Armando Cardozo, President of PRODENA, invited me to present my information on illegal exports to the Society's members. On the 7th of April, Presencia published an interview with Dr Cardozo who added his voice to those already calling for a ban on further exports. As he pointed out, adverse world opinion was likely to prejudice Bolivia's chances of attracting international funding for its conservation projects. This was one of the main points I had raised at the meeting with PRODENA members. He also proposed that the period of grace allowed the dealers be withdrawn and that the animals and birds awaiting export should be returned to the wild without further delay.

Presencia took up the idea on the 8th April. It quoted Noel's opinion: "That it was not possible to return all the birds and animals back to the wild because amongst the 40,000 involved, many couldn't fly, and there was a risk they would be eaten by predators."

Although his comments declared it unfavourable, the idea of returning the captive animals back to the wild was gaining momentum through public debate. I thought the healthy birds and animals could be rehabilitated, and I proposed that this be done in

Amboró, thereby killing several birds with one stone: propaganda for the Park and for the fight against the export issue, and by adopting such a daring scheme, good publicity for Bolivia. I say daring, because the return of so many animals and birds back to the wild had never been tried before.

The article also stated that 542,963 animals and birds had been legally exported from Bolivia between 1975 and 1983. Maybe, but with all the illegally exported ones the real number would have been substantially more.

After my meeting with Gaston Bejarano, Cheri and I decided to downgrade the danger to us by releasing our information to the public. We chose to do this by approaching the "Unión Juvenil Cruceñista", a confederation of moderate right-wingers representing the interests of young people in public affairs. Not altogether innocent of past excesses, many Cruceñans considered them to be fanatics, even fascists, and would have advised us to give them a wide berth. But, with their penchant for direct action we saw them as suitable allies to have in our fight to end animal exports. Two of their representatives, together with a journalist, came to our house one morning. At this crucial meeting we gave them an outline of what we knew about the members of the animal mafia and the illegal side of the exports, and we supplied them with copies of most of our documents.

We stipulated that they should only use this information to generate the support of the Unión as a whole, should they decide to take an active part in the campaign. We saw the theatrical approach favoured by the Unión as a way of dramatizing the situation in a way nobody else could. We asked them to consider a raid on the dealers' premises and to confiscate the birds and animals in full view of the press and television cameras. When they left our house, we were not sure how this ploy would turn out.

I don't want to give you the impression we were the whole anti-export crusade, that would be to claim too much but, we were the catalysts, and as an incorruptible influence, a force to be reckoned with in a situation where people were used to getting their own way through bribery. Apart from our efforts, using Gene's information to stir up trouble, and persuade Noel to make a stand, Reg Hardy, PRODENA, CITES, and some societies in La Paz, played a vital role in the successfully developing campaign.

We certainly didn't want any direct publicity, this was why we contacted the Unión, but we did want the word to get around that Cheri and I had nothing more to reveal, and that nobody would benefit by trying to remove us from the scenario. Rumour had it that some of the animal dealers were out to get us. In a situation like this, it's easy to become paranoid, to see evil skulking in every crevice. But fact is fact, everyone concerned had warned us: Gene, Conway in New York, Noel, and more recently Gaston Bejarano. Through him we were confirmed in our suspicion that our cover had been blown for some time, and as you will see the amorphous threats to our safety gradually took on a more frightening quality.

During Carnival, Cheri and I went with a group of our friends to a dance club called El Caballito where I was the victim of a cowardly attack. After one or two numbers, a friend of ours cut in to ask if he could dance with Cheri. Somewhat relieved, I returned to our deserted table. This gave me the opportunity to do what I like best in crowded situations, sit and watch the fun.

I had been sitting down for maybe a minute when, without warning, I was grabbed in a half nelson from behind and dragged bodily over the back of my chair to the floor. I couldn't see who had done this and for a moment I thought it to be the over-boisterous welcome of some friend or other. Then I was kicked in the back of the head and booted painfully in the ribs. While I was struggling to get to my feet and free myself from the half nelson, punches and kicks started to come from all directions. I managed to stand upright and elbow-off the man behind me; I even returned a few punches before, once again, I found myself back on the beer-puddled floor. My friends were still happily dancing, oblivious to the considerable commotion. It was like a nightmare; everyone, except me, having a good time, their faces lit by the vacant smiles of the inebriated, seemed to be mocking me. During a slight let-up in the pounding, I got to my feet again, and luckily for me, my arm was abruptly seized by a man who stepped out from the melee and pulled me clear of the fray. Whilst he hurried me to the exit, I heard one of my friends on the dance floor yell, "Look what they're doing to Robin."

When we got back to the house, I felt the need to analyse the incident. Since my companions had completely missed the action it was a shock to them all when they saw I was soaked with blood. After their inevitable protestations of remorse, we started in on the main issue: was it a bunch of drunks gringo-bashing or was it something more personal? Since none of our group had anything to contribute, and I did not recognise my assailants, our search for a motive was a rather hopeless pursuit. Reason told me that six or seven men don't suddenly get up in a group and act together, as they did, without prior planning, so planned, yes. The motive was another matter. The whole incident took place without any of the usual cries of jubilation or insulting references to gringos, which are the trademarks of xenophobia. On reflection, the attack had been business-like, impersonal, something that just had to be done.

But the precise motive for the attack still eluded me. I needed to know. If it was simply high-spirited bullying, I could forget it, but if it was meant to be another warning, an upgraded "get-out-of-town threat", this was more sinister. I was also conscious of the paranoia-creating circumstances of the situation, another reason why any uncertainty was a luxury I couldn't afford. But like a stubborn floorboard, we weren't quite able to nail the matter down.

A week or two later I ran into an ex-colleague from CIAT who said he saw the fight but regretted he had been too far away to do anything for me. "But you know what, I'm pretty sure some of those guys are mixed up with animal exports."

Apart from a number of amorphous threats like this one, I received whispered warnings of the get-out-of-town kind quite frequently, and on one or two occasions these were somewhat dramatically backed-up with a pistol muzzle in my ear. And the watchers? They still spied on us from outside the house.

After several more newspaper articles relating to the postpone-ment of the ban on further wildlife exports, the Unión Juvenil Cruceñista entered the fray in an article that appeared on the 15th of April in Presencia. It referred to the documents we had given them and printed some of the photographs I had taken of the dealers' premises. It also mentioned the large bribes paid to officials of the CDF. At the end, the article gave the names of the export companies, and more significantly, listed the owners by name. This was an important precedent that gave teeth to the article; while the exporters could remain anonymous, they were unlikely to be shamed by public opinion.

Given the circumstances, Cheri and I were both under a good deal of stress, and sadly, the swing of Father Time's scythe had scattered sharp tears on the path of our married life. Cheri wanted a temporary separation and wished to go home to the States. Tired of the endless quarrels between us, and cognizant of her need to do so, I agreed. We had become like two toucans, always spoiling for a fight. One of us had to go. Cheri left with promises to phone me at regular intervals.

On the 25th April the Association of Animal Exporters published their customary half-page of bold type. The declaration was titled, The Facts about the Export of Wildlife. It started out with the forthright statement: "None of the newspaper articles mentioned that the majority of the birds exported were parrots and parakeets, which the public knows are PESTS in this country; and those they exported were not protected by CITES agreements. It is true," the article continued, "that we export monkeys, but we must point out that these go to accredited laboratories for scientific research." Reference was made to the benefits mankind obtained from this research, "... and thanks to monkeys the production of Hepatitis B vaccinations, and in the future, vaccinations for malaria, venereal disease, etc., will be made possible." And it specifically mentioned, "The Thalidomide case which had left 50.000 defective children." It called for the postponement of the ban on exports because: "...all the experts, including Prof. Noel Kempff Mercado, agreed it was necessary." It repeated the claim that many campesinos and native persons depended upon the capture of live animals and birds for their living. It ended by saying that people criticised the export of renewable natural resources but said nothing about those that were not renewable: "...petrol, gas, tin, and minerals in general." It was signed, "Juan López. Executive. Identity Card No 1495846 Santa Cruz."

It was a clever article if it was meant to appeal to the less informed public. Apart from the lies, "We do not export cats [jaguars, ocelots, pumas, margays]," its fault lay in all the half-truths and red herrings. Statements that were only likely to annoy the Government's legal advisors and administrators. To a man, I'm sure they would have said, "Who you trying to kid, Mr Lopez? Do you really take us for a bunch of numbskulls?" Once again, foxy Señor Lopez had made a mistake. God bless him.

In spite of the doubts expressed over the plan to return the animals still in captivity back to the wild, the CDF invited me to discuss the idea with them. The meeting was a brief one since we had no information on the number of birds and animals involved, and we knew nothing about their condition. Sick and injured animals, should there be many, would greatly complicate the situation. Without any real expectation of getting approval, I did make another suggestion;

as Noel considered the return of the animals to the wild to be impracticable, why couldn't the CDF sell them and donate the proceeds to Amboró National Park? We could, I went on, have done with a reliable vehicle, that if donated to us by the CDF, would be a clear demonstration of their of interest in the future park. There was some support for this suggestion, but as Noel said to me in private, given the untrustworthy record of the CDF and the ease with which they could embezzle the profits, nobody was going to agree to the proposal. In the end it was decided that until a firm decision was made, I would go with Felix Perez, the CDF's wildlife inspector, to make an inventory of the exporters' remaining stocks.

This appealed to me because it would give me a bit more official status with the CDF, and I could roam at will in those places I'd been forced to visit with such stealth. I would feel like a victorious General returning to scenes of battles lost in the past. On the other hand this carried a risk, by coming out into the open, every exporter was going to see my face. I reminded myself the war was not yet over, we had carried the day and won a twelve-month ban. Until the ban became permanent we could not rest, nor should we take unnecessary risks. Once again, McGee was back in action.

The suggestion that I be an independent witness to the up-and-coming shenanigans may have come originally from Delfin Goitia, the head of the CDF in La Paz. We had got to know each other when Reg Hardy introduced us in La Paz. Delfin had become a keen supporter of our efforts, both for Amboró and the anti-export campaign. He had asked me to take more interest in the CDF in Santa Cruz, and he specifically asked me to prepare a parrot identification manual for the CDF staff responsible for checking export shipments.

I didn't say so, but I knew very well the CDF inspectors knew their parrots better than I did; after all, their private incomes depended upon their ability to spot the more valuable ones, the protected species. Even so, I threw myself into the task with enthusiasm. I produced a synopsis of all the species which might pass through the Bolivian export system, and I illustrated these with small coloured photographs of the corresponding plates from Forshaw's 'Parrots of the World', Onishi's present to Noel, which I had to borrow again. I sent copies to Delfin and I gave Felix one for his use in Santa Cruz. (Later, I discovered Felix owned a valuable first edition of Forshaw's book.)

The foregoing was then my initiation into the CDF, a relationship that was to become more formal over the years. Sooner or later it would have been necessary for me to keep in close contact with them because of my work for Amboró, but at the time I saw it as an opportunity to involve myself directly in the control of the wildlife trade should we fail to obtain a permanent ban. In the interim, I thought it better to leave the official side to Noel. It would have been unwise to go behind his back; not only for the sake of our growing friendship but, also to concede (as I have said) that little was achieved at the official level that did not come from him directly.

The two days doing the rounds of the dealers' premises with Felix Perez held few surprises for me. Indeed, I was more surprised that my furtive surveys of the past had provided a relatively accurate picture. Felix had already done a comprehensive survey in November 1983 when the temporary ban had first been established. He had registered 21,460 animals (mainly monkeys and parrots) awaiting export. He gave me a copy of the breakdown, the number of individuals for each species held by the then nine active dealers. It appeared that three of the twelve firms listed by Gene had gone into liquidation: Merck Sharp y Dhome, Research Animal Breeding Centre (I now knew these to be the same Onishi firm), and Gene's own Laboratory Supply. The inventory listed: 4,733 macaws, 15,146 other parrots (including 1,978 amazons), 24 toucans, 1,373 monkeys (mostly at Zoological Garden Suppliers), and a variety of other animals and birds. After Felix's survey, you will remember the exporters were allowed until the end of 1983 to clear their stock, a deadline later extended to the 1st of May 1984. It was on this day that Felix and I set out to do the new inventory. Since no further extension of the moratorium had been made we did not expect to find much left.

Onishi had a cage full of monkeys, an illegal Red-fronted Macaw, and thirty or so other macaws. His was our first port of call. Then we crossed the road to Minton's, the Dutchman, who had about eighty macaws and a pathetic Spider Monkey crammed into a cage so small the poor creature couldn't even stretch himself. Carrillo's unkempt premises, strewn amongst the orchard behind his house, had twenty-nine sick macaws and nothing more. We then travelled right across town to Romero's, the place where I had talked to the guard and seen the rifle propped up against one of the buildings. He still had a lot of Amazon parrots, a distressed looking Margay, quite a few macaws and other birds, and a diseased Mono leon, actually a Saddleback

Tamarind, probably from the Beni. Another dash across town to the north brought us to the compound of Juan Lopez, our foxy executive officer of the Association of Exporters; he had a large number of moribund macaws, mainly the blue-and-yellow sort. Claure had a few emaciated monkeys, about 150 assorted small parrots, thirty macaws, and nothing else. Antelo's cages were nearly empty; some parrots aside, the only birds of interest were eight dying Military Macaws. We did not go to Fauna Sudamericana as Felix assured me that nothing whatsoever remained there.

None of the owners we met acted in the least hostile towards us. They talked quite openly about the loss of the good old days, how things had to change, how they were going into the pig-rearing business, or whatever, poultry breeding, egg production, farming; but none of them, if they knew who I was, displayed any malice towards me. Cambas, for the most part, displayed the traditional stoicism I have alluded to before. Onishi was not there; nobody meets this elusive, legendary Japanese. He is a man who avoids social contact as a cat would a cross-channel swim. Neither did we meet Juan Lopez, Dr Rolando Romero, or the Dutchman, Joseph Minten; if we had I would have risked a sly punch to his vitals. The latter had just been accused of exporting (via Santa Cruz) a quarter of the world's surviving Golden-lion Tamarinds from Brazil. How could anyone sell into extermination one of the rarest and prettiest species of monkey just to satisfy his greed? Maybe I would have hit him twice. Lots of the larger birds bore witness to the lack of expertise of the hunters: lots with their wing feathers clipped far too short, and one cage-full of Macaws, their wings not just clipped, but cut off at the carpal edge, would never fly again. Many, those lacking tail feathers had been victimised to cater to the folk dancers, a barbarous throwback to the old days when tradition dictated that headdresses be made from macaw tail feathers. Some tradition! Others had been mutilated in violent ways, their beaks broken or their eyes missing; and a good many, with deformed feet, testimony to the illegal use of chicken-wire cages. Apart from these manifestations of cruel nescience, Mengele's Ark, about half the macaws were diseased, bald emaciated specimens which would have to be quarantined from the rest. In total we found nearly 400 macaws, a bit less than 300 smaller parrots, twenty-one monkeys, and the usual collection of less familiar items. Given time to recuperate, we thought about ten of the monkeys and 300 macaws could eventually be returned to the wild. The rest of the macaws and most of the smaller parrots looked well enough to liberate immediately. At least we now had some solid facts to go on.

But unnecessary cruelty was not limited to these criminals; during the debate about their fate, Reg Hardy suggested to Delfin Goitia that all the birds and animals left over, he thought there were 14,000 of them, be destroyed! A rather peculiar position for the founder of a society devoted to wildlife conservation.

Fortunately, as Delfin pointed out to him, the strong civic feelings in Santa Cruz would not permit any further abuses. Reg's uncharacteristic faux pas was apparently in response to a report from Delfin that it would cost $200,000 to rehabilitate the 14,000 animals and birds. As it turned out, the mathematics were about right. In fact, our survey showed we were only dealing with 700 birds and a few monkeys, about half of which could be released after a few days, which was just as well. What we would have done with 14,000, I have no idea, but kill them? Never!

The macaws were the main problem. Having been locked in small cages for seven or eight months, their flight muscles had atrophied through lack of exercise. Many others, short of essential vitamins, were suffering from malnutrition; their plumage was in poor condition, the skin of their naturally bare faces anaemic-looking, and nearly all of them were listless. If they were to recuperate the macaws would need ample space in which to exercise themselves. A full-grown macaw has a 6ft wingspan, so 300 of them would require an enormous space in which to fly. The zoo couldn't offer us their walk-through aviary, the macaws would have destroyed it and the carefully nurtured vegetation in it, and we were unlikely to get the funds to build one of suitable dimensions. The galpón in our garden was a distinct possibility, but taking all things into consideration, I thought the solution lay in releasing the birds into a controlled habitat, an isolated piece of forest where the birds could be protected and fed. This could be done relatively inexpensively and had the advantage that the birds would partly feed themselves, or at least through their natural inclination to chew on anything available, they would obtain some of the essential natural vitamins they needed. Finally, I saw this would be the solution for the majority of the parrots and our galpón could be used to quarantine the sick ones.

I was eager to see any plan to return the birds and animals to the wild be carried out in Amboró. This was logical because most of the creatures were natives to the area. But Amboró was not home to the handful of parrots which would have to be released elsewhere: the Alder Parrots, hundreds of miles to the south, and the Scarlet Macaws a good distance to the north in the Beni.

As I was preparing the project proposal the phone rang, and an excited voice told me to get myself down to the city centre, quickly! The Juvenil Cruceñista had raided the dealers' compounds and confiscated their stock. "They," Felix told me, "are now releasing all the birds and animals in the Plaza Libertad!"

****

When I got there, it was a scene of indescribable chaos. The Unión had brought all the animals to the plaza and lined up the cages outside the University's administration building. Opening all the cages, they released hundreds of terrified macaws and bewildered parakeets into the square. Wherever you looked groups of excited spectators surrounded one or more of the freed birds. Those unable to fly were crawling across the broad avenidas, or lying on their sides in the gutter, one wing waving in the air. Others, true Icarians, unable to gain height, crash landed among the crowds while attempting to escape. The traffic was brought to a lurching halt and one unfortunate macaw brained itself on the windscreen of an oncoming vehicle. At the centre of the square a small group of parrots had joined a family of Squirrel Monkeys in the comparative safety of the trees. A good many others, having commandeered the backs of the public benches, lunged with their beaks at passers-by. Yet others had left the immediate vicinity of the square and were being pursued by groups of young lads up the narrow streets. Some of the public, oblivious to the rationale behind the release, were taking the opportunity to help themselves to a pet, and I saw one efficient young man stuff a screaming Red-and-Green Macaw into the boot of his car. A scruffy looking dog was barking at a crippled one trying to take shelter inside a crowded café. And the pandemonium, the harsh raking of freed birds, the answering shrieks of caged ones, the piercing scream of a little girl with a kamikaze macaw enmeshed in her hair, was synergised by the growing panic. In every direction, overexcited people were blindly running to-and-fro. Not looking where they were going, some nearly ran into the front of passing cars; their drivers adding to the cacophony with the constant tooting of their horns. Another scream; this time a little boy turning to his mother for comfort, his hand gushing with blood where a dismayed parrot had bitten him. A cat behind the portals of an open doorway was dragging a hapless parakeet and hundreds of others, in a tight vociferous flock, were strafing the plaza with quiversful of strident cries. A group of excited kids flashed past in pursuit of an agile coatimundi. A hurry of cameramen rushed from place to place filming the best of the action. Newspaper reporters were stopping people in the streets to hold impromptu interviews; and a perplexion of policemen were consulting together under the trees in the centre of the square.

Then, realising the havoc the freed animals and excited crowds were creating: trampled flower beds, children and one old lady knocked to the ground, a series of near traffic accidents and, hounded by the dog, the crippled macaw ousting a crowd of hysterical customers from a café, the Union began their attempt to round up the startled birds.

I doubt many readers have ever tried to pick up a full-grown angry macaw. A bird on the ground will nearly always take up the same posture: lying partly on its back, half on its side, one flapping wing extended, its head ready to lunge forward to give it's very powerful beak the opportunity of removing a stray finger or thumb. It's a good defensive position fortified by the ear-splitting screams of the critical reaction (a zoologist's term for the near suicidal behaviour of a cornered wild animal). The recapture of a bird under these circumstances requires a bold adroitness available to only a few.

A less disturbed macaw might oblige you by climbing on to a stout stick, and oblivious to the protests of the policemen some of the young people were tearing branches off the trees for just this purpose. Unfortunately for them, a frightened macaw on a stick is a daunting adversary worthy of an experienced fencer. To begin with, the stick has to be a long one and the bird must be kept near the tip, a muscular feat of Olympian proportions since an adult macaw weighs several pounds, otherwise its beating wings, as it struggles to maintain balance will give the uninitiated a good many painful lashes across the face. The normal reaction in this case is to drop the stick and run, a ploy that was being adopted by most of the Unión. Then, even should the bearer of the poled parrot be strong enough to support the bird's weight, the stick will almost certainly droop. Polly's natural reaction to his lowered status will be to climb up in the world, towards the uppermost part of the pole. Under most circumstances, this will lead the angry bird directly to the bare hand of his would-be saviour. Again, the normal reaction is to drop the stick and run.

Another, and much more hazardous method, is to pick up an annoyed parrot by its tail. I say hazardous because, having witnessed this inadvisable procedure on this day, I can tell you the bird's swift response is to flap its wings, a manoeuvre which brings the avian's bill into contact with one's limbs in a very efficient manner. One young man was gaping with dismay at a large ragged hole a parrot had just made in his arm. So much for the tail-grasping ploy.

There are even less advisable ways of catching large angry parrots. Try picking one up by its feet, or if you are really insane, try grabbing one by the neck. And on no account would I advise the kind approach, picking one up with both wings. With their large wingspan a macaw spread eagled in this way will end up with its bill uncomfortably close to the jugular vein. Try it if you like. Some of the Unión tried, but didn't quite like it.

Without special equipment there is really only one way, paradoxically the most unlikely, that you, patient reader can add to your list of useful household tips. To get hold of an non-placid psittacid, grab it firmly by the tip of one wing. Believe me, one wing tip can be reasonably safe. But before you rush off to practice it on you budgerigar, and before the voices of you, pet lovers, are raised to cry "shame", I must point out that this unique solution requires experience to avoid damage to the bird. Some of the Unión were doing more damage than good.

So, every choice is a Hobsonian one, and carries with it some personal risk, the wings of disaster like the scene in front of me. One bird swinging by one wing tip, wildly flapping its free wing, lunging with its metal-shears bill has deftly penetrated the young man's thigh with one of its raptorial claws, ...uh-uh. And there he stands, Polly firmly anchored with it's claws drawing blood through the victims clothing, the young man holding equally firmly to one wing tip, an embarrassing position of stalemate. The man can't bend down to release the claws because this will bring his face very near to Polly's. He can't let go, because Polly in his fall will not only take a firm hold with his other foot, but will also take hold, and now much more seriously, with his mandibles.

If you happen to have been clumsy enough to put yourself in this predicament, there is only one thing to do, scream for help. Then, if armed with a stout stick, or maybe a broom-handle, your assistant might be able to lure the parrot on to it. If this Polly ploy doesn't work, accept your fate as calmly as you can, or as one young man from the Unión Juvenil Cruceñista was now doing, beat the bird firmly over the head.

These were, then, the incredible scenes of violence and torture being acted out by persons and parrots on that sunny May morning in the arena that used to be the peaceful central plaza of Santa Cruz. Small groups of gladiators were having trouble with obstreperous avians wherever you looked. A few, having come off worst, were parroting the screams of their uncooperative victims. One group, guilty of parroticide, were staring down at a dead one. As John Cleese would have said, an ex-parrot.

Others, having managed to carry a stricken bird back to its cage were having great difficulty trying to shove the wretched beast through the door, a task like trying to force a live lobster into a small cooking pot. Screams of rage and frustration from humans and birds, and the raucous laughter of jeering onlookers filled the air.

I saw the arrival of the Municipal lorry from the zoo; at last, a few properly equipped men who knew what to do were available. They came with the heavy leather gauntlets and stiff wire crooks the task called for. Satisfied, I turned away and made for home.

Several blocks away I had to brake sharply to avoid the Coatimundi and the same group of excited kids in hot pursuit. I noticed one of them had a golden coloured tamandua clutched under his armpit. Santa Cruz, never a dull moment.

****

Back at the house the phone was ringing. It was Cheri calling me from her hometown in Missouri. She wanted to come home; she missed me and loved me. Would I be happy if she came back soon? Yes, of course. Of course, I would meet her at the airport. No, we were not going to row anymore; things were going to be better. Even as I said the words, I knew they would not be. She had been away six weeks. I didn't have the heart to tell her how much I, too, had needed the break; how much I had enjoyed the argument-free days and nights, and my reversion to enjoying time, by which I mean long hours, even whole days, spent in my own company, free of the constant distraction of loud music and loud visitors that she brought to the house.

She had left in a state of nervous exhaustion. "I've got to take a break, R.O., I am going to have a bloody breakdown if I don't get away from here, just for a while, R.O." She always called me by my first two initials. If only I'd had the courage to tell her that we were through. But how can you tell the woman you have loved and lived with for five years something like this over the phone? I have always loathed phones anyway. I could never understand how people, and Cheri was one of them, could enjoy conversing on the bloody contraption. How could I tell a bit of black Bakelite my innermost feelings? How could I tell her that, silly, as most people would find it, contemptible, as she would consider it, that, I was already lost to her? That, and I didn't understand it any more than she would, I was falling in love with Miriam, a girl less than half my age. Oh! Why hadn't I found the courage to tell her then?

The gloom that descended over me as I struggled to justify my confused emotions, now deepening to despair. Three more days, then I would have to tell her, come straight out with it. Maybe even at the airport, before things got intimate. Should I ring her back now? "Look old girl," I could begin with enough coldness to prepare her for the realisation…No, not even that, surely she knew. Confirm? Yes, confirm our mutual coldness. No, not that either. I find this too difficult to write about; maybe I don't even know what it is I want to say. Maybe it's just too personal, too private. Ring her back? Or wait for her to arrive? She would have to come back anyway, wouldn't she? Collect her precious toucan, pack her things. Ghastly practicalities that would have to be faced.

****

Miriam? Where are you? God, what a bloody mess. Please, Miriam, come back and amuse me with your simple chatter, your lighthearted ways. Thursday. Yes, it was full of woe; and I was getting rapidly inebriated, trying to drown with alcohol the burgeoning conflicts that played havoc with my thoughts.

"Hola. Don Robin, está usted aquí?" (Hello. Don Robin, are you there?) It was her at last; my depression backed off a few large paces, then lifted as fast as a gaily coloured balloon will free itself from the grasp of a dismayed child, to rise up, up over the heads of its concerned parents. Miriam brought me a letter from Reg. She too had been to the plaza and had seen the last of the animals rounded up and taken away.

The letter. How could I take it seriously? What did it matter now? It was full of instructions for me, full of things Reg wanted me to do: "Looking at the pile of stuff I have for you," it began, "I could have done with some of the information requested in my last letter and various telexes sent to you. What has happened about those maps? Please check urgently. I need your help with item six… What ideas do you have on…? Can you activate all this for me? Obviously, any other suggestions are welcome; however, you must work on Delfin as well. Will you ask him? What is happening about the Decreto Supremo for Amboró? Let me know what goes on at your end. As I have said before, communications seem to be a problem." John Burton, David Attenborough, Armando Cardozo, and Gaston Bejarano…they said this, or they were helpful about that, or they promised this or wanted that. "Hope all is well with Cheri and that she is back in Santa Cruz." I read the penultimate line again, "Communications seem to be a problem." Then back to, "I hope to find a way of taking up your ideas and supporting your project. Stick around!" as if I didn't intend to; insinuating the only reason for me to do so would be the promises Reg held out to me. "Communications seem to be a problem," I found it repeated in the second paragraph. Yes, they bloody well were. Why can't I get through to you, old man, that I don't want my life organised by you? "Communications…"

I got a sodding letter from him, full of insensitive provocations and instructions for me, every damn week. No Sir! Not me. Find somebody else. Don't try and rush things here. Bolivia is Bolivia and I, liking the place, accepting the way it is, will not be pushed around anymore. I have my own philosophy, my own way of getting things done. Push where I can, let up when I can't. Keep a low profile, stay out of the limelight, and give no reason for jealousy. Sort myself out too. God damn it. I nearly screwed up the letter and threw it in the wastebasket along with the copy of a telex stapled to the back. More information wanted, more reports to prepare, urgently!

Putting the letter away in the file I checked out of curiosity how many letters Reg sent me in the last month. One on the 21st of March, another the next day, a longer one on the 28th. A long telex on the 8th April, and this one received today, written on the 11th. In three weeks, he had written to me four times and telexed once. Of course, I was dismayed, dismayed to find myself so remorselessly taken over, unremittingly reduced to Reg's thing. At six in the morning you would find me already at work, with much still to do twelve hours later when I collapsed, just to comply with his wishes. I had my own interests, I was working for virtually nothing; what did he think my reaction would be to his stream of commands? Communications seem to be a problem, if he could get his way he would have me attached to a 'beeper'. He was in a rush to get things done in time. But in time for what? His death? Clearly, the man was not only insensitive, but also a little bit mad.

And what were his motives? "I suggested he destroy them." I went back to study the file in search of the answer. Apart from his personal letters to me, Reg sent me copies of his letters and telexes to everybody else, and all those he received. The Hardy file was three inches thick, but if I thought, on the 3rd of May, that I was being engulfed, it was nothing in comparison to the inundation to come.

4 May : Letter ("hasty") Reg to me.

7 May : Telex Reg to me.

7 May : Comunicaciones de la PRODENA

7 May : Letter Reg to The American Bank (copy to me).

7 May : Telex Reg to Ralph Peterson (copy to me).

7 May : Letter Reg to IPPL (copy to me).

7 May : Letter Reg to NYZS (copy to me).

10 May : Telex Ralph Peterson to Reg (copy to me).

11 May : Telex Reg to me.

11 May : Telex Reg to Ralph Peterson (copy to me).

13 May : Letter IPPL to Reg (copy to me).

14 May : Letter Reg to Ralph Peterson (copy to me).

14 May : Telex Reg to WWF (copy to me)

15 May : Telex Reg to me.

15 May : Telex WWF from Reg (copy to me).

15 May : Telex WWF to Reg (copy to me).

15 May : Copy of Reg's phone bills!

15 May : Letter Reg to me.

16 May : Telex Reg to me.

17 May : Telex Reg to me.

17 May : Letter Reg to Cheri and me.

17 May : Telex Reg to NYZS (copy to me).

19 May : Letter Reg to The American Bank (copy to me).

21 May : Letter NYZS to Reg (copy to me).

23 May : Telex Reg to me.

28 May : Letter Reg to NYZS (copy to me).

28 May : Letter Reg to Bernie Peyton (copy to me)

28 May : Letter Reg to WWF (copy to me)

28 May : Letter Reg to IPPL (copy to me)

28 May : Letter Reg to British Ambassador (copy to me)

28th May: Telex Reg to me, "Suggest you fly up to La Paz Sat/Sun 9/10 June as we should jointly see a number of people there before I head for Santa Cruz. Regards."

King Canute would have been hard put to stem this flood.

Then there were the phone calls from him usually between five and six a.m. on account of his indifference to the five (GMT) or four (BST) hour difference in time. Half awake, three-quarters asleep, I would have to contend with a barrage of intricate questions: "How many monkeys are there to release? How many parrots? What other animals? What is Kempff doing about all this? What are the authorities planning? Have they accepted your proposal? You must keep me informed. How can I help if you won't tell me what's needed? You will have to improve your lifestyle if you want to get things going." Fuck me, the man was giving me nightmares. I felt as cornered as those poor birds in the plaza.

All the shops were closed, the petrol pumps shut off. Miriam, the other two girls, and I spent a good part of the day searching for something to eat and drink; looking for cigarettes, skulking about in the hope some profiteer would sell us petrol, preparing Clyde for the road, queuing for cooking-gas cylinders, trying to pay the bills, persuading the vet to attend the puppies, finding a way around the street barricades. Yes, Santa Cruz was in chaos again, and inflation constrained the populace to acts of disobedience. Nothing could be achieved without four times the usual time and effort. Dunning's money was coming to an end. Reg was holding out promises of more, but not yet: "Stick around," he would say. Meanwhile our commitments were being financed with the last of our meagre funds supplemented by the little money I had saved over the years.

Ring-ring, ring-ring. It was Reg. "What's going on? The lion monkey you reported, what species is it? What's happening with the other animals and birds? Why don't you send me a telex if you don't have time to write? I must know what…" I put the receiver on the table, lit a cigarette and went to fetch my bottle of Amontillado. After taking a satisfying slurp, I picked the phone up again. "…The CDF. I think I can get you the money needed for the parrots. The New York Zoo Society says it will help. But I have to keep them informed. You really must try to keep me…" I put the receiver back on the table. When I picked it up again, after taking a long look through the window in search of composure, the line was dead.

Ring-ring, ring-ring. This time it was Noel. "Robin, como está, hombre? Mira, tenemos una reunión en una hora. Donde? Aqui, en mi oficina. Por favor tiene que asistir. Deseamos charlar con usted, Robin. Ya, muy bien, pues. Hasta pronto." He rang off. He didn't need to tell me what it was about, he knew I knew already. In one hour did he say? We were to discuss the fate of the animals and birds that the "Unión Juvenil" had confiscated on behalf of the people of Bolivia, and which were now crowding the quarantine station at the zoo. Noel would be worried about the situation because the zoo ran on an insufficient budget, and another thousand or so hungry mouths to feed, extra cages to clean, and vet bills to pay, would be to overburden his budget at the cost of the zoo's own valuable collection.

The meeting was really a foregone conclusion. Nobody was in a position to oppose my plan to return the confiscated creatures back to the wild. Noel voiced his reservations about the plan but did not see any alternative; his assistant agreed and added that the zoo wanted some of them. The Dean of the University's Veterinary Department wanted some of the sick birds for study, and all of the monkeys. Esteban Cardona (Director CDF Santa Cruz) formally proposed the rehabilitation project be accepted. Oscar Llanque (Wildlife Chief, La Paz), who had already discussed the issue with Esteban and me, seconded the motion. After these gentlemen had expressed their opinions, everybody turned to the boyish looking representatives of the "Unión Juvenil Cruceñista", clearly they were in charge.

Since they had arrived late, and had not heard the proposal before, I was asked to outline the main aspects. I wanted, I told them, the thirty monkeys and all of the 300 large parrots. I thought the parakeets and parrotlets would recover quickly, and they should stay in quarantine for another week, and then be set free. Since Noel wanted the remaining mammals for the zoo, they were already spoken for. Taking up the discourse, Noel said he wanted most of the primates: all of the Spider Monkeys, some of the Brown Capuchins and Squirrel Monkeys, and the lone Mono leon, if it survived its present illness. I had to have some of the monkeys. I suggested five of the Brown Capuchins would serve my purpose, a bit of publicity for the International Primate Protection League, who Reg had informed me were willing to donate $500 towards the rehabilitation scheme.

I went on to tell them I would like to take all the birds to El Carmen (a small settlement in the Park), and under the supervision of our two workers who lived there, involve the whole village in the care of the birds. The others at the meeting considered this suggestion to be further evidence of gringo guilelessness, they wouldn't hear of it. I had to concede, but to sacrifice this romantic notion for the sake of pragmatism was a disappointment I had already prepared myself for. Bolivia was not ready for it.

After their first refusal, I then outlined my alternative plan that now had a better chance of being accepted. I intended to take all the large birds to Buena Vista for rehabilitation. There they would be released into a patch of woodland on the quinta owned by our Dutch friends, Leonard and Chela Muyzenburg, and tended by Pieter and Els Brekelmans (our other Dutch friends).

The sick ones would be taken back to my galpón and kept there until they, too, were fit enough to go to Buena Vista. For those that did not naturally inhabit there, arrangements would have to be made for their transport to areas where they did.

Of the 400 large parrots (nearly all macaws) we thought about one hundred were in good enough condition to fly off immediately. The remainder would stay in the trees until they were strong enough to do the same. Meanwhile, we would feed and protect them, and provide them with the necessary medicines. Depending upon expert advice, in other words whatever Noel recommended, we would supplement their artificial diet by gathering wild fruits and nuts. Our two employees from the Park would take care of the birds and Pieter and Els would be responsible for their day-to-day management. We estimated it would take three months for the majority of the birds to recover, or die, as may be the case. (As it turned out, this was optimistic.) After this any birds remaining would have to be returned to the CDF for disposal as they saw fit; maybe these should be passed on to the University.

In order to implement the scheme, we would require enough money, technical advice and practical support. As of yet we did not have a single peso, but I reassured them we were sure of getting the money, here I was sticking my neck out. The Dutch were willing to shoulder much of the responsibility but wanted a 500-dollar deposit to cover damage to their banana plantation, should the birds cause any. I needed wages for the workers, money to feed the birds and buy them their prescribed drugs, and running expenses for the vehicle, the dreaded Clyde. And, I went on, the galpón required some minor alterations.

We also wanted publicity for the Park, possibly the preparation of a video for transmission on TV, and publication of a monthly progress report in the local newspapers. My motive for stipulating the latter was to neutralise the spiteful rumours that were bound to circulate, like those saying we were selling the birds on the quiet. It all had to be above board. We intended to keep daily records so that we might learn from the experience. The CDF vet should take full responsibility for the health of the birds, a duty that would require regular visits to the galpón and Buena Vista. Finally, we needed to sit down and work out exactly how many animals were involved and how much it would cost to feed them and treat them with the drugs they needed. I would work out the other costs involved, and, I finished off, together with the CDF I would put in a formal request for the necessary funding.

Nobody at the meeting demurred, and without further preamble the scheme as proposed was accepted, pending the money being made available. There were however a number of practical points which had to be discussed. Since there was a two-week-long petrol strike I wanted the CDF to supply me with fuel from their reserve tank. I also wanted the CDF carpenter to carry out the alterations to the galpón: make it bird-proof, and put in a small side door. These conditions were agreed to.

One urgent problem remained. The dealers had isolated the sick birds from the healthy ones. When the "Unión Juvenil Cruceñista" confiscated them, they had mixed them all together. The vets needed to sort them all out again and make their prescriptions without any further delay. The contagious disease was apparently due to a fungus that attacked the lungs. The recommended drug could be readily administered with the bird's food, or by mixing it with their drinking water. I was none too confident of the vets, although it should have been possible to find one good one among the fifteen hundred said to be professionally registered in Santa Cruz. (None of those we knew had managed to keep our puppies alive, something about our failure to give them their distemper vaccinations in time, something about our having given them their distemper vaccinations too early. You couldn't win.) Either way, I viewed the veterinary side of things as the weak link in the chain, and so it was.

Immediately after the meeting I sent off a telex to Reg requesting an initial $1,000 to get the Rehabilitation Project on the move. I got a reply back a couple of days later in which Reg said he was sure we would get $2,000, mostly from the New York Zoological Society. "In the meantime, I am mailing $1,000 today." This was good news; our project was on its way and given that the Amboró Park boundary was only two miles from Buena Vista, I felt my original idea, that of moving all the birds to El Carmen, had not been wholly compromised.

Reg also referred (with considerable confusion) to the lone Mono leon from the Pando (which Felix and I found at Romero's premises): "This is what the Bolivians (he must have meant Brazilians) call the Golden Lion Tamarind (Leontidius rosalia) that comes from S.E. Brazil (not Pando) and very rare. Some 30 were recently smuggled to Belgium. If you have one of these, you will have made a lot of people's day. Please check and advise. Anyway, it looks good."

This was Reg jumping the gun in his enthusiasm to make everything we were doing appear larger than life, an attitude which got my goat because I disliked any hint of the 'circus' to enter our work. I did not expect him to convert a perfectly ordinary monkey, common in northern Bolivia, into one of the world's rarest and most endangered species. When reference was made to the discovery by his PRODENA representative of a rare Golden Lion Tamarind in Reg's next Comunicaciones, and also in TRAFFIC's official bulletin, long after I had clarified the monkey's true identification, I was incensed; nobody, I thought, was going to take us and our Amboró project seriously, because whichever way you looked at it we would be labelled as bunglers or charlatans.

As I mentioned earlier, I knew all about the Golden Lion Tamarind through the BBC World Service, their arrival in Belgium having created quite a stir, and because Minten (of Minfauna) had been involved in their re-exportation from Santa Cruz. Once again, as Noel maintained, Bolivia's name was being dragged through the mud to the profit of a crooked gringo.

And talking about crooks, another scandal (which had been perpetrated a few months earlier) had finally seen the light of day. The affair, like the report of a missing person, only attracted the attention of the authorities when the gas-filled corpse rose through the murky water to the surface, and where in its turgid corruption it now laid exposed to view. It was reported in a La Paz magazine called "Ultima Hora" on the 4th May. The crime concerned CITES documents, or to be more precise, permit number 00153. The permit gave Santa Cruz Pet Farm, this was a new one to me, authorisation to export 375 macaws, all fully protected by law. It bore the signature of Gaston Bejarano, and the signatures (that CITES thought might be false) of our old friend Benigno Rodriguez and Noel Kempff Mercado. The fact that the birds were protected species wouldn't have caused more than ripples of censure. Far more important was the permit itself. You see International Organisations like CITES have little tricks to make it hard for people to falsify their documents. One of the methods they use is to build little mistakes into them, minute errors so hard to detect that a forger is likely to overlook them. In the case of 00153 somebody in Germany, where the shipment was to be imported, had with true Teutonic efficiency rumbled the deception. He, or maybe she, had sent the document to CITES headquarters in Gland to confirm their suspicions.

CITES, having confirmed the authenticity of Bejarano's signature, had finally and irrevocably caught Gaston with his pants down, or as Reg wrote me, "All covered in bird shit."

"Onishi: forges Certificates of Origin (CITES permits) and Gaston Bejarano is Onishi's man." I couldn't help but recall Gene's prophecy, and Gaston's own words to me: "Let me tell you something funny, they showed me the lot." At least I felt I had been vindicated.

Now that the shipment had been irrevocably exposed, my confidence in Gene's information was justified yet again. TRAFFIC's terrible breach of confidence now returned to them with as much welcome as Rumpelstiltskin's arrival at a christening party.

The next few months found me entirely preoccupied with three things: the Parrot Rehabilitation Project, my disintegrating marriage, and the swift erosion of my confidence in Reg. Now the export issue was over, at least for a year, I didn't need to worry about it, though just now and again I caught sight of the watchers who now spied on us from a blue Toyota. The "Unión Juvenil" had proven to be stalwart allies, and the newspapers had enjoyed a field day now that the dam had burst. I maintained my friendly association with Noel and I hoped he would survive the latest infamy in which he was implicated. I had my own troubles.

The parrot rehabilitation project had necessitated a number of meetings and clarifications to get it going. An annoying article appeared in El Mundo on the 5th May, which said, "Clarke was ready to donate $15.000 to save the birds and to cooperate with studies at the University (of Santa Cruz), together with the University of Texas."

****

Cheri returned from her hegira looking well and as beautiful as ever. So much had happened in the days preceding her return I had not had the time, time I should have made, to consider our future relationship. For one thing or another, exhaustion robbed me of any desire I had to resolve the problem, and taking the easy way out, I welcomed her back with a genuine gladness in my mind and a good deal of confusion in my heart. Having failed to come clean at once, I let my conscience dictate my actions and stifled my heart's subterfuge with true Punic faith. Or in other words, if you don't like these Flaubertian sentiments, I let things slide.

Our world now came to be dominated by the parrots in the galpón; the harsh sound of their voices, in apparent recognition of my wayward heart, screamed insults at me from dawn to dusk. To get away from the noise I took to visiting a kiosko some distance from the house. Even there, sitting over a cold beer, their screams floated across the field and over the roofs to reach my ear.

I had often wondered who in their right mind would want to keep such strident pets. Now we had sixty of them. "Yaak", you silly bugger. "Raa'ch", you macho. "Raak", you rascal, was all I heard, day in day out. Even the nights were not free of their abuse, for in my dreams their rebellious cries took on anthropomorphic significance and their gaudy plumage the flamboyant vestments of condemned courtesans:

As we dance around the guillotines
Lining the Petit Trianon
In the gardens of Versailles
Now, all together:
As we dance around the guillotines,
Lining the Petit Trianon,
In the gardens of Versailles.
Thank you, Ian Rhythm-Stick. Don't call us, we'll call you.
Raaak.

Cheri had become obsessed with the parrots, given her own way she would have made a pet of each and every one. I had to keep reminding her, when I caught her hugging one to her breast, that these parrots were destined to return to the wild. So much did she love the beasts that even in the light of her understanding she was constantly slinking off to the galpón, love glinting in her eye. She only stopped, or nearly, when I pointed out to her most of these were sick birds and that her loving attentions might easily contaminate Cyrana, her blessed toucan, which still tyrannised the household.

Every day the CDF sent one of their redundant guards over to clean out the galpón, help one of the girls carry the fruit and corn back from the market, and prepare the food for the parrots. It was not so much that we needed an extra hand as my wish for the CDF to participate officially; after all, the majority of the birds were in Buena Vista beyond their help, and assisting us was the least they could do. Mind you, it was surprising how much work was involved: preparing the galpón for use and the daily ritual of scrubbing it down with disinfected water, sterilising the food bowls, chopping up the bananas and other fruit, boiling the maize, which the birds ate in prodigious quantities, dispensing the drugs, repairing the perches, burying the rubbish and, sadly, dead birds, and so on, our work was never done.

Our long-suffering friends in Buena Vista were, if anything, even more overwhelmed than us. It had been necessary to build a palm leaf shelter because of the cold windy weather. And, though they had an extra assistant, they had a lot more birds to take care of. With the low temperatures their own bananas didn't ripen, and without a large marketplace to serve their needs they spent much of their time looking for alternative sources. Then, there was the considerable organisation required for the collection of wild fruits and nuts.

It was also surprising how much all this was costing. Nearly two-hundred-dollars-worth of materials to convert the galpón; cash for wire, food bowls, bins and other sundry items; forty, fifty, thirty, sixty-dollars-worth of disinfectants and medicines; thirty dollars for food each day; petrol and repairs to Clyde, which was now put to hard use ferrying birds and supplies up and down the deteriorating road to Buena Vista; wages for the assistants and the small band of daily labourers who gleaned the forest for palm nuts and other suitable provender. Then there was the $400 to pay for the production of the video.

The film was made by a Yugoslav with an unpronounceable name, Matias? Matiash? Matyash?, take your pick. An unbelievable character, Matiash (as I will call him) looked like a throwback from the Old Testament. His head, including most of his face, was covered with straggly blonde hair; his large blue eyes had a taunting glint, which together with the derisive curl of his full lips, his image a parody of the Mad Messiah, a likeness he intensified by his partiality for exotic garments and his mocking manner.

Matiash was a pain in the neck but a competent maker of films. He was eminently likeable when you got to know him, at least some of the time. Matiash did all the filming, cleverly scoring it with music and parrot screams, and I wrote the commentary. The film opened with a scene of us collecting the first shipment of birds from the zoo; and closed with their release, which, as some could fly, showed images of a happy flock circling over the woodland at Buena Vista. The script, which I tried to read in a serious David Attenborough voice, was mainly criticism of the animal exporters and their customers. Matiash wove the verbal indictment of pet-lovers with disturbing images of mutilated birds. The video was a great success and was eventually shown on TV screens throughout South America and, I heard, in the States. Even today, many years later, it is sometimes repeated in Santa Cruz, or I see cuts from it used in advertisements. It was a pity it was to become the cause of more trouble between Reg and myself.

Come to mention it, it was a pity the whole Parrot Rehabilitation Project was to cause so much trouble between us. One of the problems started with Reg's penchant for finding a scapegoat when things went wrong. As I have explained, the animals and birds under our care were those that had not been exported in time to beat the ban. Felix Perez, as you will remember, had done an inventory of all the stock in November 1983 when the ban on animal collecting had come into effect. The seven hundred or so animals and birds we got for rehabilitation were the remnants of the 21,460 he registered at that time; the ones too sick, or too damaged, to export. In addition, we had liberated the healthiest among them. So it was not surprising that some of the birds that came under our care died.

The sickest of the parrots had been installed in our galpón, and in spite of the medicines we gave to them, two or three were dying each day. At Buena Vista the situation was better: of 250 parrots in their care our Dutch friends only found one or two dead each morning. Distressing as this was, nobody could work miracles, and we accepted the losses in the knowledge that we were doing our best.

Later, when Reg came to inspect the project, he accused us of not taking sufficient precautions, and of not caring. Coming from the man who had recommended that 14,000 animals and birds be destroyed, this made me pretty angry. It looked to me as if he was more concerned with justifying the money he had raised for the project than he was for the fate of the birds. Then, as I have said, the project was turning out to be really hard work, especially for our Dutch friends in Buena Vista. Nobody was being paid, we were all volunteers. The injustice of Reg's accusations, on top of all the other vexations, became too much to bear.

Here (in tune with the aroused feelings my memory rekindles) I find myself running ahead of events, for the showdown did not take place quite yet. In accordance with Reg's instructions, I met him in his office in La Paz on the 10th June. Neither of us was feeling friendly towards the other. Trying to organise the Rehabilitation Project (to get all the parties concerned to cooperate, and to cope with the constant changes to the original agreement) had worn me down. Reg, equally frustrated by the vicissitudes of the operation, was not in a good mood either. For example, he simply could not understand why we were unable to tell him exactly how many birds we were caring for. For my part I had ceased to care; the parrots were counted and recounted the day before we collected them from the zoo. Overnight some died, some went to the galpón and some went to Buena Vista. Counted again (with difficulty at Buena Vista because of the dense vegetation) we would find we had twenty more than before. Later, I would learn some more had arrived from somewhere else and, without informing us, had been put in with the rest. Or I would find out that the zoo had decided they would, after all, like a few more of these or those, or whatever; and of course, some were able to fly, gone in a group of thirty yesterday, back in a group of ten today, birds were coming and going in all directions.

After I explained all these things to him, Reg, who had uncharacteristically not done so already offered me a cold beer. Basically, he wanted me to agree to his figures should anybody enquire. I didn't see why we couldn't be straightforward about it. The situation had already been aggravated by a gremlin in the telex (Reg thought we had started with 400 Macaws), whereas we only had 300. And, to add fuel to the flames, while Reg had obtained the $500 from the International Primate Protection League (in return for some film of us releasing monkeys), Noel had decided he wanted them all for the zoo. I cabled Reg to tell him there were no monkeys to release; he telexed me back, "Go find some!"

Reg wanted me to supply him with precise information. I was too busy trying to organise the project. The communications problem had led to misunderstandings and unfounded suspicions; we were both annoyed by the frustrating circumstances. To give you an idea of our mutual aggravation here is a copy of a telex from Reg to me, and my reply:

For: Robin Clarke. 16 May 1984

From Reg Hardy

1. Worried night so sought advice of friend ex-WWF now with ICBP Cambridge.

2. She advises strongly against passing info in your telex to NYZS as you have made it sound as though all volunteers have pulled out and that you are paying someone for a pricey service to get it off your back, and why should one have to give cash guarantee anyway?

3. She says NYZS will want to know precisely whom the money is being paid to, more or less precisely what for, and for how long. What happens after 'next two months' — you previously told me it would take four months.

4. Also why $200 renovate galpón if now not being used? Justification and cost breakdown would be required.

5. She says WWF, NYZS, etc. work to the book and having spent some time with them recently I can but agree with her. To get off on a wrong foot now would jeopardise Amboró's chances. If we give amount, quantities and time we must stick with it and not change every week.

6. So sit down quietly and draw up detailed project acceptable to NYZS and telex to me soonest. Meantime can only suggest you tell whoever to hold $1,500 cheque as that kind of cash not in bank and won't be for some time, and I can't advance anymore. Or why not go back to galpón?

Regards. Reg.

****

To: Reg Hardy. 16th May 1984

From: Robin Clarke

NYZS and anybody else interested in the Parrot Rehabilitation Project must realise that this is an emergency project involving logistical problems under Bolivian conditions, and all the tees will not be crossed, nor the eyes dotted, before all these birds die. So they must help us or not, but not waste our time and money.

Can they send us one post-graduate student with vet experience, his keep, and a liberal supply of drugs for the birds? If so we will do this project and they can come and see it when they like.

TV video about complete, held up by petrol strike. NYZS should be very happy to receive very good publicity for $2,000.

Cheers Reggy. Robin.

****

So there we now sat, two ruffled people in Reg's hotel room in La Paz, memories of the piddling quarrel obstructing our conversation, Reg struggling to mask the condescension in his voice, I, uncaring and indifferent to his threats of pulling out altogether. One of the niggling issues concerned money, not only the shortfall we might have to face in order to conclude the parrot project, but also my own financial situation.

The money John Dunning had promised me, the elusive $8,000, was nearly finished. He had Promised me he would cover our personal living expenses for one year, starting from September 1983. By May 1984 we had only received $1,000 from him. We'd done a good deal more for the Amboró project than we had originally planned to. We were running Clyde, albeit with Reg's contribution for its reconditioning; we employed the two men in El Carmen to look after the horse and mule we had been donated, and to help us with surveys in the Park; we'd covered some of the expenses for a TV documentary about Amboró; we had printed our own project stationery; we'd given quite a few public lectures, and we'd done a fair amount of entertaining on behalf of the Park. Confident we would eventually receive the promised funds we were happy to dip into our own meagre savings to complete the project's aim: to get Amboró National Park officially declared.

It was no wonder, then, that by the end of June 1984, ten months after we started the project, we were coming to the end of our money. So, imagine our consternation when John Dunning started to blow hot and cold:

12th December 1983. 'Did you get the first $1,000 I deposited in Cheri's account?'

18th January 1984. 'I am amazed and disappointed at the amount of opposition to the project... WWF feels it would be a waste of time and money to promote or finance it...'

21st February 1984. 'I am very gratified to learn that I had the wrong impression of the situation ...'

8th March 1984. 'I am holding the next $3,000 to send to Cheri's account for you as soon as I get the word that WWF will pass the money to you.'

25th April 1984. 'Congratulations. WWF now agrees to take my contributions for Amboró. [By this time he had given us $500 cash, and so went on]...then I will plan on the remaining $3,500 I promised for your first year's effort on Amboró.'

With these last words in mind, things didn't look too bad when I went to meet Reg in La Paz. But we were still in deficit at the time, and though not exactly fighting to keep the wolf from the door, we were worried about covering the rent and the bills. I'd decided we couldn't afford to stay in Santa Cruz where rent for the house was swallowing up a third of our budget. The solution was for me to resolve my marriage difficulties (by which I meant the final break) and to move to Buena Vista where houses could be rented cheaply and where I could spend much more time in the Park, as I desperately wanted to do. It would also be possible for me to help our Dutch friends with the Rehabilitation Project. When I told Reg about my plans, he was adamant that I stay on in Santa Cruz until the decree to establish the Park was signed by the Senate. Since the Dunning money would run out by the end of the month, he offered to give me $500 a month for July, August and September.

Confident in the knowledge we would get paid I did not ask him to deposit the money into our account until the 20th September. In a telex to me of the 2nd October he replied, "However I must point out right now that I made no promise of a further $1,500. My notes are quite clear on this and you are confusing it with the money you said was owed to you by Dunning."

I did not stay long in La Paz. After clearing up some personal business (residents' visas for Cheri and myself) on Monday, I took the early morning flight back to Santa Cruz on Tuesday, directly this time. I didn't see Reg after our first meeting; neither of us being very happy in each other's company, I passed my spare time walking the streets of La Paz. Since Reg had some business to attend to, he was to come down to Santa Cruz a day or two later.

As it was necessary for me to be in Cuzco on the 1st July to learn how to do Spectacled Bear surveys I was anxious to get back home to prepare our affairs for my absence. The field course in Peru was important to me because Noel always claimed there were bears in Amboró, and I wanted to learn how to detect their presence. We had been toying with the idea of doing a bear survey in the Park for some time, but our complete ignorance of the matter had been a stumbling block as large as Amboró itself. Spectacled Bears are very secretive animals and even a dedicated bear man rarely manages to see one. However, the expert can spot their presence by the telltale signs they leave in their passing. Footprints are the most obvious, the tattered remains of bromeliads, one of their staple food plants, less so. He can also make a positive identification from their faeces, yes, bears shit in the woods, and anywhere else, or from the claw marks they leave on the trunks of fruit-bearing trees. From a few "bear" facts an expert can build up a fairly accurate picture, even predict how many individuals occupy any particular area. I was hoping my trip to Peru would give me the opportunity to persuade Bernie Peyton, the course teacher, to pay a visit to Amboró.

Bernie was the acknowledged authority on the Spectacled Bear. As he worked for the New York Zoological Society that had connections with the World Wildlife Fund, I saw an opportunity of killing three birds with one stone. He would certainly find bears in the Park if they were there, he could make an official report on the Parrot Rehabilitation Project — which as you will remember was largely financed by the New York Zoological Society, and he could act as locus tenens for Curtis Freese, the World Wildlife Fund representative, who after numerous postponements now looked as though he would not come at all.

The overbearing Mr Hardy arrived in Santa Cruz on a cold and misty morning. It was the day of the showdown. The pot of vexation having simmered too long now boiled over and extinguished the flame, it only needed one small spark to blow the roof off.

Reg arrived with his friend Ralph Peterson, a man of endless patience and unassuming ways. The first thing Reg did was to look at the birds in the galpón. Reg, his red face now beetroot from the cold, came back looking, if it were possible, even redder. "Those parrots are cold," he started blustering. "You are not looking after them properly. Why the hell haven't you put a heater in there? Don't you realise how cold it is in there? It's like a bloody icebox. Of course they are dying, what would you expect, that they like being frozen? My God, it's a good thing we came to see, Ralph."

Reg was right, it was cold. Santa Cruz was always cold for a few weeks each year. Maybe Reg did not realise this for, as far as I knew, he had never been to the lowlands during the winter months. When, a week ago our Dutch friends had told me they, too, were concerned for the parrots sitting in the open woods, I rang up the zoo's vet. He told me not to worry about it and, in response to my suggestion that the galpón be sealed up to keep it a bit warmer, he said: "None of the zoo's birds ever die from the cold. Why should yours? The birds are suffering from "Pneumoconiosis", a fungal disease of the lungs, and they are best left as they are. And, don't forget, the birds in the wild aren't all dying either, they're quite hardy."

Hardy. Yes. When I explained all this to him, he simply refused to listen. He could swear like a trooper when he was annoyed, and this he started to do, "If there isn't a fucking heater in that bloody galpón by the time I get back this evening, I'll quit. You can forget your bloody project. I'll be through." With that he turned on his heels and left, taking Cheri (who agreed with Reg) with him. Ralph lingered a moment to tell me not to get upset. Don't worry, Robin, was the way he started nearly every conversation with me, and patting me on the shoulder, turned and followed the others to his jeep.

I suggested he destroy them.

I stood there staring at the fifty-dollar note Reg had thrust into my hand and then slowly tore it into tiny pieces. I needed to smash something. I should have waited to smash the heater the fifty dollars was supposed to buy. But even this was not the showdown; the day was to be long, and that little event took place in the evening. The only consolation the day held for me was to come later; my chance to unwind at a friend's wedding supper.

By the time lunchtime came round my petulance had subsided. I'd even managed to persuade myself that for the sake of the Park I had to learn to endure such torments, and I could make a start with Reg on his return, show humility, humble myself. I went out and bought a heater, and with Miriam's help spent the afternoon filling the holes in a course of latticed bricks with newspapers. Standing on top of the ladder, stuffing the papers into the crevices, I really did feel out in the cold. The others had gone off to Buena Vista to see the rest of the parrots, visit our Dutch friends (of the "Dat moet een tuinkabouter zijn" fame) and talk to the owner of the Estancia San Rafael de Amboró, Rafael Mendez.

It was Ruben Poma who first took me to the Estancia. We were on our way back from filming in the Park. We had been up the Rio Mucuñucu when a violent storm broke out to put an end to our plans. As we had failed to get a good panoramic shot of Cerro Amboró, Ruben judged the Estancia as being the place to offer the opportunity. Turning into the Estancia, I remember being very hungry indeed, and at the time I could not understand why Ruben had chosen to delay our longed-for breakfast by stopping off to film the mountains. I shouldn't have worried. Pragmatic to the last, Ruben knew we would be offered the traditional Camba hospitality, and very soon we were slurping cups of strong coffee and tucking into sizeable platefuls of food.

Rafael Mendez's house stood in the middle of a flat meadow spread out like a green carpet on the top of an embankment. The meadow was flanked on two sides by forest, from the edge of which clumps of bamboo sprouted here and there to break the leafy walls in a charming manner. At the back of the house the land fell away to a winding stream; with its string of weed-enamelled pools it looked like one of Atahualpa's emerald necklaces. Beyond the stream the folds of the hill were drawn together into a steep knoll, which looked man-made — recalling to mind the one at Glastonbury, and excited the imagination with suggestions of buried treasure. While Ruben was chatting to Rafael and his unusually blonde wife (she could have been a "Glastonburyian" herself), I took the opportunity to climb the knoll. Looking down, past the stream and over the roof of the house, I could see a shallow lake. Raising binoculars to my eyes I saw it was inhabited by waterfowl and rose-coloured spoonbills. The whole setting had a park-like appearance reminiscent of a Kentish manor house I once visited on the North Downs. Even without the backdrop, few people could resist falling in love with the place, and I did, head over heels!

The land beyond the lake, as far as the foot of the Andes ten miles to the west, rose in a series of ancient river terraces, apparently covered by uninterrupted forest. Until now I had resisted the urge to look at the mountains and focused my attention on the unobtrusive allure of this pastoral scene, but like a mighty king commanding attention by his grandeur — indifferent to his absolute authority, sat the "Big Chief", the green velvet of his cloak falling aside in a series of sweeping curves and deep folds.

I had never seen the peak from such a vantage. Much nearer now, the rock itself was no longer the Chief's what-you-ma'call-it (as it had appeared to me on my first evening outside Buena Vista), it was El Cacique himself. "What I would give to live here," I said to myself. And no sooner had this dream suggested itself to me than I made plans to do just that.

Rafael Mendez was not an easy man to deal with, he was foxy. Had he been more prepossessing he could have passed for a wolf. During my visits to Buena Vista, I nearly always drove the extra twenty miles to talk to him. Soon enough he admitted to me he was thinking of selling the Estancia, at a price. The wily old man knew he had me hooked, but at the price he was talking about it would always remain my dream and nothing more. This was my strongest card for at first, I didn't take the negotiations at all seriously. Fish or fisherman? I'm not sure which one of us was which, but it was not long before we were discussing the price. I gradually pulled him in: fifty thousand dollars became forty, rapidly fell to twenty-five with all the cattle and the horses thrown in, became fifteen for the house and half the land, but no livestock. I was working towards eight thousand dollars with a third of the land (about 120 acres) and the two horses. I still couldn't quite afford it, but Reg could!

When Noel and I submitted our Amboró project to the Senate, I had already anticipated Rafael's intention of selling the Estancia at a realistic price, and had included the cost of the entire Estancia in my part: "The Practical Development of Amboró National Park" of our joint report. When we went to look at the place, Noel thought the whole lot was worth twenty thousand dollars. I saw it as an ideal centre for the proposed Park: Warden's house and office, visitor's centre, guard station and paddock for their horses. The house was rustic and large by local standards, and its living space amplified on three sides by a broad veranda. I knew I was destined to be Warden, and prior to Reg's visit, our negotiations had got to the point where I knew Rafael would conclude the deal on my next visit.

"I have in mind the time when you have the Estancia as a base," Reg had written to me back in February. He knew a hook when he held one, and like Don Quixote's horse, Rocinante, I would have cheerfully supported him in any of his vainglorious quests for just that one carrot.

And this is where the matter stood on the day of the showdown. My promise to knuckle down to Reg's overbearing manner was then moulded by the object of my desire, the Estancia; not the Park, for that was something Reg could never take from me.

The moment Reg and the others left for Buena Vista without me, I knew, don't ask me how, but I knew I would never live at the Estancia, a cold realisation that added to my shivering body as I filled up the holes in the galpon's bricks.

I hadn't told Reg how close I was to getting Señor Mendez to agree to my offer. It was my intention to surprise him with my success, and thereby consolidate my already considerable right to live there as Warden. By involving Ralph, the professional realtor, it was likely all my efforts to clinch the deal, even buy it for ourselves if necessary, would be wasted. Two naive gringo gentlemen, Rafael would chew them up.

Without deception I had convinced Rafael that we (meaning the Park) had very little money, and as was true of all Bolivian projects we had to operate on a shoestring. Faced by two well-to-do gringos Rafael would drive a hard bargain, and probably outfox them on all the small (but to me important) issues as well. Worse, he might push his luck and put Reg off for good. However, there was a fresh development to prod Reg's interest, though in all other respects I considered it negative.

Reg had persuaded Operation Raleigh to reinstate Bolivia on their itinerary (the British Ambassador had wisely put them off in the first place) by convincing them they would help the Park by their efforts. He now needed a base to offer them and the Estancia, I hoped, would prove too irresistible for him to turn down — especially for $8,000.

Reg, Ralph, and Cheri got back from Buena Vista quite late. Their obvious gaiety and lack of sobriety was heightened by the contrast of my own sober mood. I was waiting to go to my friend's wedding party in a state of repressed expectation. I'd sat around smoking endless cigarettes, longing for the relief the party would provide me after the rigours of the last few weeks. I imagine my own solemn disposition must have appeared highly amusing to this inebriated trio; and with a reversal of roles as played by Felipe and I at the River Chonta, they would have been better off letting me lie. But in the security of their drunken camaraderie, they didn't.

Cheri was the first to start, "Hey, R.O., we've had such a good time. As you can see we've had a few drinks. Reggy bought them all, I mean. Hey, and guess what, R.O.? Reggy is going to buy the Estancia and I'm going to live there and breed toucans. Reggy really likes the idea, R.O., and I'm so excited!"

Ralph also took up the business of the Estancia, "Don't worry, Robin. I think Reg is convinced we have to buy the Estancia and Rafael has more or less agreed to sell for $20,000. There are still a few matters to clear up, but don't worry, Robin, I think you will get your beautiful home. I know how much it means to you, Robin. So don't worry."

Reg, flushed of face, weaving around the furniture on his way back from the lavatory, was more direct. "Did you do as I told you, Robin? Is the heater in place? Have you put plastic sheet over the holes in the bricks?" I just nodded at him, my mouth turned down, my eyes turned up in patient resignation. He and Ralph went outside to take a look.

"And guess what R.O.?" Cheri started up again. "Reg was really impressed by the parrot thingummy in B.V. What do you call it, you know the reblitational thing? He really got on with Pieter and Els and he wants Pieter and Els to help us out with the Park, R.O. Reg says he can find the money, R.O., and everything'll be great, and Leonard and Chela are leaving, R.O., and Pete and Els are going to stay on, R.O., and look after the parrots and the horse, so we won't be alone. Everything is turning out great, it'll all be so great, R.O. Wow, I'm so happy."

I cannot remember exactly how the showdown's closing stage began, but if you consider its beginnings were planted in the events of the morning, then the evening was the end. I know we were all in the room together, now joined by Matiash the video maker. I think the trouble started over Reg's plans for Operation Raleigh.

Raleigh, you may recall, was a worldwide scheme organised by the explorer Colonel Blashford-Snell (in Ethiopia we called him "Colonel Pong") and a team of experts. It was proposed by Prince Charles as a follow-up to the successful Operation Drake, which had given thousands of young men and women an opportunity for adventure and the chance to rub shoulders with people of other cultures. The Venturers, as the young people were called, were expected to participate in amateur research projects or in community work, and each team was supposed to be organised by suitably qualified leaders. I did not see in what way Operation Raleigh was likely to help with Amboró, with the advantage of hindsight, correctly so, and I resented the large amounts of money the scheme would squander, given that its aim had been unnecessarily misrepresented by Reg (by which I mean he sold it to the Bolivians as a prestigious scientific expedition, which it was never meant to be). Raleigh's proposed budget for its Bolivian phase, about $180,000, would have provided sufficient funds to keep Amboró running for eight years, and at the time, as I have told you, Amboró was almost entirely dependent upon my own private funds. Now Reg was proposing to raise $20,000 to buy the Estancia for Raleigh, and dearly as I wanted to live there, in terms of priorities for the Park and the shortfall in funding, it was a waste of money. The cost of the Estancia was originally supposed to come out of the $250,000 Reg assured me he would get for the Amboró project; it was not supposed to be the be-all and end-all of our plans. The Park desperately needed a Warden, guards, an office, a vehicle, and about half a dozen large notice boards.

When Reg and Ralph informed me they were going ahead with the purchase of the Estancia, and with the same breath admitted that no other funds would now be available for me to keep the Park going, my peevishness of the morning turned into anger. Eight thousand dollars was one thing, twenty thousand quite another. Eight thousand was acceptable, as it would permit me, as Warden, to live close to the Park, from where I could get on with the project. Twenty thousand, so that a group of young people could have an adventure, as I saw it, was completely unfair. Fair and unfair, words which always came to my lips in my dealings with Reg, but aware of their childish connotations, words I rarely expressed. But to get to the heart of my feelings, I thought it unfair that he was now planning to spend money, money he told me he didn't have, buying approbation in England at the expense of Amboró and myself.

When Reg launched into his standard pep talk about the benefits of Operation Raleigh's proposed visit — how good it would be for the Park, what an opportunity for me to make my name — I resented the unwitting patronisation and transferred this resentment to a diatribe about the waste of money and the snobbery that I saw lying behind the scheme.

Unable to keep my earlier resolution to maintain the peace, goaded as I was by their lack of concern for the shortfall of funds for Amboró, I condemned the Raleigh proposal (that prestigious scientific expedition) with the palpable derision I thought it deserved. To somebody like Reg, bent on making his mark with the Old Boys back home, I was not only demonstrating my contempt for him, but also showing intolerable disrespect for Prince Charles and patriotism in general.

Gripped by paroxysms of rage he started yelling at me, "I should have listened to the bloody Embassy. They said you were a fucking nuisance. They said that you were anti-establishment and a bloody troublemaker. I have had enough of you. Why for Christ's sake did I take you on?"

Stamping the floor, spilling his gin all over his front, bright scarlet of face, Reg looked so ridiculous I couldn't help but start laughing, even jeering, "Well, there you are, Reg. You should have bloody listened to your precious Miss Coombs at the effing Embassy, you miserly misshapen little man." Later, I did not feel proud of myself for this exchange, but at the time — my feelings having been too deeply gouged by Reg's reference to such a sensitive issue — I poured scorn on top of insult, and continued, "What I would like to know is why you're doing all this, Reg? What are you bloody trying to get out of it? You don't have the slightest interest in Conservation. You don't even like animals. You with your bloody 'well, kill all the parrots' and your effing Operation Raleigh. Go on, tell me. What is it you want? No, wait. I'll tell you. You just want a bloody O.B.E., don't you? You want to kneel in front of the Queen, don't you? Arise Sir Hardy. That's bloody well it, isn't it? Well fuck you, Sir, I've got a wedding to go to."

I turned away from him and started on my way to the door. I'm not sure what happened next; with my back to him I didn't see what he did. All I heard was, at first, complete (stunned?) silence, then a tremendous CRAAASH! The sound of breaking glass followed by the thump of his body falling to the floor. I turned to see Reg lying in a puddle of beer and gin, his eyes turned upwards into his skull, silently mouthing words nobody could hear. My God! It looked as if he had just had a fit, or maybe a heart attack. It seemed Reg's stab at my most sensitive feelings had been amply repaid by my thrust to the heart of his.

We all rushed over to help. Somebody said they thought he had hit the floor with his head; others that he had hit the edge of the table in his fall. Full of remorse now, I wanted to ring for a doctor, but the others thought Reg would be all right if we laid him in his bed, which we did. Then, typical of his friends, they decided there was nothing more needed of them and they started to leave. Ralph first, "I'm sorry, Robin, but I have to go to a dinner engagement. Don't worry, I think he'll be all right." Then, of all the bloody cheek, after having amused herself all day, Cheri announced she was off to the wedding with Matiash and that I'd better stay and look after Reg.

I hadn't intended doing anything else. But Cheri's complete abdication of her duty — she was once a nurse—leaving her good friend Reg in the lurch, and her selfish disregard for the seriousness of the circumstances shocked me with its cold disdain. When they had all gone, I went in to check on Reg. He was awake. He had a nasty looking bump on his forehead and was complaining about the pain that we alleviated with a bag of ice cubes. Half unconscious, half awake, Reg mumbled to me he would be okay, that I should leave him to sleep.

By the next morning he was well enough to get up and, surprisingly, apart from the swelling on his forehead he didn't seem any worse for last night's incident. Both of us, feeling a little ashamed of ourselves, but still resentful, said little to the other. Ralph came by, and after Reg had drunk some coffee and breakfasted, he went off with him to catch his plane. I knew if he ever returned our relationship was going to be on a new footing.

A few days later he sent me a letter, and all credit to him, it was written with business-like efficiency, and contained no reference to the showdown or any hint of malice. It was a brief summary of all our discussions over the Park, and the good news that it looked as if the Decreto Supremo would soon be approved by the Senate. The only thing missing was reference to his promise of the $1,500 we were going to need for the next three months.

I suppose I should have written to him immediately, but I decided not to bother; that one way or another, with or without him, I was going to get Amboró going—fulfil my promise to the dead otter, comply with the demands of my daemon, throw all caution to the wind, do or die, and move up to Buena Vista as soon as I could. No longer, I promised myself, was I going to tailor my plans to suit the false hopes held out by others. I was going to make Amboró National Park my Park, and to hell with everybody else. I sat down, lit a cigarette, and revised the current situation.

Reg had started intimating, even before the night of the showdown, that he would like to find somebody else to work with. Dunning had finally sent us some more money. The Parrot Rehabilitation Project was running smoothly and an end was in sight. Our Dutch friends in Buena Vista told me there would be a whole house available at El Hombre Nuevo at the end of September since Leonard and Chela had made the decision to leave Bolivia for good. It really did look like Amboró was going to get its Decree to make it an official National Park. The CDF had once again changed its entire staff, and the new faces, having gotten to know more about Amboró through the parrot project, were making the right noises — more involvement in the Park. They suggested I work for them, albeit on a semi-voluntary basis. Noel's bird book was soon to be published, and, busy as he was with Huanchaca, was happy to leave Amboró in my hands. The Unión Juvenil had become my allies and I was accepted as honorary member and advisor. The animal and bird exports were on their way to being closed down and a permanent ban had been rumoured. Apart from the now occasional watcher at the gate, I had no reason to expect any serious trouble from that quarter.

But my relationship with Cheri had taken another turn for the worse and we now occupied separate rooms in the house. She spent most of her day listening to music or practising her flute. In the evening she went out with her friends or brought them back to the house, the sound of revelry and laughter often continuing late into the night. Oblivious to the squawking of the toucan that now shared her room, she slept late into the morning. For all intents and purposes we had separated.

My spare moments were spent in reading and writing or watching the TV. Miriam was the only person I saw regularly. It was during this time that I at last made the effort to brush up my Spanish. I say "at last" because it was evident to me my language shortcomings were an obstacle in my negotiations over Amboró. Miriam's first English lessons enabled me to revise my Spanish from scratch, and perforce correct many bad grammatical habits I had picked up. Since I usually turned in early and rose with the dawn, it was not surprising I didn't see much of my wife. When we were alone there was no real rancour between us. We would talk freely together and sometimes spent time in each other's rooms. Our mutual friends would visit us together or separately. It was an unconventional arrangement, but under the circumstances worked surprisingly well.

So apart from my financial circumstances everything looked as though it would be resolved. For the first time, I saw I would soon be free of all those responsibilities that kept me tied down in Santa Cruz. It was towards this end that I now worked.

By the end of July I was back from my trip to Peru. In spite of the heater, the sick parrots had continued to die; in fact, the number dying each week had increased. We had decided to give the rest a chance of survival by moving them up to Buena Vista. Reg had taken up an option to buy the Estancia that would expire in February 1985. Rafael, the wily owner, had somehow managed to sell off sixty acres in the meantime.

The Estancia would be purchased in the name of PRODENA-U.S. Inc. Originally there had been just one society — PRODENA Bolivia, with its headquarters in La Paz. Its activities were controlled by its President and the members, together with Reg. By the end of 1984 there were three more societies: PRODENA-U.S., PRODENA-U.K., and PRODENA-INTERNATIONAL. All were registered charities with Reg in control. Each had a board of directors made up of his friends and business contacts; none of these people lived in Bolivia, some had never visited it. I do not know if Reg had to do all these things for legal reasons, or whether, as I suspected, he did them to ensure he had complete control of the money – both, probably. In any case, it smacked of high-handedness and led to the belief among the Bolivians that they were being outflanked, which of course they were. Reg had always wanted to run Bolivia's conservation programme through PRODENA, and on more than one occasion had stated that only he, one or two nationals, and I were of any use to the Society and that the rest were not worth consulting. This was another bone of contention between us. I saw the future of conservation as being dependent upon the Bolivian authorities, principally the CDF. I did not like to see the executive power vested in outsiders, and Reg never suggested one of the local Bolivians (or needless to say, myself) be on one of his boards. [It may surprise many readers that these notions were even possible, but at the time of these events, conservation was still a primordial concept among Bolivia's intellectuals and politicians.]

Considering the seriousness of the showdown, my relationship with Reg continued much as before. We were still receiving a steady stream of telexes which he usually signed off with, "TTFN, Regards, Saludos, Love, or, alles van die beste." Our replies were invariably signed, "Love Robin and Cheri."

On the 21st of August the news we had been waiting for arrived in a telex from Reg, "You may have heard inside information from the palace that the President signed Decreto last Friday, August 16th. Have a drink on me."

Finally, and yet in record time, only eleven months after we started with Dunning's money, we had the Decreto Supremo: No 20423! "John," I thought, "will be proud of us."

John had been trying to make another visit for some time, but on every occasion something occurred to prevent him. We had just received a letter from him dated 20th August, "Some family developments here are going to postpone our trip to Bolivia but rest assured we will be there just as soon as we can." Hardly had I finished reading it then the telephone rang. It was John, to say he was arriving in two days, "Can you get my room booked at the Hotel Asturias? Will you arrange a chauffeur driven vehicle and some people to work with me?"

They arrived on the 1st of September. John wanted me to work for him and generously offered me fifty dollars a day to do so. Unfortunately, I was not in a position to take fourteen days off, but I arranged for three of our gringo friends to help him, one of whom had done so in 1983. He also brought with him a cheque for $2,000, which for tax purposes he had passed through the World Wildlife Fund. This was good news; in order to legally pass money to us they had given us an official project number and title: WWF Project Number: US-351, Establishment of Amboró National Park.

I was a little bit disappointed by John's reaction to the good news about the Supreme Decree. He kept saying, "That's wonderful. Congratulations," words he often used for quite trifling matters. And a few days later he asked me if I thought we would be able to get the Park officially declared.

It was, as I said, unfortunate that I could not spend more time with the Dunnings, especially as we were so short of money at the time. Even with the cheque John had just given me, he still owed me $2,500, and Reg continued to disagree with me over the $500 he promised us. I think John took umbrage at my rejection of his offer of employment, and this may have had something to do with his eventual refusal (after he sent me a further $1,000) to finally settle the matter: "I admire your persistence in trying to get Amboró in a permanent, secure status. After all I have invested $6,500 in your attempt. That original offer was based on substantial help from other sources...so I agreed to make up the difference between your other sources and $8,000. I never expected my share to be over half of it."

Since my trip to Peru, I had been constantly on the go. As I had persuaded Bernie Peyton, the Spectacled Bear expert to visit Amboró in October, no sooner had John left than I was again occupied with visitors to the Park. I had already taken four Argentinian biologists, and a valiant Englishman called Bob Le Souer (who came on behalf of Operation Raleigh) on field trips to Amboró. Clyde was continually in and out of the garage, the parrot project still required my attention, and I was trying to find the time to prepare a report and do the accounts for the New York Zoological Society as Reg had requested. Not only did I have to get everything ready for Dunning's last-minute arrival, but to cap it all, John brought a bulky correspondence from the World Wildlife Fund that required me to prepare an Annual Report on the Amboró project, and to account for all the monies spent so far. The latter was unreasonable, since the money I received from John was a personal gift, and I had had no reason to record our expenditure, nor save the many receipts that were now being demanded. On top of all this I still had to contend with a steady stream of telexes and letters from Reg, and I had many urgent things to do, some for the CDF, after the Decree of the 16th August.

Needless to say, the situation in Santa Cruz had not improved, and the constant failure of even the simplest errand was adding to my extreme frustration. I no longer had any spare time to read or watch TV and hardly ever felt energetic enough to deal with my own personal problems: sort out my marriage, write to my family, look for a source of funding, and so on. My sleep was interrupted by visions of persecution: horrendous half-humans chased me around the streets while my family and friends called out to me as though from far away, "Robin. Robin, where are you? Why don't you answer us, Robin?" I was near to collapse.

Bernie Peyton rang me up on the morning of the 7th of October. He was calling from his hotel room in Santa Cruz. Having arrived with a friend of his late the previous evening, they had gone out to find somewhere to have supper and a quick drink in one of the local bars. Apparently, they had run into trouble almost immediately, "Bienvenidos a Santa Cruz", over something to do with bird exports. They had, he went on excitedly, fallen into conversation with some men in a bar when my name cropped up over the animal export business. Bernie and his friend, Randall, had jumped to my defence when one of them started making disparaging comments about my part in it. Then without warning, another group of men got up from an adjacent table and surrounded the two Americans as they sat at the bar. What had started as an innocent conversation now turned into an ugly confrontation. Jeering and insults progressed to shoving, and then someone threw a punch. Outnumbered three to one, Bernie and Randall took the hint, quickly backed out of the bar, and away into the street.

"What the hell's going on here, Robin? I think we had better get together immediately. Can you come to the Hotel Asturias? Now?" I put the phone down. "Here we go again," I thought. "So it's not over yet."

We spent the whole day preparing for our visit to the Park. Randall was a part-time journalist living in Peru and Bernie had invited him to join him on the trip. He had accompanied Bernie on several trips before. Randall wrote the odd article for a magazine called "South

American Explorer", and he was sure he could place an article on Amboró in one of the next issues. "My ambition is to write a book on him," he said, indicating Bernie with his slightly mocking manner. They were obviously good friends and I looked forward to their company for the next two weeks.

Their unpleasant confrontation in the bar had left them more puzzled than annoyed. Their second evening in Santa Cruz was to fill them with horror.

This particular evening started in our house. The heavens were filled with millions of stars as bright as only they can be in the unpolluted skies of far-flung places. Immediately above, Scorpio followed the inscrutable pattern of Leo to the east, and to the west, Orion the Hunter aimed a steady bow at the head of Taurus. Set in the blackness, the Seven Sisters stood out like the cockade on a bull's leg. The stillness of the night was only broken by the sounds of the city, our passionate conversation, and the murmur of small owls as they spoke to one another of their desires.

I had run around all afternoon trying to put Clyde into good enough condition for their visit. The two Americans had visited Noel and the CDF, and they had dealt with all the irksome chores that are the lot of the traveller in foreign parts. By evening everything was ready. Cheri wanted to go out on the town, and eventually it was decided she would accompany Bernie and Randall, while I stayed behind to look after the house, this night being one of the rare times all three girls had gone out. After reading for a while in the peaceful silence after their departure, I fell into an exhausted sleep.

At three-thirty in the morning I was woken by a low whimpering from the passage. I lay there for a moment trying to make out what it was. Then it dawned on me that it sounded like Cheri crying, but eerily, the cry of a violated spirit. It was not the uncontrolled whimper of the wronged, the angry, or the frustrated; what I was listening to was the sound of a tormented soul, inhuman in its heart-piercing forlornness, a noise compounded from all the world's saddest moments, the wail of a supernatural grief.

I lay there a fraction of a second longer, the hair on the back of my neck tingled with alarm. Suddenly I knew something awful had happened. I flew out of bed to the door and on opening it saw Cheri slumped to a squatting position on the floor. She had her arms about herself, like a soldier, who having been shot in the stomach tries to hold himself physically together. Her clothes were all torn; she had lost one shoe, her hair hung down in muddy coils. She was trembling with uncontrolled emotion, her eyes were tightly closed, and through their swollen lids huge teardrops gathered and fell to her lap.

She could not speak. Shudders shook her whole body, and as I bent to lift her from the floor, another loud wail, almost a scream, came from her lips. I saw she had been hit about the face, one cheek was swollen and there was blood at the corner of her mouth and visible on her teeth. I lay her down on the bed and with difficulty got her to sip a little whisky, hugging her to me for comfort.

It was a long time before I managed to get a word out of her. Then it came out so slowly, like the blood seeping from her mouth, those ghastly, inevitable three words, "I...was...raped. They...raped me, R.O., I couldn't do...anything. There were t-t-two of them...R.O."

I took her into the shower and helped her wash all the evil scum away. I soaped her and shampooed her hair. She kept asking me to do it again and again; she wanted to clean herself to her very core, to wash not only her skin but also to purge her soul; wash the memory from her thoughts, expunge the incident from her mind. The more the warm water soothed her, the stronger came her tears. Now words started pouring from her in an endless stream as steadily as the water from the shower head. "I know I did a stupid thing, R.O. I know I shouldn't have come home in a taxi alone, but you see, R.O., we all went in the taxi to Bernie's hotel and I thought it would be all right, R.O., but it wasn't, the bastards, the bastards, the mother-fucking bastards. They took me, they had guns, R.O., they...they...Oh God...oh my God. I couldn't do anything. I didn't know where we were, R.O".

"They...they...took me to a field, I don't know where it was, somewhere near here, not far from here, they...No, one wasn't so bad, but the other one, R.O., he was a real bastard, R.O., I was trying to fight him off, but he s-s-said if I didn't...he was going to kill me, there was nothing I could do, R.O., then...then the other one...he wasn't as bad, I kept trying to fight them off but I couldn't. One held me...down, R.O. Then I got free and I ran, I was screaming for help, for h-h-help but there was no one there, no lights there, nobody heard".

"Then they came after me in their taxi, I...I think it was a blue one, a...a To-Toy-Toyota, R.O., not a real taxi, you know. They tried to kill me then, they...I was running as fast as I could, they...they tried to run me down. The bastards, the fucking murdering bastards.

They RAPED me and they tried to KILL me, R.O. Why couldn't they have just let me go, R.O.? Why did they have to tr-try to run me down? Oh my GOD, my GOD...," but her words were cut off by a long scream and a renewed spell of whimpering.

Sobbing, she had fallen into a delirious unconsciousness. I sat beside her stroking her hair, my thoughts full of the most dangerous resolutions. How I was going to find the bastards, torture them, and kill them...slowly. That car, could it have been the same one the watchers used? A blue Toyota...was it THE blue Toyota? Was it THEM? The animal mafia? "He will have you killed. You will be eliminated." Blue Toyota...blue Toy...blue..., I fell slowly, very slowly to sleep, whilst equally slowly the sun rose in the east.

The light, I thought to myself, to light up the scene, that field, those two men, their guns and my poor, poor ravished wife.

5. Buena Vista

I put Fawcett's book down and pressed my face to the window, one of those absurd aircraft ones, small and deeply inset in its white plastic moulding. It was like sticking one's head into a potty. "Why," I wondered, "does everything to do with modern air travel have to be so humiliating?"

It was now light enough to see something of the Amazonian Rain Forest far below us, a light mist hanging above the trees, an endless sea of foamy green here and there broken by the serpentine course of a slowly moving river, its convolutions like the writhing of a dying snake, loop after loop, each coil's throe recorded as an ox-bow lake.

My first glimpse of South America; the ocean of trees evoked memories of everything I had so fervently read about it. Though nothing else relieved the monotony, the effect was mesmerising. I could have gazed at the scene forever, but the porcelain-coloured clouds soon returned and blotted it out. I returned to my book. Fawcett was describing his ascent of the Rio Verde flanking what he called the Hills of Ricardo Franco, what the Bolivians call "Huanchaca", and what Conan Doyle came to call "The Lost World", that in his imagination he populated with prehistoric monsters. If only he had known, for now as I sat reading, monsters of quite another sort were inhabiting the place.

I looked up from my book into the sleep-worn features of the ageing airhostess and gratefully accepted the cup of black coffee she offered me. We were, the intercom informed us, soon to begin our descent towards "El Trompillo", the airport serving Santa Cruz de la Sierra. While sipping my coffee I maintained a steady watch in the hope of seeing the forest once more, but the sheet of cloud smothered the view on all sides.

Then quite suddenly a hole appeared and rushed towards us, and squinting downwards I saw a cluster of buildings straddling a hill. From every side, like the strands of a spider's web, dusty white paths converged on a red brick church that sat like a well-fed spider at the centre, the shrivelled skins of its discarded meals, the uneven rows of

164

houses astride the tracks. At the foot of the church a green tree-shaded square was placed like a Thanksgiving offering. A line of piebald cows was walking down one of the tracks; their progress marked by dust that hung in the air like a vapour trail behind them. A glimpse of a tall microwave tower, a quick slash of eroded red soil, and the scene was gone as quickly as it had come.

Writing about this years later I recall the scene as clearly as if it had been yesterday, for it was burnt into my memory with the sharpness of a branding iron. And as fate would have it the scene and the circumstances in which I saw it were to take on an almost occult significance for me. It was my first glimpse of Bolivia and I remember quite clearly how the sight filled me with a curious, inexplicable longing to visit the hilltop village. I'd fallen immediately and entirely unreasonably in love with it, a "coup de foudre" as the French would say.

The village was, of course, Buena Vista; it was destined to be, because the Gods had planned it so. Buena Vista, the good view through a hole in the clouds.

"We'll have him sit on the right. The plane will have to make a slight detour to the west, a thunderstorm due north of Santa Cruz should do the trick. Okay? Now, we will have to time it just right you know, we can't work miracles. So careful, you fellows, we need a little hole in the cloud at just the right place."

"We believe he will be engrossed in a book."

"Oh hell! Okay. A distraction, err.. I know! Get the stewardess to bring him a cup of coffee. Put it on a tray; he can't hold his book and the tray at the same time, can he? I know his sort. He's a bit shy, bound to look away out of the window as soon as he's accepted his coffee. Go ahead. I'm sure it will work."

I made no immediate attempt to identify the hilltop village, "coup de foudre" or not, for there were other towns it could have been; or so I thought. As soon as we had settled into our provisional home in Santa Cruz we were moved to go exploring. I recalled the little town and studied the maps. It didn't take us long to realise it was probably Buena Vista. When, eventually, I went there I recognised it for the very same place. Even now, when the normal flight path to Santa Cruz is cut off by storms, I hear the Miami flight passing overhead.

What a lot of fuss about nothing! Maybe. It depends on your beliefs. When it comes to such subjects, spiritualism, clairvoyance, I believe the parameters must be very subtle, for if not we would know more about them; or, of course, you're probably right, they don't exist.

My gut tells me they do. Once, on my way to Buena Vista, driving Clyde at full speed along the main road, I suddenly had a strong feeling I should slow down. A few hundred yards further on a vehicle approaching from a side road skidded at the junction and came to rest across my path. I just had time to slow even further, glare pointedly at the white-faced driver, and sweep safely by. Premonition? Coincidence? Chance?

****

I left for my new life in Buena Vista on the 1st December 1984. I packed all my precious possessions into the back of Clyde and left Santa Cruz. Although I say all of my possessions, I didn't have much: my bed and mattress, a table and four chairs, the typewriter, two large boxes of books and clothes; everything I owned was piled into the pickup's open back. Steptoe's nag could have easily coped with the lot: "Harold, you're too soft on that 'orse". Most of our furniture had been sold and the remainder lent to my friend, Nick, to fill his half-empty house.

Cheri had moved in with some friends of hers, taking the dreadful Cyrana with her. At last! The house stood bare and empty.

God had blessed me with a dry day; the move would have been inexpedient had the unabated rain of the two previous days continued. The only dark cloud in the sky that morning was Mrs Tartar who came to make sure I wasn't stealing the plumbing. She insisted I hand over the cash to replace two cracked panes of glass, and then refused to return our deposit.

I said a last farewell to the girls, the remaining "Paquio" tree, and the little colony of Thornbirds chittering from their nest of twigs hanging over the gate. Miriam's parents had given her permission to move with me to Buena Vista, but as she was to follow me in two weeks, this day I set off alone.

The landscape was wearisome: flat grasslands, bare ploughed fields, and vast areas of sugar cane, the only feature of interest being the cloud-shrouded mountains of Amboró thirty miles to the west. After half an hour of monotonous repetition past the cotton-ginning plant, the cooking-oil plant, the milk-processing plant and the offensive black smoke of the plywood-processing plant, I arrived at Warnes where I stopped to fill up with petrol. The heavy night's rain had left the courtyard of the filling station under a foot of water, but rather than waste time queuing for service at the busy town of Montero, I waded through the flood, and was soon on my way again.

The drive as far as Montero reminded me of the endless commuter trips I had made whilst working for the British Mission. The flooded ditches to each side of the road attracted large numbers of Snail Kites. The top of nearly every fence post was graced by one of these interesting birds staring down into the weed-choked water. Curiously, though birds of prey, they feed exclusively on water snails, winkling them out with their sharply hooked bills. The heap of discarded shells lying below each post reminded me of a "Thrush's Anvil".

I was taking more notice of the bird life because I had decided to make an inventory of all the species to be found in Amboró. Noel had told me the Park had a very diverse avifauna; a statement borne out by a collector who knew the area well. During my visits there I had also gotten this impression but, in those early days my trips to Amboró were taken up with more mundane matters. I wasn't expert enough to identify the great variety that could be heard, and more rarely seen. "When I live there I'll have time", I said to myself. And so it was to be.

After Montero the road stops thrusting north, or rather, from there I turned west towards Buena Vista and the low rolling hills that presage the rise of the Andes. After a few minutes I slowed to a crawl, obedient to the notice which told me "Velocidad Maxima 15km", and leery of the white-helmeted sentries outside the headquarters of the regiment, the "Panteras", who many years ago captured Ché Guevara.

Then I went on through the river-road of Puesto Mendez, the very troublesome section I described to you when I first made my way to Buena Vista. Today the water was shallow and none of the tractors or boy-pilots stood about to offer their services; only one snow white Egret was there in their stead. Set beside the road on a low concrete plinth was a small black cross and a fresh bouquet of red flowers. As is the custom here, this forlorn little shrine marked the spot where somebody had died. In some places, "Accident Black Spots", dozens of these shrines crowd the edge of the road. Each cross, seldom more than a foot or two high, had the name of the deceased and the date inscribed on the crosspiece. This particular shrine, to Richard Parada, was amazing in two respects: in spite of the flood waters (which sweep away tree trunks and rocks) and the activity of heavy earth-moving equipment to keep the road open, the shrine survived it all and still stands there today. It was the only one I ever saw invariably decorated with flowers. What story could it tell? Whose beloved ended his days here? A hopeful proud father? A grieving mother? A loving wife?

After Portacheulo, another agricultural town (actually it was better known for the amount of "pichicata", cocaine base processed there), the road passed by spacious pastures in which hundreds of zebu cattle grazed. Here the asphalt surface of the road was severely damaged, and for the next fifteen minutes I had to creep along from one set of potholes to another, being careful not to unship my possessions from Clyde's open back.

The tropical forests of Bolivia begin about here. Although little is left now, large patches stood to either side of the road on that day. Along this stretch I often saw Cyrana-like toucans but, today I saw only the bleached bill and tattered feathers of one killed by a passing vehicle. The sad little heap reminding me of my marriage to Cheri, the colourful beauty faded away, the intimate warmth of its plumage squashed in the dust; one wing with a few primaries still attached lifted with the wind to wave a last farewell. 'Goodbye. Adios, my gay pretty bird'

****

Now a new bird was entering my life; no toucan, rather a vibrant hummingbird, as sweet as the nectar those little birds sought, as capricious as their darting flights of fancy, throbbing, pulsing with "Joie de vivre". Tremendous short bursts of energy when the mood took her, content to idle away long hours when it didn't. How would Miriam settle down in Buena Vista? Would she be lonely without the ebullient distractions of her many brothers and sisters? What about my Dutch friends, Pieter and Els, would they take to her too?

Lost in thought, full as they were with such imponderables, the last ten miles passed without me being aware of them. Then just before arriving at Buena Vista, at the bottom of "farting-bus hill", I had a flat tyre. I had been dreading this because somewhere under the untidy heap at the back lay the spare wheel and jack. Had Clyde's tailgate not been permanently jammed, I might have been able to slide the wheel out from under the bed but, as it was I had to unpack half the load before I could get at it. It, too, was flat!

Now I had real problems. With an open-backed truck on the side of a main road, without a garage, or even a house, or any trustworthy help in sight, I could do nothing. So, I reloaded my precious cargo and limped into town like a tramp with one boot. As before, I arrived thirsty and rather hangry. Guess who? Yes. Good ol' Melon Breasts with the same old slouch brought me a beer in the same small glass. As they say, nothing ever changes, we just get older. Bolder too, for this time Melon Breasts gave me a leer, maybe she did like me after all. After the beer I picked up the reinflated tyre and followed the narrow sandy track that led to El Cairo. Past the ever-broken water pipe (for a long time I believed it to be a spring), past the airstrip, the basketball court and the soccer pitch, and out into the fields again. Then on for another mile, and there I was, "El Hombre Nuevo" carved on the gate.

****

So here I stood, on the threshold of my new home and my new life; like a shipwrecked mariner, or come to think of it, more like the truck-wrecked driver of Puesto Mendez. My boxes, books, and bed cast down on the sandy space around me. Els and Pieter Brekelmans went out of their way to make me feel welcome, laughing cheerfully, kissing me, shaking my hand, and thumping me on the back.

Pieter, broaching a bottle of red wine: "Cheers Robin. Welcome to your new home. Welcome."

Els: "Oh, ya. Wilcome Rrobin. Wilcome. Ant your Miriamcita?"

My Miriamcita. Did it really show that much? Els could see it; she knew before me, I was in love with that hummingbird. I missed her. How many more days would it be? Had she said the 12th or the 14th?

It was easy to get along with these two amiable Netherlanders. Have the English always found it so? Certainly, throughout my travels I have counted Dutch people among the best of my friends. In their unassuming classlessness they lack two-facedness, the faithful attendant of the English character. No-nonsense folk, the Dutch take everything in their stride; and maybe because of the sea which threatens their odd little country, they are a light hearted, fatalistic, fun-loving people; free of pretensions and rarely snide. The ease of the relationship was disencumbered by language problems, since most of them seem to speak some English.

If there is a common character flaw in their make-up, it was Pieter who put his finger on it. "Neffer forgot, Robin, busy ness is busy ness and I hem a busy ness man", his attractive droopy moustache lifting in self-amusement, the twinkle in his green eyes calling into question the implied self-criticism, yes, Pieter was a bit of a wag. His florid face would split with laughter, and turning around, off he would go in the manner of a man used to tramping furrowed fields. He was the perfect foil to my own serious nature. Els always said we were like "brodders". She was right, I loved Pieter like a brother.

This was the first of many happy days spent in their company. We sat on the grass drinking wine and beer, getting a little high on the "buena vista" the hilltop farm afforded us. The same view I described on my first visit, that now seemed so far in the past but, was in reality only three years ago. Spread out at our feet laid the familiar scene: the scrubby marshlands, patches of remnant forest and green pastures, interrupted here and there by uncertain glimpses of the River Surutú's sandy beaches. Then, lifting our eyes, we could look straight down the eastern wall of the Andes to where they blued away sixty miles to the south.

Drunk with happiness at my escape from the big city, intoxicated by the wine, the beer and the bright cheerfulness of the waning moon, secure in our mutual friendship, free of past failures, and now with the promise of fulfilling my dreams, I slid inexorably into those dreams that were beyond my control and, which I only rarely remembered.

"Pieter iss in neet of a goot friend, Rrobin. Ant you are like his brodder. We well be so heppy togetter, wit you ant your Miriamcita. You well see. You well see. You well s...."

It was morning, the room already bright with sunshine streaming through the large windows of my bedroom. I lay there listening to the sounds outside; sounds which in their familiarity would soon go unnoticed by me. The mango trees were in fruit, and I heard the "woik-woik" of contented Oropendolas, attracted by the bright yellow of the ripening crop, as they stuck their ivory-coloured bills into the flesh of the overripe ones. The cry of large and small birds followed the daylight into my room. The harsh "caark-caark" of Purplish Jays and the unoiled hinges of the Araçaris' excited squeaks was now smothered by the grating arrogance of Chachalacas. The softer "tssit-ssit" of the Bananaquit and the low whistles of the "Sayubú's" (Sayaca Tanagers), all these avian sounds were syncopated by the "tock-tock" of lethargic cicadas.

Heaven or Hell? Heaven or hell? Heaven.... I drifted into sleep once more, a warm contented feeling replacing the anxiety of the last few years: Reg Hardy, Bernie Peyton, Cheri's rape, bird exports, blue

Toyotas, Italian terrorists, Mrs Tartar, cracked windows, Noel, the zoo, Amboró, Curassows with blue-knobbed bills, Raleigh, Rolando Paz, denuncias, Bejarano and Onishi, Oilbirds, Dunning, ecological societies, CIAT and the British Mission, El Belgrano, pineapples, Cyrana and parrot projects. The WWF, promises, money, funds, guns, cages, immigration officials and thick policemen, Cuzco. "They will kill you, Robin" of Gene and Chachi, decretos, newspaper articles, Illimani from the Plaza Hotel, "Pichicateros", angry Germans, threats in bars, "El Caballito", soft breasts and lips, Paquió trees, Don Juan's widling trick, flood waters, drunken Collas, "El Porton" and its Screaming Pihas, a dead otter. Heaven or Hell?

"He neets a brodder. He neets a brodder, a brodder, a brodder, Rrobin," Els's soft tones. "Woik-woik. Woik-woik".

"Are you sleep, Rrobin? I hof som' coffee for you".

Coffee! One day I will patent an alarm clock which says: "Coffee. Coffee, dear sweet, is ready. Coffee. Wake oh wonderful one."

"Coffee, I hof som' coffee for you." And there she stood. If Pieter was my brodder, here was my sister; with her large soft brown eyes, her too pointed nose, her angular cheeks, her inviting lips, now parted in a friendly grin to emphasise her prepossessing nature. "Blek coffee, Rrobin. One sooga, ya?"

"Tssit-tssit. Tssit-tssit. Tssit-tssit".

****

We now had to decide the fate of the few parrots leftover from the Rehabilitation Project. Pieter and Els had devoted a great deal of time and effort, "a real labour of love", to ensure the majority of them were able to return to the wild. Their success was evident with dozens of healthy birds circling over the copse where we had kept them; bands of thirty, forty, Blue-and-yellow Macaws, smaller flocks of Red-and-green Macaws and Military Macaws returning to their roosts below the house each evening. There is scarcely a more beautiful sight than looking down on a flock of Macaws as they wheel and turn above the dark green of the trees; flashes of brilliant red, yellow, blue and green, like living jewels, coming together, separating again, then landing like a shower of sparks in the canopy.

But not all can be harmony, and ten birds remained; incurably sick, broken-winged, or just too old. It was obvious to us these could never go back to the wild. They, we decided, would be returned to the CDF who could find them homes, or more likely hand them over to the University's vets.

Another problem was lack of money. As I have said, it was surprising how expensive it was to feed and otherwise provide for dozens of large parrots. Every day, maize, bananas and other fruit had to be bought, and twice a week we paid for sacks of nuts that the local people gathered from the "Motacú" and "Chonta" palms that grew in the fields and forests bordering the quinta.

It was the food that drew the Macaws back each evening, noisy groups squabbling over what we were able to provide. Now we made the only possible decision. The project would be wound up; each day we would provide less for the birds to eat until no more came. It was a hard but necessary decision. Pieter and Els had been unable to make more money from the farm than had the owners. They were poor, we were poor, the parrot project was out of funds. It was the beginning of summer, and with plenty of wild fruits in the nearby forests, it had to be now or never. Els cried.

****

So did I, but with rage. Clyde, oh, the blue monster, now refused to go. I was limping back to Santa Cruz to collect Miriam; she had passed a message via the bush telegraph that only a Bolivian can fathom how to use, to say she would come four days earlier than expected. Did I want her to? Clyde was spewing oil over the engine from where it spattered on to the windscreen. This was the second time in a month. I had just paid $500 to have the engine overhauled and the piston rings replaced. Since there was no money coming in, and as I really did not need a vehicle now, I decided then and there to fix her up once more and return her to her owner, Guy Deuel. Miriam and I would have to return to Buena Vista by bus or hitchhike our way back.

When I got to Montero, I found the road to Santa Cruz closed for a motor rally. Then, to put a cap on it, the road into town was closed as well, a large crowd of people having collected to watch the vain efforts of two men trying to extinguish the flames consuming their car. I took the opportunity to go to the nearest service station and buy two gallons of used engine oil, enough to get Clyde to Santa Cruz.

By the time I got back, the burnt-out vehicle had been pushed off the road and the first rally cars were screaming round the bend. Throngs of ill-disciplined people lined the road, and a few adolescents were playing chicken; one rally car had to take sharp avoiding action, and momentarily out of control, almost ploughed into the rank of spectators. People fled in all directions. Some fell into a dirty ditch. One mother grabbed her toddler in an effort to avoid the on-coming machine that sprayed us with gravel as it clawed to regain control. Seconds later, two more cars approached the corner now blotted out by a pall of dust. I ran for the safety of a gateway, pushing my way through the crowd of laughing townspeople returning to watch the action. I was frantically trying to turn them around, screaming at them to turn back; nobody took any notice of me, and fortunately, the two cars passed by, horns blaring, lights flashing without further incident.

I was beside myself with anger; the complete lack of official supervision, the artless simplicity of the people, the reckless dash of rally drivers flashing past in pursuit of dubious honours, the expensive waste of it all. High-tech machines thundering past the penniless people standing beside their unkempt hovels; my own chagrin with the closure of Santa Cruz's busiest highway so that a score of wealthy playboys could go motor racing. Galling as it was, there was nothing I could do but wait for the road to be opened once more.

Somebody told me some of the rally drivers use the occasion to ferry cocaine base to finance their avocation. Whether or not this is true I don't know, maybe it was simply my informant's sour grapes. But sour grapes or not, nearly every year scenes as I had just witnessed led to the demise of innocent bystanders. In the boondocks, it's the rural folk misled by the pace of high-speed cars who become the victims.

****

When I got to Santa Cruz the oil warning light told me what I already knew, and it was with sad relief that I drove Clyde into a small yard adjacent to the mechanic's house. Guy, better known as "Sir Guy Deuel, Guru of Slack", generously offered to reimburse me with the money I had spent reconditioning Clyde, with the provision that I arrange for the sale of the vehicle.

When I went to the post office the next morning it was closed; a hastily chalked announcement explained that the workers were on strike: "Por motivo de HUELGA". I didn't bother to read the rest, I'd read it before. The place was, and still is, a disgrace; my mail was frequently opened, presumably in search of cheques or valuables, and not long ago the authorities announced that due to the overwhelming quantity of mail that had accumulated during the current troubles, they had been forced to burn ten tons of it. 'Thank God,' I thought, 'I had sent all my important stuff with that American.' At least my Annual Report to the World Wildlife Fund (with copies to the New York Zoological Society and John Dunning) had got away safely. What was he called? Gary? No, Jerry? Yes, Jerry!

I didn't get very much done, as nearly all the shops were closed. I managed to cash a Barclays Bank' cheque (only done for residents, and always at an extortionate exchange rate, 12%!), before I met Miriam at the bus station. She had too much baggage for us to hitchhike, so we boarded the Colectivo for Buena Vista. The Colectivos are run-down charabancs that catered to the rural population. They will carry everything from pigs to bananas and sacks of rice, destined for the Santa Cruz markets and, in the reverse direction, to stores for the village shop: cement, lengths of steel, mattresses, sugar, flour, etc. If one wishes to travel in more comfort and get from place to place in a hurry, the "Micros", minibuses that offer an efficient service but, won't carry more than personal luggage are better. They are also the favoured form of travel among single females, school children (who can travel at a subsidised price) and the elderly.

The poor, the drunks, the locos, the unclean, the overburdened and the standing passengers all got on the "Colectivo" with us. I preferred to travel on these buses for the good-natured bonhomie of the mixed fare, rather than the commuter-like, no-nonsense atmosphere of the "Micro". United, maybe through their sense of being underprivileged, the passengers on the Colectivo tend to behave with the same loose familiarity as would factory hands on a Work's Outing to Skegness. It only needed one wag on the bus, and there invariably was one, to create the right mood.

No sooner had we started than somebody at the back yelled out that he had forgotten his wife.

"That's all right as long as she doesn't 'ave the money" answered another. This simple repartee was met with much laughter.

Then the first man again: "No. I've got the money, she'll have to walk home." More laughter. "And carrying everything too," he added to a riffle of sniggers.

"That's all right as long as it's not far," said the second man.

"No. Only to El Comando [80 miles]," said the wag; and to screams of delighted guffaws added, "Trouble is, she's got a wooden leg."

Then, when the loudest of the laughter had died down, in a hardly audible voice, he finished off with: "The real trouble is I got her leg with me!"

By the time the passengers had settled down again, I blush to mention how the conversation went on. Some woman had asked to see the leg; another advised her to keep her legs crossed if he showed it to her, and on, and worse.

Then we were driving into Montero; the dreariest part of the journey nearly over.

After a beer and a chicken "salteña" while we waited for the bus to fill up, we were off again. The frivolity of the mood did not return until we had passed the Ranger's H.Q. with its white-helmeted sentries in place. Always a potential trouble spot, I noticed a change of mood: a discernible tenseness among our fellow passengers as we approached the sentries.

The bus had to stop here and one of the soldiers climbed the steps to push his head around the threshold of the door. "Policia Militar" was clearly stamped in large black letters on his helmet.

Today, satisfied with a cursory glance, he allowed us to get on our way again. The sound of heavy boots descending the steps stayed with me for a while. In a few more weeks I knew things would be different because; it would be "Conscript Time". In early January, the PM's white helmet will come aboard and scan the passengers, while his comrade stands in the doorway with a machine pistol. They will be searching for any likely looking lads, boys of military age; and woe betide any who can't show them his "Libreta de Servicio Militar", proving he has already done his duty. Without it, he will be summarily marched off the bus and in through the barrack's door. Here he will be stripped to the waist to look for his military tattoo on his arm or chest. Without this, too, he will be subjected to a formal interrogation. If he now fails to satisfy them with details of his military service, he will be led immediately to the barber, have his hair cropped, and be drafted into the army for eleven months.

But in Bolivia there is always a way out. If you have the money, that is. You can legally buy your "Libreta". The better-off do, the poor can't afford to. The price seems to depend upon the circumstances but is likely to be about $400.

With our return to the stretch of river-road (soon after the barracks), a flood of relief passes through the bus. The crossing was still relatively dry, and a lively conversation sprang up about other, more difficult, times. We pass Richard Parada's shrine with its bouquet of flowers and churn our way through one stretch of dry sand that set the bus a-shuddering as the wheels scrabbled for a hold on the shifting surface.

At Portachuelo, the Colectivo filled to capacity, and there was some delay whilst the roof rack was piled high with more assorted bundles. With seating capacity for forty people, the bus now carried sixty, and anybody like myself with a seat adjacent to the aisle found himself assailed by a variety of petty annoyances.

The bus left the middle of the road to avoid the worst of the ruts and holes in the crumbling tarmac, and now travelled along at an uncomfortable angle in sympathy with the 10° camber. It took on the disposition of a roller coaster, and the standing passengers a group of Samba dancers with no place to Samba. Short of handholds, they are forced to cling to anything within reach. A large Colla woman reached across my face to grab the rail on the back of my seat; I had to bend my head to the side to avoid her voluminous breast. Her daughter took a firm hold of my shoulder with her warty little hand, and in these uncomfortable positions we bounced along on our way. The unabashed intimacy between strangers that the situation obliged provoked another flurry of suggestive comments from the good-natured peasants. One young woman, her blush-red face framed by the braids gathered to the top of her head, changed places with her grandmother, presumably to avoid the embarrassing proximity to one old man who looked rather drunk. Further down the bus a baby started to cry and its mother, an unlovely-looking woman, reached down inside her blouse and flipped out a milkless-looking flat breast. Nobody moved to offer her a seat, and free of any apparent resentment she patiently nursed her child back to sleep.

Unaware of the rapid change in the weather, I was jolted out of my daydreams by an ear-splitting crash of thunder. This was immediately followed by a rain-laden gust of wind that sideswiped the bus with noticeable effect. The mood on the bus swung back again to one of apprehension, and as the heavy drops of rain raked the bus from stem to stern many of the passengers cast worried glances towards their pathetically valuable bundles on the roof-rack.

In the ensuing distraction, one young man, whose features were typical of one suffering from mental disabilities insisted on getting out then and there, since he realised he had overshot his destination, contrary to the good advice of some of the passengers who urged him to do so at the next village, where he could shelter from the storm. Apart from the soaking, he could now expect his predicament was made more serious because he had two sacks of flour on the roof rack. Finding difficulty in making a decision, he took refuge in a torrent of tears as we left him sitting on his sacks by the side of the road: no doubt anticipating the beating his father would hand out to him for his error.

I doubt there was a single person on the bus who did not feel sorry for him, everybody realised the inevitable damage to the flour would be a calamity of no small account to the family's fortunes. And, whether in sympathetic accord with him, or aware of the threat to their own belongings, several old women began to beat their chests in lament.

The bus was forced to take to the middle of the road because of the swirling brown flood at the sides. The rainstorm cut the visibility down to a few yards. The slow, lazy beat of the single windscreen wiper fought a losing battle with the cascades of water pouring off the roof. The tenseness of the driver, who now hunched forward in his seat to peer through the opaque glass, was matched by the looks of agitation on the faces of the passengers. To add to our troubles, the rain now started to beat in through the many windows lacking glass, and steady trickles of water fell from a dozen cracks in the roof. Some of the passengers took out scraps of plastic sheet with which to cover themselves. The old woman in front of me was helpless to do anything about the steady dribble of water which fell down the back of her neck. Several others, having similar difficulties, now abandoned their seats to join the dejected crowd standing in the aisle. Packed like sardines, nobody rose to the occasion by trying to make jokes. Still nursing her infant, the mother stood there with rainwater, or maybe tears, coursing down her cheeks.

Relatively well off in our shower-proof jackets, Miriam and I huddled closer together for comfort. "But isn't it always so?" We gringos and Europeans are always relatively well off". Munching these thoughts reminded me of something Mrs. Thatcher had said on the BBC World Service: "The poorer nations of the World really have to learn to help themselves," made me sad and irritable.

Yes, Mrs. Thatcher, your cold advice is no doubt correct but, please explain to me and the other people on this bus what we can do about it. Will you explain to your comfortable, selfish, greedy fellow-Europeans that the poor countries are being denied the possibility by the self-serving machinations of , just to take one example the European Community's Agricultural Policy. Why have they, through their extravagant subsidies of meat and dairy produce, almost totally destroyed much of South America's export potential? Why is it that a common or garden factory worker in Europe, stamping out washers for thirty-eight hours a week earns thirty times more than a common or garden coffee picker in South America gets for his sixty-hour week?

Do you really believe, Mrs. Thatcher, the desperate people on this bus should pull their socks up? And even had they any to pull up, do you really think they would be rewarded by the fruits of their efforts? Of course not! They don't have the time, or the energy or the education or the health or the clout or the backing of you and your lot to do it. What will become of that young, flour-carrying, man and his family? The rainstorm will destroy his family's hopes of following your monetary principles as surely as it is now washing the rice seedlings from their fields. Life here, Mrs. Thatcher, is made up of these small tragedies. Do you think the family insurance policy will protect them? Do you think there is a Community Chest to help them out? Do you think life here is one long game of Monopoly, Mrs. Thatcher, everybody with huge beefsteaks to lose on the roll of a dice, to dabble in the Property Market, or the Stock Exchange, or invest in the misery that Foreign Exchange Rates cause the poor of the World?

Do I hear you say I am being unfair? Do I hear you say that monetarism, belt-tightening, must be given a chance to work? You may well be right. But in the meantime, Mrs. Thatcher, who is there to care for all the millions, their belts so tight it near cuts them in two, until the benefits trickle slowly down to them? And who do you think is going to pay the price?

All of us, Mrs Thatcher. ALL OF US.

"Here I sit under my banyan tree, scrutinising my navel, listening to your words. So I have to work harder, 'tis truth, nobody has, after all, filled my proffered bowl with tattered peso notes. 'She's right!' I am heard to say to myself. So up I jump: 'Make an effort,' I think, and off I go to chop down another forest, to grow more rice, to make more money. If I can cut down thirty acres of those wretched trees I can become rich (weather, health, pests, weeds, market forces permitting) I can make another five hundred dollars a year. Then in twenty years' time, I can go out and buy myself a British-made tractor to help me clear another forest. 'But no,' I say to myself, 'First I'll get me a chain saw, a good reliable American one, you know to help the poor Yank Economy along. Now with that I can really set to work, Mrs Thatcher, fell another fucking forest.

"So I do, Mrs Thatcher, so I do. But lo and behold, here comes that raving nitwit from Amboró Park. Do I believe my ears? What is he telling me? It is silly to cut down trees. There is no future in it. The soil is too poor. Well, I know all that, but what does he expect me to do about it? "Right", I say, "I see that, but how am I expected to make a living whilst I do as you say?" Plant coffee, plant cocoa, plant spices, plant timber trees, you have told us. "Where am I going to get all the plants from? Who will lend me the money to buy them? Where do they sell them? As I cannot eat them, who will feed me and my family while we wait eight years for the harvest?"

"Yes, I know I'm cutting down the forest. No, I don't know how many millions of others are doing the same thing. What's it to me? What do I care that the World is being destroyed? It's not my fault. I'm not destroying it. I'm just trying to scrape a living. Caramba! Do you think I am working my guts out because I like it? If he's so bloody concerned about it why doesn't he get his ruddy British Mission to give us a hand then? You tell me.

"It's all right for them, the bastards, with their swishy shoes and socks on their feet. Look at me, hands a rhinoceros would be ashamed of, the skin so thick I can't tell if I am wiping my arse or not; my feet short of a toe or two, slip of the axe; cataracts in my eyes, I don't know a pretty girl when I see one. Full of sores and spines; bitten by insects day and night; my wife so full of Chagas she's hardly got the energy to draw water from the well.

"Then you get that bloody Reagan calling us all Communists. What a laugh. Who the hell has got the time to be a Communist? What is a Communist, for fuck's sake? Am I a Communist because I'm poor? Because I have to work hard to fill my belly? Provide for my wife and children? Jesus, I wish somebody would tell me, put me right. I wish there really was someone to help me. To help me…"

****

"What?" God, I hadn't even noticed. Miriam was asking me to help her get the heaviest things down from the luggage rack, no soaking roof-rack for us! We'd arrived in Buena Vista. The rain had stopped, but now the sun was very hot causing steam to rise from the wet gravel of the road. The whole place smelt of tarpaulins, not that there were any about.

As luck would have it, spoilt foreigners again, Pieter and Els were sitting drinking a beer at Blanca's, their brown Chevy truck parked handily in front. After we piled our things into the back, we joined the others for a welcome cold Ducal (a German brewed lager). It was four o'clock, the temperature a pleasant eighty, the worst of the humidity's enervating effect carried away by a discernible breeze which caressed the plaza. Sitting in plastic garden chairs on the pavement outside Blanca's became our Club. It was a good location from which to study the village folk as they strolled leisurely past on their way to fulfil trivial errands. (I nearly wrote hustle and bustle, there was never any hustle and bustle in Buena Vista). Even though the raised sidewalk of the piazza was wide enough to leave sufficient space, the respectful people stepped down into the road to pass around us. Those unable to would hesitate a moment and with a "permiso" ("excuse me"), would hurry through the gap.

Until the people got used to us, our habitual beer outside Blanca's, that on hot days became a twice-daily ritual, was regarded as a typical example of gringo decadence. Drinking was something reserved for weekends, and when the locals drank they did so with the intention of getting drunk. Hardly anybody drank for the sheer pleasure a cold beer offers on a hot day. When they drank, they sat inside; they didn't hide themselves, the doors and windows would be left open for anybody to see, but the gesture was there, if you didn't look, you didn't see.

Lowry sums it all up when he records Sigbjørn's conversation with a barman.

Lowry: "How do you mean they don't like to see you drink down there?"

Barman: "Not every day they don't. They like to drink themselves. But they don't like to see you drink. Makes them mad."

But tolerance is a mark of the local people and only once, and quite rightly, did I hear of any complaint. This was years later when two of my young colleagues were foolish enough to sit drinking beer in the plaza while the whole village enacted the ceremony called "The

Stations of the Cross", the most sacrosanct of the village's annual festivals. Sacrilege indeed!

The subtlety of many of the social codes makes it difficult for us foreigners to detect them. The customs related to drinking alcohol is worth pursuing. We have already established that, unobtrusive, or not, a glass of beer might upset the sensibilities of the people; but drinking in the plaza is offensive! Because there is nothing else to do in a small village much of the social life is concerned with events relating to the Church calendar; friends, family and religion come together on these occasions. In most towns of any size, a church dominates the plaza, and polite socialising takes place in the plaza under the watchful eye of the Church, so to speak. Ergo, the plaza is close to being hallowed ground, and strictly speaking, our innocent refreshments were unethical because of the implied disrespect for the Church and its congregation.

A roll of thunder somewhere behind the pueblo told us the rain would not be long in returning, and mindful of our damp belongings retrieved from the "Collectivo", and now in the back of Pieter's truck, we prudently made for home. The narrow track was slippery, and every now and then we had to edge our way around deep muddy holes. Thick drops of rain greeted our arrival at the quinta. We just got the pick-up unloaded as the main blast of the storm hit us. The rainy season had arrived.

Home. Every man needs a place to go home. But what is home? When one is young, home is where your parents live. When one gets older, home is where you live with your wife and kids. But there are many people, like myself, who have never had a real home. Before I came to Buena Vista I'd never lived in the same house for more than five years, nor in the same town for more than ten. I carried my home around with me like a tortoise, wherever I happened to stop was home. If I am staying with a friend and I want to go back to his house I say "Let's go home". I rarely say "Let's go back to the house" or "Lets go back to your place". So even though I knew the little cottage at "El Hombre Nuevo" was only to be my quarters for a few months, pending our move to the Estancia, I thought of it as our home.

And if home is where the heart is, then my heart agreed with me; this pretty thatched house, thirty by twelve feet, now contained my heart's desire: my hummingbird, my "picaflor".

****

The next four months passed in joyful routine. With the first light of day I would set out to go bird watching. By ten, when the day was already hot, I would return home. There wasn't a bird book for Bolivia, but with the help of those that were available I spent an hour carefully checking the provisional identifications I had made. Any doubtful records would be discarded and the rest entered into my field notebook. Miriam would then go with me to put up the bird nets ready for the evening. Then off we went to the village with Pieter and Els, and whilst the girls did the shopping, Pieter and I would discuss the morning's events over a cold beer at Blanca's. After the girls came back and also refreshed themselves, we returned home, had a light lunch and whiled away the worst of the afternoon heat reading and sleeping, and me checking my morning's field notes.

At four o'clock Miriam would return with me to the bird nets, that we now opened. We worked with only three nets because it took us a long time to record the details of each bird and then identify them. Each one had to be carefully described and measured: total length, length: of bill, head, body and tail, colour of bill, gape, eye, legs and plumage. Each and every part had to be described: forehead, lores, cheeks, crown, nape, mantle, back, rump, chin, throat, upper breast, lower breast, belly, crissum, wing coverts, secondaries, primaries, under wing coverts, under tail coverts, outer tail feathers (above and below), inner tail feathers (above and below), and, finally, any special markings. With this done we let the birds go.

Everything had to be done as fast as possible because in the heat of the tropics, birds tire quickly and die. We inspected the nets every twenty minutes. On some days there would be four or five birds each time, on others none at all. Miriam, being especially nimble fingered, would set to work on the smaller and more difficult ones, those entangled in the fine mesh of the net. Sometimes the birds were so caught up it was necessary for us to cut them free. I would deal with the easier ones and start recording their details with the help of a tape recorder. Those that couldn't be processed immediately were placed in cloth bags. When the weather was hot, we wore fine gloves to help keep the birds cool, and before releasing them we gave each one a drop of dilute honey from a glass dropper. All these precautions paid off and only once did a bird die from exhaustion.

We did have some minor mishaps. The first was with a large Nightjar we caught. When captured, they usually open their mouths to such an extent that the gape is large enough to insert a billiard ball.

When this one accompanied this surprising ability with a hiss like a puff adder, Miriam was so startled she let go of its body, and the bird flew off leaving its tail feathers still clutched in her hand.

Another time a dog shot out from the bushes to ensnarl itself in the net, that it quickly demolished before extricating itself.

Parrots were the most difficult birds to deal with because they were aggressive. I don't know what we would have done had we caught a flock of them. Imagine it. It's getting on for dusk, suddenly a flock of thirty parrots swoops into the net. Frightened, they clutch the fine mesh with their long toes, and attack fingers with strong sharp beaks. One by one we start cutting them out of the net; it's getting darker, forming inky black pools under the thickest part of the canopy and then, like smoke drifting rapidly outwards fills the whole forest.

Such things have happened, but fortunately not to us. We always closed the nets long before sunset, especially when working in the forest. After taking them down (because we couldn't risk having them stolen) we would hurry back and go with our friends to Blanca's. Then, back at home, I would transcribe the information on tape to my notebook while Miriam prepared our simple supper. If I had time, I would spend an hour or two on the identification manual I was compiling for Bolivian ornithologists, not that there were any. On most evenings we went over to the main house and passed the time chatting with our friends before going to bed.

If the day was wet, I spent most of my time on the identification manual, sneaking out for a while if the rain lessened. Sometimes, when it was really hot, we would walk down to the Surutú and sit in the river's cool water, or like children splash water at one another. Sooner or later the biting flies or the rain would drive us home again.

Whatever I did, wherever I went, my binoculars were always with me. I soon got to know the commonest birds and rarely did a day go by without something new. At first the variety was astonishing, and one family of birds, the Flycatchers, bewildering, with 180 species registered in Bolivia.

An average 120-acre farm comprising a patch of mature forest, a fruit tree plantation, thirty acres of scrub, and a cattle pasture, might support more than four hundred different species; nearly twice as many as are regularly found in the United Kingdom. Bolivia has more than fourteen hundred different birds, six hundred more than the whole of North America.

And yet birds, and wildlife in general, can be difficult to detect, especially in a mature forest. The stereotype of a tropical forest is one that resounds to the noise of birds and insects from dawn to dusk and the nights are filled with the strange cries of unknown beasts; behind every tree lurks a four-footed denizen, and the ground is littered with poisonous serpents. Why this romantic notion should be so commonplace is difficult to explain. Early Spanish explorers (who so often exaggerated the individual attributes of South America's animals) can hardly be blamed in this instance, for their own narratives are full of starving men wandering the vast empty forests in search of something (anything!) to eat. Paradoxically, the idea may have been fostered by the famous naturalist-scholars of the nineteenth century: Wallace, Bates, Spruce, Waterton and others who wrote scientific accounts of the incredible variety of animals. That was what the public in its cheerful complacency came to interpret as an incredible abundance of animals, a false impression endorsed by engravings of all the sylvan beasts to be found in the jungle condensed into a single woodcut.

Let me not, then, allow the same impression. As I have said, the variety of birds is great, but more often than not one could walk all day through a tropical forest without coming across a mammal or snake, without hearing a single tree frog, and without seeing or hearing more than a handful of birds. The very essence of a tropical forest is its absolute quiet; at times a silence so profound the sound of a falling leaf is remarkable. Only during the early morning, and less so in the evening, does a tropical forest even suggest otherwise.

This is what Henry Bates had to say: "We often read in books of travels of the silence and gloom of the Brazilian forests. They are realities, and the impression deepens on longer acquaintance. The few sounds of the birds are of the pensive or mysterious character that intensifies the feeling of solitude rather than imparts a sense of life and cheerfulness.

Sometimes in the midst of the stillness, a sudden yell or scream will startle one; this comes from some defenceless fruit-eating animal, which is pounced upon by a tiger cat or a stealthy boa constrictor. There are, besides, many sounds that it is impossible to account for. Sometimes a sound is heard like the clang of an iron bar against a hard, hollow tree (this is the call of the Great Tinamou), or a piercing cry rends the air; these are not repeated, and the succeeding silence tends to heighten the unpleasant impression that they make on the mind".

Reading through this again I find Bates may have gone too far, for there are times, most often during the breeding season, when the forest comes alive, especially after prolonged heavy rain as the warmth of the sun coaxes the wildlife into activity.

Sometimes the pervading silence is disturbed by the arrival of a mixed flock of birds containing as many as forty completely different species. These large mixed flocks are a peculiarity of South American forests and pose many intriguing questions. What attracts the birds together? How do they start off? Do they remain together all day, roost together at night, and stay together the next day? Where are they going? Is there a plan? Which bird leads the band? What benefit do they derive from their odd association?

After many years watching them, I have, I think, answered some of these questions. Sitting very quietly in the forest, these busy flocks sometimes pass closely by without noticing me, engrossed as they are in feeding. And yes, they come together to feed because the activity of one species is a help to another. There is no hard and fast rule, but here in the woodlands of Buena Vista, a typical performance goes something like this: It's dawn. Among the first birds to wake are Woodpeckers, species that largely depend upon hearing their prey rather than seeing them; a group of seven or eight start working the upper parts of the trees, hammering away at all the rotten little branches, removing the bark from larger ones, and pecking at infested fruit in their search for grubs. The material prized loose by their activity falls to the vegetation below and disturbs the insects found there. This attracts other insectivorous species. Canopy Woodcreepers and Flycatchers chasing the disturbed cicadas, crickets and moths through the air causes further commotion, and dislodges even more insects.

Then more specialised insectivores (like Xenops and other Woodcreepers) go to work and are joined by those more catholic in their choice of food: Vireos, many Flycatchers, Tyrannulets and some Tanagers, birds that will hunt for crawling insects and spiders amongst the leaves, or quite happily stop to peck at a few ripe berries. All these birds squabbling and flitting to and fro in the upper canopy (shaking leaves, dropping berries) adds to the general commotion, and puts even the most cryptic insects to flight.

The wave of activity then passes to the middle canopy where in the growing daylight other species wait to catch the fleeing insects: Antwrens, more Flycatchers, Warblers, Thrushes and Manakins, and so on down through the different levels of the forest's system, the rain of debris (falling arthropods and dropped fruit) reaching the Wrens, Antbirds and other semi-terrestrial species waiting below.

All this is perfectly understandable, but why those that mostly eat fruit or nectar (some Tanagers, Honeycreepers and Euphonias) should habitually join this motley crew is more difficult to explain. Among the theories put forward, the winner has remained the old standard, "the-more-eyes-the-better" theory of predator detection. The protection given to an individual by joining a group of mutual back-scratchers. How nice!

The trouble is I don't believe this to be so simple. To explain my theory, I will refer to the family of Fruiteaters, not the fruit eaters in general. Fruiteaters join mixed flocks for the benefit of group protection but in a way much more sinister than the "more-eyes-the-better" theory.

You see, the Fruiteaters are more ponderous, less excitable in their activity, their food is not trying to flee their attentions, so they have more time to be wary, more time to watch out for that hawk or snake than do their busy insectivorous colleagues, whose mad dashes in pursuit of highly mobile prey among the foliage, or even from tree to tree, not only make them conspicuous to a hunting falcon, but also vulnerable to something nasty hidden among the foliage. I believe that Fruiteaters gain an advantage by mixing with birds more vulnerable to predation than themselves.

As the light fades at the end of the day, the flock disintegrates. The Fruiteaters are the first to drop out, content as they are to pass the night regurgitating the less rapidly metabolised berries in their crop. The Flycatchers, dependent upon good lighting for the hunt and warmth to maintain their prey in an active state, also soon retire. Much of the driving force having wound down, the omnivores also drop out and only the more crepuscular species: Woodpeckers and Woodcreepers, continue for a while, but more and more individuals drop out of the restive band in search of a good roost for the night.

With the light of the new day, the insectivores with their more urgent needs (because of the faster metabolism of a high protein diet) are the first to awake, and the flock begins to form anew. The complete stillness of dawn is now broken by the Woodpeckers as they chip away at grub-filled limbs, and behind them the growing clamour (as if a party of excited schoolchildren were approaching) of the other birds darting hither and thither, pushing amongst the vegetation, or leaping into the air in pursuit of a Tree cricket, or fighting for possession of an unwise Katydid, or a Tenebrionid beetle, that has fallen to the forest floor.

And just to complicate matters further, I should mention the participation of more sedentary, territorially conscious birds like Thrushes, Trogons, Puffbirds, Manakins and Hummingbirds. These may accompany the flock for short distances, but leave them as the main body passes on.

Mixed flocks provide enchanting moments for those of us with the time and patience to wait and watch, and though I believe the sort of model I have described to be at the heart of the mixed flock phenomenon, there are still many facets of their "modus operandi" that require investigation.

Some birds never join mixed flocks, reconciled to a life of solitude like me. Lonely Tinamous with their flute-like calls creeping warily through the undergrowth in search of grubs and seeds. Jacamars are content to sit motionless on a favourite perch, but steel themselves for the momentary action needed to capture a passing dragonfly. Others, cliquey types like Parrots, Jays, Oropendolas and Toucans, stick together, foraging through the trees and bushes in the company of their own kind. How like humans they are with their prejudices!

And talking about humans, many birds seem to be creatures of habit too, going their rounds each day they pass from one favoured fruit-bearing tree to the other, or flying off to visit a vine they know of with the regularity of a Chelsea Pensioner. Some of them are so punctual in their routine one could set a watch by them.

Or as in my case. I wanted photographs of Chestnut-eared Araçaris together with the smaller Lettered Araçaris (named for the calli-graphic marks on their bills), that often accompany them. I found out that a mixed group regularly visited a bush heavy with round green berries, known to the locals as "Huevos de gato", a common bush among the many trees called balls of something or other.

So there I sat in the morning. The Araçaris arrived on the dot at 10:45 a.m., hung around for fifteen minutes or so, and then promptly departed. Taking photographs of birds in the dim light of the forest is notoriously difficult. And so it is unless you know their habits well. Arriving at ten o'clock I set up my camera and flash on the tripod and hid myself as best I could. Precisely forty-five minutes later, having arrived without a sound, I detected a bright blue eyebrow below watching my equipment with suspicion, the first of the Lettered Araçaris was approaching. If at that moment I had reached out to focus the camera on it, I would have had to pack up my equipment and go home, for even the slightest movement would have scared the bird away; and screaming in panic the rest of them would have gone with it.

Birds like these approach their food in the same way as a deer will a watering hole, with extreme care, but once they have committed themselves they seem to throw caution to the winds and take little further notice of their surroundings. So, I waited. Tantalisingly slowly, the bird gradually relaxed, flew down to the bush and started to pick at the berries, only occasionally stopping to peer in my direction. Then three more birds arrived, the first one's mate, and a pair of Chestnut-eared Araçaris. Relaxed by the casual behaviour of the first to come, the rest soon settled down to feed. Now was the right moment for me to very slowly focus the camera on the four birds. Flash!. Then, while they were still trying to decide whether or not they had seen something, a second photo, Flash!, and pushing my luck, a third, Flash!. By now they were showing signs of uneasiness, the flash is of such short duration they hardly notice it, but together with the click of the shutter mechanism, and the slight movement of my fingers as I wound on the film, the birds stopped feeding to eye my temporary hide with growing suspicion. I was quite close to them (about fifteen feet away) and for a minute or two I had to remain absolutely still. A horsefly settled on my nose, and as I focused the camera on one particular bird, I felt the sharp pain of its bite. A cloud of mosquitos hovered around me, and some, sampling my blood, were oblivious to the highly recommended repellent I was using. Flash!, Flash!, Flash! I managed to get three more shots. Then one more horsefly, and yet another, pierced the back of my neck. I tried to ignore them Flash!, Flash!, the birds were no longer concerned. I knew I had them now.

Then quite unexpectedly, a Dull-capped Attila arrived to peck at the same fruit. I'd often heard its call, a strange "weet weet weet weet" (the weet's increasing in pitch until suddenly cut off by the low, sad, drawn-out "wooooo") make it a melancholy, disappointed one (the lament of a football supporter as his favourite centre forward puts the ball over the bar of an open goal).

I had never seen Atilla so close, and I must say, that except to an ornithologist, there is nothing very special about it, it's a large cinnamon-coloured flycatcher with a piercing white eye. I'd just had time to snap off a hurried shot of it when a dead branch, having come loose from above, struck me a resounding blow on the top of my head; and in its fall carried away most of my improvised hide. All the birds fled. I remained there for a few more minutes fighting down the pain and the shock of the incident. Why? Why did the branch have to fall just at that moment?

The throbbing ache spread from the back of my head to the sides of my face. I rolled over on to my side and wiped away the tears that sprang involuntarily to my eyes. Disturbed ants were biting the sensitive skin between my fingers. I let them continue as their sharp bites distracted me from the growing pain in my head. Attracted by the commotion the horseflies had returned in strength and were stabbing me with renewed vigour.

A few yards from where I lay a tiny hummingbird, even with its slender bill and white tipped tail (no longer than my index finger), was hovering a foot above the forest floor, seemingly intent on feeding at some strange flower. Then it whipped away with lightning speed and hovered a foot from my face: "Are you all right mate? Got a nasty one there didn't you? Well, I must be off".

With the dusky band crossing its breast I knew it to be a Reddish Hermit. A minute later it was back again paying attention to the same flower. This aroused my curiosity, for they rarely do this (they too go their rounds), and forgetting my recent misfortune, I reached out for my binoculars. Even though I had adapted them (by unscrewing the objectives to increase their focal length) the bird was still too near for me to focus on it. But, through the blurred image I saw something, something about the strange flower that excited me. Each time the tiny bird hovered in front of the flower a minuscule beak was extruded from its centre. Great Scott! At last, I had found a Hermit's nest.

This was an important breakthrough for me. These little humming birds are quite common here and yet I had never before found a nest. The books state that they make them of cobwebs that they attach to the tips of palm leaves. How many times had I stared through my binoculars at the tips of palm leaves? Staring awkwardly upwards at the swaying fronds my only reward had been a cricked neck.

I scrambled to my feet and went over to examine the nest. It was, as the books confirmed, made of cobwebs, and yes, the nest was attached to the tip of the leaf. But I'd never imagined the books meant the leaf of a palm seedling. Both in form and colour, the nest resembled a miniature ice-cream cone, its weight pulling the tip of the leaf over to protect the nest from rain. It was remarkably soft to the touch, as light as gossamer, and in the tiny, fingertip-sized cup a diminutive naked chick thrust its open beak at me. It was obvious the tiny creature had just hatched for, like a snow-white pearl, an egg lay beside it. I dragged my camera equipment over and took pictures of these miniature marvels.

What I really wanted was a photograph of the Hermits feeding their chicks, but having disturbed them enough for the moment, I decided to return the following day. At least I now knew why the branch had fallen on my head; if it hadn't, I would never have found the nest. Before I left, I took a jam jar and some string out of my pack, and tied the jar to a stick which I stuck in the ground two feet from the nest. This was my false camera; I used it to accustom birds to the presence of strange objects near their nests, a ploy that on my return allowed me to take pictures without having to wait for the birds to settle down.

The next day it rained all morning, and the afternoon was overcast and dark. When I went back to the nest the following day the second chick had also hatched. I set up my tripod and camera, attached a long extension release, and moving some distance away, sat down to wait. After a short time one of the adult birds approached, but it was surprisingly shy of the camera equipment and refused to fly to the nest. I must have waited for two hours before I succeeded in taking the picture I wanted. The bird reacted strongly to the Flash! and disappeared immediately. I waited another hour, but intent as they were on making a detailed examination of my strange equipment neither parent returned to the nest.

Again, I decided to leave the birds in peace, and when I returned a few days later, sadly, the chicks had gone, maybe eaten by ants, or some other predator. When I eventually got the slides back the photograph was disappointing. The shot had caught the bird just as its near wing concealed its head. A fraction of a second sooner or later (the books say seventy wing beats per second) I would have had an incomparable picture. Such is the luck of nature photography. Even so, the pictures of the egg and chick and those of the Araçaris more than compensated me for the blow on the head.

It is strange how individual birds differ in their reactions, for some time later I found another nest. The incubating bird allowed me to approach within a foot of it, set up my photographic equipment and flash-off as many pictures as I wished. But again, having thought I had found the ideal subject, one night the eggs disappeared, and I was deprived of the picture I really wanted.

Many people believe that predators are attracted to a nest once it has been touched by somebody. They say the predator will follow the human's scent trail. One might think that most animals would be wary of a man's scent, and even if this were not so, how many men go in search of birds' nests? How many human scent trails would lead them to food?

If, instead of referring to human scent, we refer to an animal's scent (human or not), I think we will be closer to the truth. In my garden, when filling up holes in my lawns, I know that tomorrow I will discover that some animal or other (most likely a fox) will have undone my work. I hope my observations have resolved the issue for any "Doubting Thomas".

My own records show that in the area of Buena Vista about eighty per cent of birds lose their eggs or young to predators, amongst which ants are the chief ones. This figure is in line with findings of workers in Central America who showed there was no correlation between nest predation and human disturbance. Furthermore, they claim that naturally high levels of predation account for the fact that tropical birds rarely lay more than two or three eggs, while their temperate cousins, that suffer much less predation, lay four, or often more. And, in contrast to their kin, tropical birds are capable of replacing lost broods very quickly. One could say: tropical birds expect to lose their eggs and don't put them all in one basket.

'Expect!' Do I hear accusations of anthropomorphism? 'Los hechos son hechos, hay que reconocer la verdad!' In other words: 'Get away with you!'

****

With the passing of the weeks, I got to know the birds in the immediate area of my new home. The use of nets had shown us there were many species living around us that we rarely saw, and judging by the variety of unknown calls, yet others I wouldn't get to know unless I developed new techniques.

Among the most vociferous of these were the formicarids, Antbirds and Antthrushes. They are a fascinating family, more secretive than shy, and nervous of anything except the gloomy twilight conditions of the forest floor. They are small, most of them no larger than a Wren, and they get their name from their habit of following columns of army ants. They normally feed on the escaping insects that the ants flush from their hiding places, but on one occasion I watched a White-backed Fire-eye stealing its provender from the ants themselves. When ants are not about, the birds revert to normal foraging activity, creeping across the ground, looking (like Christopher Robin) at all those places where beetles like to hide.

Intent in their search for food, it is easy to imagine that a bird and its mate could become separated amongst the undergrowth and perhaps never meet again. In order to keep in contact with one another, antbirds maintain an incessant vocalisation; usually the males with a pleasant series of sharp descending notes, and the females answering more quietly and less often. For one common species their contact calls went like this, Male: "Wife, what are you doing now?" Female: "Nothing, I'm here". "Male: Well don't go too far", pause, "don't go off on your own", pause, "why don't you answer me?" Female: "Don't worry, I'm here", and so on.

There were six or seven different types of Antbirds calling to one another in the forest behind El Hombre Nuevo. At first, I spent hours crawling after them on my hands and knees trying to get close enough to identify them. Apart from the badly lit conditions, they were restive creatures, and in the fleeting moments when they came to rest they merged so well with the background of dead leaves I still couldn't identify them.

After many frustrating hours worming my way after them (an unwise procedure with cryptically coloured vipers around) I employed an idea that gave me instant success. Abracadabra! I recorded their calls with the tape recorder, and then played their own voices back to them. God! What a laugh. As soon as the little macho males heard their own voices, up they would come, hopping over the floor and fluffing up their plumage with ludicrous self-importance, all the while yelling at the imagined intruder as if there were no tomorrow. They really did look cross, and as long as I remained still, these tiny balls of fury came right up to me in their eagerness to see the opposition off. One even had the temerity to sit on the toe of my boot. They bobbed up and down, twisting from side to side and flicking up the feathers of their backs to display their hidden white interscapular patches. The angry little birds made it easy for me, and I was able to identify them at last.

But even to this day, there remain calls I don't know. There is one I find particularly perplexing, an incessant long series of calls made up of trilled notes lasting about three seconds. I hear this song every day during the summer months, and I have sat for hours trying to head it off since it won't respond to a recording of its own voice. Another, a sort of low bubbling, emanating from a bush in front of me. And calls from the top of the forest canopy. Piercing reminders that I still had much to learn.

You may have noticed I rarely refer to a bird's song, but call or vocalisation, Yes. Very few tropical birds can be said to sing. The few exceptions: the familiar Thrushes, Saltators and a few Finches are probably of northern origin.

On the other hand, many tropical species produce the most wonderful calls: the bell-like tones of the Oropendolas, the police-car siren of the Black-capped Donacobius, Tinamou flutes and the heartrendingly beautiful soliloquy of the Musician Wren. Surely there are fairies in the woods?

And who could fail to be captivated by the hollow, resonant throb of the Spectacled Owl. Commonly refered to as "El Taladero" ("The Coffin maker") as he nails the lid down: "bub bub bububububub bbbbbbbbbbbbbbbbbb". Or the harsh roar of the Great Potoo, so like that of a jaguar many people have fled in terror from it. These and many others are music to my ears.

I could go on ad infinitum, but I must turn away from these wonderful things, and after four blissful months leave my beloved world of nature, so guileless, trusting and unequivocal, to pick up my mantle once more and return to the main story; back to the mean world of men, as it seemed to be in Bolivia.

Author's note: Using tape recorders to lure birds towards the observer was already standard practice; that I had to discover this method myself, is a measure of how cut-off we were from the rest of the World.

6. "This is My Town"

"Es mi pueblo, Señor." How those words ring in my ears.

We are sitting in the "Alcaldia", the Mayor's office and the town's dance hall. It's late Saturday night. The Mayor, Don Aquile, had invited us to be present at the festivities, his election as the town's new Mayor. Sitting together with him was a dubious honour that had been accorded us because we had just opened an official Amboró National Park office on the opposite side of the Buena Vista plaza. I had the feeling I was going to be the sacrificial lamb, "Larry the Lamb?", for truly this was like "Toy Town".

Machine-gunned by the gusty gamut of electric guitar music, we sat in discomfort, the painfully distorted sound made it impossible to think, yet alone talk but, talk is what Aquile wanted to do.

Aquile: "Si. Es mi pueblo, Don Robin. Conozco mi gente."

Me: "Claro, Don Aquile." Everything was "claro" ("of course", and everybody was a "Don". It made relations easier. To show respect, and to agree is to make friends. And yes it was his town. And yes, he did know his people.

One-eyed Aquile was telling me that I had better listen to him because he knew his town and its people, what they would accept, and what not. Aquile looked like your archetypal, not pirate (he was too morose), Archbishop! A monophthalmic Makarios, fatter faced, even darker featured, and with the ubiquitous glint of gold from his teeth. Did Makarios have gold in his teeth?

Aquile: "People here will not understand your Park."

Me: "Not mine, Don Aquile, not mine."

Aquile: "People here will not like this Park." Thick moustache over thick fleshy lips, a permanent puffy smile, the creases so deep bats could safely roost in them. Bushy black eyebrows over one black patch and one black eye, the eye lazy in movement matching the lips, Don Aquile tended to mumble. This, given the volume of the dance music coming from the six feet tall loudspeakers made it difficult for me to listen to him, as he advised me to do. And by now we were a bit drunk.

Aquile: "Claro. The Cholas".

Me: 'Are you one?' I wondered.

Aquile: "They will never understand. It will be too difficult to move them out of your Park. Here it will be very difficult to do a thing like that. "Dificil, sumamente dificil. La politica. Dificil".

The crowd of village folk dancing to the hot beat of the music overflowed the concrete dance floor. Apart from a few spotty youths none of the men were dancing, mostly young girls and middle-aged women swaying together. It reminded me of a Parent's Day Hop. High-whitewashed walls stained green with mildew and a roofed-in veranda surrounded the floor, large enough to accommodate a tennis court. As the dance floor was open to the sky, I could see a few of the brightest stars.

Don Aquile was already tiring of the conversation. The people here can only concentrate for a very short time. I was looking for the opportunity to disengage myself and was on the point of getting up when two of Buena Vista's confirmed drunks came over to us.

First drunk: "Buenas noches, Don Aquile".

Second drunk to me: "Buenas noches, Meester." How I hated being called that. They were hardly able to stand and bore the determined look that drunks have when they are intent on picking a bone. I searched among the dancers in the hope of attracting Miriam's, or Pieter's, or Els's attention. We had a prearranged signal, theirs a wink, mine a head over left shoulder ('let's go') for just these occasions. And I smelt trouble. Again.

Don Aquile didn't bother with such refinements; he got up and left, barely taking the time to bid me good night or shake my hand. One of the drunks sat on his chair that he put directly in front of me, his face a bare six inches from my own.

Then he started: "Somebody told me that you are in charge of the Parque Industrial. Es cierto?"

Me: "No, Parque Nacional. No es Parque Industrial", I corrected him.

Him: "Claro. Parque Nacional," he mouthed as he nearly fell off his chair into my lap, at the same time dispensing a good third of his "trago" (a small glass of alcohol) down my trouser leg.

Him: "You gringos, you want to take my land away. You want to take our lands away and give it to the other gringos, no?"

Me: "Well, no. You see a National Park belongs to the people of Bolivia, it's a Park for the Nation, it's not got anything to do with gringos".

I was making a mess of it. He didn't need to ask me why, in that case, was I here? He said as much by the way he screwed up one corner of his mouth, a movement that closed the eye above.

Me: "What? Oh, a National Park? Well, you see, it's governed by International Conventions. It's a place where nobody can touch anything. No logging. No hunting. No agriculture". The second drunk had taken hold of my sweater just below the collar. Hastily, I continued, "It's not a gringo Park, it's a Bolivian..."

The second drunk: "Si. Y yo soy Boliviano. Y quienes le dieron permiso para que venga a decirnos que podemos y que no podemos hacer? Es mi pueblo, Señor! Gringo de mierda. Gringo Carajo!". ("Yes, and I'm Bolivian. And who gave you permission to come here and tell us what we can and cannot do? It's my town, Mister! Fucking Gringo. Gringo, Fuck you!")

Thank God. Pieter was coming my way, his attention attracted by the lacuna of ill will around me, and now that the two drunks were shouting together, the rising consternation among the bystanders.

I quickly rose and backed off, suggesting they call by the office on Monday morning to pursue the matter if they wished. As we collected the girls and left, we could hear a barrage of insults coming from the vicinity of our recently vacated table.

Later we were to hear the fiesta had turned into a bottle fight. Over a hundred glasses were smashed in the melee and a good many people hurt into the bargain. This had nothing to do with the Park; it was an on-going feud between Buena Vista and the neighbouring village of San Carlos. The next time San Carlos held its festival party a "Buenavisteñian", who had five wives, was killed in a brawl when somebody shot him with his own pistol.

Although these accounts are true it would be wrong of me to give you the impression they happened all the time, they didn't. The villagers were peace-loving people by nature, but when drunk they often lost their natural timidity and became insulting. I attracted trouble because of the Park. I attracted trouble wherever I went, and it was foolish of me to have gone to the fiesta in the first place. I could never go to a bar or party without having to contend with the inebriated. So, we were largely forced to keep to ourselves.

Needless to say, the two drunks didn't turn up on Monday. They never did.

****

If you stuck a knitting needle neatly through the middle of a globe, starting at Amboró, the tip would come out in Vietnam. I tell you this because there are a lot of people, who like myself, feel uncomfortable unless they know exactly where they are; who's waking up when they're going to sleep, who's busily killing one another when they're dreaming peacefully.

Amboró National Park (that, for simplicity, I often refer to as Amboró) lies a little south of the seventeenth parallel and close to the 64th meridian. It is as far south as the Victoria Falls in Zimbabwe or Atherton in Australia, and shares a time zone with New York. It is a transition zone between the sticky tropical zone to the north, and the cool temperate lands to the south. The River Amazon sticks what must be one of its longest fingers into the guts of the area, a finger nearly 3,000 miles long. To the southeast the land is drained by the Rio Paraná, which becomes the Rio de la Plata as it flows into the ocean at Buenos Aires.

The Andean Cordillera emerges from the sea at the Straits of Magellan, and runs like a wall 3,000 miles due north to Amboró. From there it makes a sharp bend to run off in a north-westerly direction to Peru, and then makes a long steady curve to finish in the Gulf of Venezuela. The western part of the Amazon basin, like the reservoir of a funnel, is contained within this curve of mountains. The elbow formed by the bend of the Andes at Amboró corresponds to the neck of the funnel, a very significant feature I will refer to below. In eastern Bolivia the lowlands give way to the elevated lands of the Mato Grosso, representing the other side of the funnel so to speak.

Recent studies of satellite information conducted at the Mullard Space Laboratories in England suggest the lowlands of Bolivia may be sinking, a phenomenon which could be linked to the rise of the Andes themselves. True, as yet the evidence is flimsy, but I will stick my neck out and predict that ten thousand years from now the subsidence will cause the River Amazon to flow, once again, in its old direction: towards the west until it reaches the foot of the Andes. This is not a wholly wild idea given that the fall of the land from the great river's watershed to its mouth, 2,000 miles to the east, is a mere two hundred metres. I said ten thousand years, I could be wrong; it could be a lot sooner, for we should not forget that current thinking also predicts a rise in sea level due to the Greenhouse Effect. This could accelerate the process by restricting the free drainage of the Amazon into the Atlantic Ocean.

If I were Nostradamus, I would see the following scenario: "Under the gaze of Scorpio, across the Great Sea, a Serpent will turn to bite its tail, an event both stupefying and marvellous to cause great trouble in the air and on land; and the New Continent will be cleft by a great flood even unto eternity".

Seriously though, if the eastern lowlands of Bolivia are sinking, the upper reaches of the Rio Madeira will cease to flow and will form a seething bog of swampy forests and bad lands, that, in its turn, will give way to a vast inland lake flanking the Andean foothills. The lake will not be long before it overflows into the River Paraná, which will swell in size and depth to become a broad strait, as the water carries away the light sandy soils of which the whole region is composed. South America could become cut in two. Most of Brazil, Uruguay, northeast Argentina, and half of Paraguay could become an island (Atlanteans might even say: "again") separated from the rest of the subcontinent by these events. (Atlantiansmight even say "again.) The outlet to the sea that Bolivia has for so long craved would be provided, somewhat unexpectedly, on its own territory.

"Bah! Ridiculous fantasy", do I hear you say? Maybe, but there are a few motes of evidence that might be food for thought. The boffins at Mullard have found a large discrepancy between observed satellite readings and altitudes given on existing maps. Their figures place the Bolivian lowlands fifty metres lower than those maps indictae.

During the last decade South America, especially the southern half of Brazil, northern Argentina, Paraguay and the lowlands of Bolivia, has suffered the worst floods on record, and even here records go back quite a long way. Talking with the locals in my neck of the woods, (the Bolivian lowlands), they say the good grazing lands of the past have become permanent marshlands. They say the huge tracts of forest, where once they roamed at will to hunt Peccary and Deer, have now become slippery, mud covered matorrals impossible to traverse. They say thirty years ago they dug deep wells to find the water that now lies as lakes on all sides. So yes, according to them the Beni and northern Santa Cruz are inexorably being drowned.

Whatever the truth, I can think of only one group of creatures that would surely benefit from such an occurrence. The Pink River Dolphins of the Bolivian Amazon, an entirely isolated population of these animals that has been trapped by the Cachuela de Esperanza: a formidable series of rapids in the upper reaches of the Rio Madeira. This population has been prevented from mixing with their own kin downstream for so long that some zoologists now consider the Bolivian dolphins a separate species, Inia boliviensis. What a joyous day for them when they are freed to go home and mix with their long-lost loved ones.

These dolphins may offer another mote of evidence in support of Nostradamus's prediction, for how was it that they became trapped in the first place? The last decade, as I have said, has witnessed floods without precedent, or put another way the rivers have become deeper than ever before, and yet Bolivia's Dolphins remain trapped. I can think of only one explanation for this mystery: the lands they inhabit have sunk, the rock beds of the Cachuela de Esperanza have become the rim of a bowl, a future lake, over which the dolphins may never escape, their own "Berlin Wall".

But let's keep to the facts and return to the geo-ecology of Amboró. You may have discerned from the foregoing that Amboró can be divided into three distinct geographical zones: mountainous (or Andean), lowland tropical (or Amazonian) and lowland temperate (non-Amazonian). The latter can be divided into two subzones: Chaco-like (with an Argentinean flavour) and cerrado (with a Brazilian flavour). These diverse ecological zones, each with its characteristic flora and fauna, make Amboró a true melting pot of different species, a biological "Tower of Babel". If, as many modern field biologists do, you were to count all the different plants and animals in a football-field sized plot you would not find as many in Amboró as you would further north in the true Amazonia. This is because tropical forests contain the greatest number of species and Amboró is not strictly tropical. Based on this method of recording its biodiversity Amboró would come a poor second.

This had something to do with our difficulty obtaining the support Amboró truly deserved, and a lot do with my determination to get at least, the birds registered. With a dearth of money for conservation many funding agencies select their priorities according to an area's known species richness, and with shortfalls of funding for longer-term biological studies the football-field method is favoured.

The truth of the matter is that Amboró, with its mixture of plant and animals from three very different geographical zones, is one of the World's richest deposits of biodiversity if assessed by the sum of its parts and not just one of them.

Because Amboró is the meeting point of three geographically distinct zones: the southernmost Amazonian jungle, the northernmost forests of the Chaco, the eastern most part of the Andes, it stands to reason that many of the plants and animals found there are at the extreme limits of their distribution. Biologically speaking the fauna and flora of Amboró is made up of pioneers, and being pioneers, they struggle to survive. Where there is a struggle to survive, as Darwin was the first to point out, there is evolution. Natural selection will favour those plants and animals that are the most adaptable. The most adaptable will be those that have inherited the greatest elasticity in their genetic makeup. Struggling to survive, they will produce the greatest number of mutations: the essential nonconformism with which Mr. Hereditary beats out new species on his "Anvil of Evolution."

Amboró is, then, a living laboratory of Ecological Genetics. As such it has a priceless gift to offer anybody willing to study its potential. If you were a genetic engineer or a plant breeder, or simply interested in multiplying animals for a biological release programme, or for a zoo, the plants and animals from Amboró should prove to be the hardiest and the most malleable, because that is what they have to be, given the vicissitudes of a climate that keeps Amboró's wildlife on the hop, so to speak. Raked by cold gales from the south one moment, enervated by hot north winds the next, dried by the sun shining from a clear sky today, and shrouded by dripping wet clouds tomorrow, nature does not allow her species to slip into evolutionary apathy.

If, to take one example, somebody wanted to cultivate an Amazonian palm tree in England, they would require a cold-resistant species. If they brought seed from Manaus (in the cradle of the Amazonian rain forest) the attempt would almost certainly fail because the seeds come from trees permanently bathed by the hot humid vapours of the Amazon's unchanging climate, a recipe for genetic conformism, an "Evolutionary Straight Jacket". Amboró's pioneering palms would be much more likely to provide the Arecacophilist's requirements. And the same argument would apply to any of Amboró's 15,000 plant species waiting to fill the corner of your solarium or the green houses of the Pharmaceutical Industry.

Without, I hope, the risk of driving you back to the Bruce Forsythe Show, or whoever it is that panders to the masses today, I should put Amboró's diversity of habitat in perspective. A Dr. Holdridge made his name by undertaking the laborious task of cataloguing the World's Ecological Variety, dividing and subdividing our planet into numerous Life Zones based on their altitude and climate, and each with its characteristic vegetation and animal life. In Bolivia he recorded forty-eight different Life Zones; from the snow covered mountain peaks and salt lakes of the altiplano to the humid tropical forests, savannas, and swamps of the lowlands. The boundaries of Amboró National Park embrace 1,000 square miles and contains eleven of Dr. Holdridge's Life Zones. In comparison the 23,000 square miles that comprise the whole of Costa Rica contains only nine. So as you can see, as biologist for Amboró I really had my work cut out.

****

Unfortunately, Amboró presented a much less delightful problem: the lowlands that consisted of two Holdridge Life Zones: tropical and subtropical humid forests, were occupied by about six hundred campesino families dedicated to the removal of as many trees as they could in the shortest time possible. And as Don Aquile avowed (you can always trust a one-eyed Archbishop) their removal to areas outside the Park was going to be "sumamente dificil". So did Noel, who had by now abandoned any hope of establishing Amboró as a going concern. "Bulldog Drumskull" (that's me) was convinced that not only could the Park survive but it might also be freed of its tree-felling itinerants.

Eager Beaver I was, but I had two factors in my favour: most of the colonists were illegally settled there, and some of them had asked me to help them get out. With a wee bit of support from Government (an official request for funding), and a million or two (in the shape of dollars) from the good old World Bank, or any of the other prestigious conservation organisations, it could have been done. So I was eagerly awaiting the arrival of Curtis Freese from WWF's offices in Washington. A two-year wait that had prevented me from leaving Bolivia to visit my family, or go on holiday with my sister to the

Galapagos Islands, or sell my tumbled-down cottage in France; in other words, a wait Godot would have approved of.

24th August 1983: Russell Mittermeier, Director, WWF-US to Mr Clarke:

I recently spoke to Mr John Dunning about your efforts to conserve a large tract of land in Bolivia. From what he told me, your project is a very exciting one, and one that might be of interest to our organisation. If you could let us have further details on it, perhaps we could be of some assistance in your efforts. I assume that you are in contact with Dr Gaston Bejarano since [he] would probably be very interested in your work.

23rd September 1983: Curtis Freese, Senior Wildlife Biologist, WWF-US to Mr & Mrs Clarke:

The staff of WWF-US has reviewed with interest your letter, and the proposal for the establishment of Amboró National Park. I am considering the possibility of visiting Bolivia in January or February 1984. I would hope, of course, to have the opportunity to meet both of you to discuss the Amboró project.

28th November 1983: Curtis Freese, WWF-US to Mr & Mrs Clarke:

Thank you for your letter of October 13 and for the additional information on Amboró. I wish to propose the dates of January 23-29 for a visit to Bolivia and to [you] the 25-29th. I would, of course, like to make a brief reconnaissance of the proposed park during that time. An overflight, as you propose, would be helpful.

5th January 1984: Telex from Curtis Freese, WWF-US to Clarkes:

Unavoidable conflicts require postponement of trip. I propose alternative dates in March for visit. Regret inconvenience.

6th January 1984: Curtis Freese, WWF-US to Mr & Mrs Clarke:

As my recent cable indicated, it is not possible for me or another WWF-US representative to visit your Amboró project during the last week of January. We therefore propose a visit to Bolivia sometime in March.

25th January 1984: Telex Reg Hardy to Curtis Freese, WWF-US:

Ref: your visit to Bolivia...feel that visit should be timed for May...neither G Bejarano nor myself will be in Bolivia March/April.

30th January 1984: Reg Hardy to Curtis Freese, WWF-US:

Ref: Amboró National Park project. I have been handed a copy of a letter written on 18 January by John Dunning to Robin Clarke. Therein he states that other WWF people have told him that there is strong opposition on the part of Bolivian Government to this project. So much so that he doubts that WWF would support it. I would respectfully advise that the information given to him is totally and utterly incorrect. The Amboró project has the full support of [everybody, as listed]. I think I am right in saying that the (incorrect) information came from Gaston Bejarano as a result of his personal dislike of Noel Kempff Mercado...I am doing my level best here to get everyone to work and pull together conservation wise and believe me, this is no easy task. However, I am meeting with a great deal of success.

31st January 1984: Alberto Navarro, Chief of Wildlife and Parks Department and Gaston Bejarano, Scientific Assessor, Chamber of Deputies, La Paz to Curtis Freese WWF-US:

Considering intervention of PRODENA and Mr Hardy as well as Mr Clarke, Bolivian Government agrees to collaborate with Amboró National Park ...

31st January 1984: Delfin Goitia, Director, CDF-La Paz to Curtis Freese, WWF-US:

Would like you to know that the CDF is totally in favour of the Amboró project and that we are processing all the legal documents...to declare it a National Park. I hope that you will be able to visit us in May.

2nd February 1984: Telex Curtis Freese, WWF-US to Delfin Goitia, CDF-La Paz (and ditto, to Navarro, Bejarano and Reg Hardy):

Thank you for your telex concerning proposed visit by WWF-US to review establishment of Amboró N.P. and other areas in Bolivia. May visit proposed by Delfin Goitia should be fine. Will advise of dates soon as possible.

29th April 1984: Reg Hardy to Robin and Cheri Clarke:

Freese assures me would visit Bolivia early June.

15th May 1984: Reg Hardy to Robin Clarke:

Big WWF policy meeting...[Freese] told me could not give a date until that is over ...

23rd May 1984: Telex Reg Hardy to Robin Clarke:

Freese...thinks better they ask one of their Latin American consultants (Dr Ponce del Prado of Lima) to come...but unable to come before August.

30th July 1984: Reg Hardy to Robin Clarke:

WWF-US/Prado visit: I haven't heard either ...

10+27th August 1984: Telexes Reg Hardy to Robin and Cheri Clarke:
Will chase Freese next week. Prado will not be coming. Instead [WWF] will rely on Peyton report after his October visit.

28th October 1984: Reg Hardy to Robin Clarke:
In connection with Amboró which IUCN/WWF say they are prepared to fund. This would entail them appointing an approved consultant to visit Amboró ...

... Dear Lisa, Dear Lisa
There's a hole in my bucket
Dear Lisa, A HOLE!
You can only have a good laugh. Can't you?

                    ****

The Agrarian Reform of 2nd August 1953, as I have stated before, declared that: "La tierra es de quien la trabaja": "The land belongs to he who works it". In other words, a man could occupy a piece of land owned by somebody else and claim Squatter's Rights. The law did not intend that somebody was free to climb over the wall protecting your house and garden, build himself a cardboard shack, and dig up your lawn to grow vegetables and rice. And yet this is not far from the way in which many immigrant campesinos interpret the law. These people, the "loteadores", have become a serious source of social friction; illegally occupying private grounds and public parks within the city limits of Santa Cruz. In the rural areas the loteadores formed themselves into syndicates comprising scores of families who arbitrarily took over hundreds of square miles of land.

With Supreme Decree 11254 in 1973 the Bolivian Government set aside the mountains of Amboró and their adjacent lowlands as a protected area, the "Reserva Natural Tncl. German-Busch", that as you will recall was to become Amboró National Park in 1984. Both Supreme Decrees recognised the rights of the people already living there, though the 1984 Decree (Article IV) was a bit more specific: "Los campesinos (autóctonos) con asentamientos tradicionales gozarán de trato especial y a su vez desempearán como elementos activos en la conservación del área".

In English "The farmers (natives of the area) with traditional land holdings will have special rights and in their turn will form an active element in the conservation of the area".

Fine. Nobody wished to deprive the traditional settlers of their rights to remain in the Park especially as the Decree stipulated that they would help defend it. The problem was the wording was not specific: who was to be considered traditional, and who not? And what did the word autóctono mean?

It looked as if the Decree had singled out the older Camba communities and the handful of Guarani Indians, the only true "Autóctonos" as legal residents of the Park, and that all the others, the few recent Camba settlers and the mass of Colla colonists, had no right to remain there.

Certainly Article 3 in both Decrees is very definite on the subject of further colonisation: y no siendo permitido bajo ningún concepto nuevos colonizadores. ("and on no account are new settlements to be allowed").

In 1981, three years before the area was declared a National Park, I spoke to the Padre Pablo, our American priest at Buena Vista whose diocese included Amboró. According to his own census Amboró contained two hundred and eighty families distributed among ten villages.

One month before Amboró National Park was declared the "Federación Nacional de Campesinos" sent a thousand families to settle there as a protest against the conservation movement in Bolivia. Of these about three hundred families took up permanent residence.

By 1986 when we conducted our first census under the auspices of the CDF the number of people claiming to hold land rights in the Park had risen to seven hundred families. In 1989 when a fully-fledged census was carried out with funding by US-AID, the total number of families registered stood at nine hundred and eight. In addition, about two hundred other colonists had claims to land in the Park but had not taken up their option to occupy it. Finally, let it be said (and with true regard for a situation that had been complicated by the movement of people in and out of the area), of the nine hundred and eight families claiming Squatter's Rights only about one hundred of them could, if we stretched the law, claim to have legal status.

Even had the Instituto de Colonización kept proper records (and they hadn't) it would not have been a simple matter to sort out which families were legal and which not, and even had this been possible, Government would not have allowed the eviction of people from the Park. Like any law prejudicial to the peasants of Latin American countries, Amboró's decree was a "Paper Tiger"; designed to pay lip service to the Conservation movement, while placing those charged with implementing it in an impossible, and potentially dangerous, situation.

As I have stated, too often, I wanted to get the colonists out of Amboró and I did not hesitate to wave the Decree as proof of Government intentions; though this was more a ploy to unsettle them and render them less willing to invest their money and effort entrenching themselves, than it was a legal threat. No, the only way to achieve my aim was to start by moving those who wanted to be moved.

But why did some of them want to move? First, I had better say that as much as I love Amboró I wouldn't want to live there. With one hundred and twenty inches of rain a year it's a very wet place. The area occupied by campesinos is cut off from civilisation by mountains to the south and by two rivers that meet in the north (the Surutú in the east and the Yapacaní in the west), rivers that during the wet months cannot be crossed by boat or during these times by foot (some Campesinos die every year trying to do so.) The rivers are wide and untrustworthy, and it would be a major engineering project to span them with bridges. In other words, quite a few Campesinos are tired of the lack of access to the Park, and consequently the lack of roads within the Park itself. If you had to walk six miles to market, carrying your produce on your back, as many of them do, you might feel the same way.

I have seen them shouldering a forty-pound stem of cooking bananas (worth 50 US cents) over such distances. And, cut off as they are, woe betide any of them who fall seriously ill, for there are no doctors within the Park. There are, however, plenty of biting insects that carry a range of diseases: malaria, yellow fever, leishmaniasis, and poisonous snakes (as I have already related).

Then add to these ailments the attendant risk when machetes, axes and chainsaws are employed to achieve their aim: cutting all the forest down.

And, as many of them have eventually found out, most of the soils in Amboró are so sandy and devoid of nutrients that according to Bolivian law it is illegal to settle colonists on them.

"Okay," do I hear you say? "Why do they want to stay?" Primarily for the same reason as anybody else, it has become their home. And I know most of them would not trust the Government were it to offer to move them; and in the absence of such a scheme they have nowhere else to go. Some no doubt enjoy the life the area offers: clear unpolluted rivers with plenty of fish, independence and privacy (including, as I have told you, escape from the long arm of the law), lots of land (120 acres each) and for some, aesthetic reasons too. It is also true the more established among them would suffer economically if they were moved, for they are the ones who have made the biggest investment in time, work, and money improving their lands, and they know this. They are also the most politically motivated and encourage the others (usually by whipping up notions of patriotism and fear of Government offers) to stay. They believe, and maybe with good reason, that the Government will sooner or later accede to their demands and change the boundaries of the Park in their favour; freeing them to sell their lands if they wish, and at a good price if they can persuade their clients to believe them when they say the rivers are full of gold, and their lands pock-marked with buried Inca treasure.

Apart from the residents in Amboró National Park, there were many selfish outsiders who for one reason or another supported the campesinos: hunters, gold hunters, loggers and, as you already know, Animal Exporters; these are just to name a few.

So, as you now know and I thought I did too, there was yet a larger threat to come. About this I will tell you later.

****

To return to what I was saying about the Amboró National Park office we had opened in Buena Vista. The office was the fruit of a contract I had signed on the 1st of February with the Bolivian Wildlife Society (PRODENA-US) who agreed to pay me a monthly fee of $500 for a period of six months. The contract was a small miracle given that my relations with Reg Hardy had reached a new ebb. He had agreed because he saw me as an indispensable part of Operation Raleigh. Alan Hutchinson, whose idea it was, thought it high time that PRODENA put their money where their mouths were and helped me to get Amboró going.

I think it necessary to summarise where Reg and I stood. I was quite sure he intended to take over Amboró National Park as his own "pigeon", and with this in mind would try to find some way of removing me as soon as I could be replaced. He had prepared a project proposal for the World Wildlife Fund and the International Union for Nature Conservation without my knowledge, but based on my work and my past reports to him and the WWF. Although he had identified me as a project leader, he now included (to their own astonishment) Pieter and Els Brekelmans as project leaders also. Reg had already started corresponding directly with them over, at first, the Parrot Rehabilitation Project, and later the Park and the planned visit of Operation Raleigh. He needed to proceed with care because he knew we were very good friends. He pooh-poohed my official status as Park Warden (which I had been ceded by the CDF-Santa Cruz) stating that nobody in La Paz knew anything about it. To quote from his letter of the 24th December to the Dutch couple, "I also attach a copy of a letter from Robin in which he states he is Acting Warden and that he works for the CDF. This is rather stupid of him as you will see from the enclosed exchange of telexes that neither Oscar Llanque [Wildlife Chief, La Paz] nor the CDF in La Paz [i.e. ditto] know anything about it." Then, curiously, he goes on: "It seems to be an attempt to discredit Ralph Peterson and myself for some reason best known to him, probably an attempt to obtain some money from Operation Raleigh which will not be successful." Then further down the letter, referring to a pre-Raleigh visit report prepared by Bob Le Souer, Reg says: "You will see that [Le Souer] considers Pieter a likely warden for the Estancia San Rafael de Amboró (not the Park as misread by Robin), would that interest the two of you?"

I failed to see why my official appointment by the CDF as Warden (on a voluntary basis for the time being), and my confirmation of this publicly should have provoked Reg's ingenuous comments. All I wanted was to get on with the implementation of the Park and to involve the CDF in its protection; after all, they were the Government Department responsible for National Parks. Given the very nature of the bureaucracy in Bolivia, its extreme inefficiency in particular, there was no reason why the CDF in La Paz should know everything their Santa Cruz counterparts were doing. Here I should mention that although the CDF in La Paz was the Head Office it only comprised the Director, Wildlife Chief, and a couple of technicians; whereas the CDF-Santa Cruz employed more than eighty people, thirty technicians and fifty CDF guards, most of whom spent their time collecting the duties from the logging companies.

In any case, I now carried a CDF Identity Card naming me as Warden of the Park; and if anybody doubted the intended solemnity of this they only had to read the reverse side, that stated: "Likewise, we request that Political Administrators [Sheriffs, Fiscal Agents, etc], the Military and the Police, give the carnet holder their collaboration and necessary backing so that he may pursue his duties, in respect of the Renewable Natural Resources of the Nation, under the current Laws for their protection. Paragraph XVII (Article No 91) in the Regulations covering the National Forest Guard [states that] CDF forest guards and technicians are authorised to carry firearms in the exercise of their duties."

Heavy Stuff. Licensed to kill.

Of course, the reference to Pieter being Warden of the Estancia nettled me. The slight change in wording, Estancia for Park, given that the Park Warden was going to live at the Estancia, was a typical Reg Hardy deception, one he hoped would cause trouble between the Brekelmans and myself. Reg's attempt to divide and rule.

Then, when Reg offered to give me half the money for a one-way plane ticket to England (maybe in the hope that if I went I might not return), I became convinced that Reg was not only trying to get rid of me (he didn't know Bulldog Drumskull); but was also intentionally keeping me short of funds. He certainly was spending a good deal of money on other things: his Comunicaciones still came thick and fast, Operation Raleigh and the purchase of the Estancia; and now he had money to offer my Dutch friends a salary. My contract with PRODENA-US, which I was yet to be offered, was to be proof of my conviction.

So paradoxically the $500 a month that PRODENA now offered me was cause for further annoyance. With somebody more reasonable than Reg our tacit agreement — that I would get on with the Park whilst he made this possible by obtaining the funds — would have been an ideal arrangement. But as it was, I could not help but think the project would be better off without him. Some organisations did not like him, WWF for instance; about whom, to quote Reg's own letter [23rd Dec 1983] to me, he had said, "I don't know how I am viewed by WWF-US as I had a set-to with WWF-International in Switzerland some years ago."

And, I was receiving disturbing news from well-placed friends, like Bernie Peyton, that Reg was not appreciated by some of the funding agencies. Was I being told I should find a way to get rid of him?

With these developments, and the background of past troubles, you can appreciate why I say my contract with PRODENA-US was a sort of miracle, it was. Reg still needed me to assist with Operation

Raleigh, but as he had made quite clear in a telex (12/12/84) to Ralph Peterson, assist: "Please make it clear to all that tasks to be performed by Op. Raleigh will be planned by and decided by PRODENA in conjunction with Op. Raleigh Bolivian Committee and other interested scientific bodies in collaboration with the Bolivian Government. Nobody else." Ralph (I'm sorry, Robin) obviously thought that the sole object of this telex was to tell me personally to keep my nose out of Operation Raleigh's business, for the dear man passed me a copy on which he added in his own hand: "Robin, for your info but PLEASE IGNORE. I will sort it out!"

Desperately tired of all this needless subterfuge, and with my newfound wealth, the $500 a month I set about trying to make a good impression on Reg, Alan and Ralph, the CDF, the people of Buena Vista and any would-be donors to the Amboró project. Though the money was meant to be my salary I set aside enough to run the office and split the remainder between Pieter and Els, and Miriam and myself, with the understanding we would work as a team to get the Park going and to make Operation Raleigh's visit a success. We made the office look as formal as possible: two desks, a hired telephone, a filing cabinet, an electric typewriter (the first the villagers had seen), and maps of the Park and conservation posters on the wall. In such a small village it really did look sufficiently official to convince the people of Buena Vista and the colonists in the Park that Amboró was something to be taken seriously.

When the office was organised, we officially inaugurated it by inviting representatives of the CDF and Buena Vista's bigwigs: Don Aquile (as Mayor), the Sub Prefecto (the Sheriff), the Men's Civic Committee and the Lady's Civic Committee (that was more influential), the Police, the Public Notary, priests from the Church and so on. Without doing so our office would have had no status. We opened the proceedings with a slide show about Amboró, what it represented and why it was needed. The CDF made a short, somewhat coy speech about their part in Amboró (i.e. me as Warden). Aquile placed the Mayor's stamp of approval on our efforts; but as usual with some reservation said, "We will have to see how it turns out".

We closed the proceedings with wine and cheese biscuits, and an invitation to the villagers outside to attend a repeat performance of the slide show in the evening.

We all thought it had been a great success. The evening's activities were even more so, the place was packed, and a considerable crowd had to stand in the road squinting through the windows at the projected slides. I say great success, albeit in Buena Vista, where there was absolutely nothing else to do, we had a captive audience.

Els and Miriam bore the brunt of the office work and stoically dealt with the (eventual) flood of campesinos who came to enquire about the Park, and in what way they were to be affected. We had placed copies of the two Supreme Decrees (that for 1973, and the 1984 one declaring the Park) on the wall, and the people were asked to read them carefully before putting their questions to the girls. We knew we would have trouble with the colonists in the Park because, as I have told you, the decrees specifically declared their presence there illegal.

What we failed to realise at the time was the extent to which the people of Buena Vista depended on free access to the Park's resources for their living. The wealthiest among them had trucks and machinery for the removal of the Park's mahogany, tajibo, cedro, verdolago, and other valuable hardwood trees; and they employed a further thirty or so men to help them. Some of the villagers made their living by hunting the peccary, deer and tapirs in the Park and others to hunt or fish for pleasure. Later, after we had completed the 1986 census, it became clear to us that nearly everyone in Buena Vista claimed to hold titles to land in the Park, and those who didn't personally work their holdings rented their land to others who did. All these people, just about the whole village, were very suspicious of us and the Park, and some were busily undermining our efforts by supporting the campesinos in their campaign to have the Park annulled. We really weren't very popular, though there were a few who understood what we were doing and they expressed their support for our efforts. All the Cambas, with or without vested interests in the Park, grudgingly admired our stated aim to remove the illegal colonists from Amboró but, for reasons that had nothing to do with conservation, and a lot to do with their dislike of the Colonists.

Oblivious to all these different factions and the corresponding jiggery-pokery of the local political scene, we were riding high; four friends carried by our wave of enthusiasm over the coral reef of mounting opposition. Opposition that was soon to express itself in a distinctly alarming manner. But at the time we really were excited with everything to be gained, and six months free from the worry over cash. And best of all, the intensification of our friendship, united as we were by our common cause. It was too good to last.

The quinta, El Hombre Nuevo, where we all lived was (as you may remember) owned by the Muyzenburgs, who in their absence leased the property to the Brekelmans who in turn got a nominal rent from me for the thatched roofed house where I lived with Miriam. On their return to Holland the Muyzenburgs had separated, and Chela had sent word that Pieter should try to sell the quinta along with its contents — sell at any price. We would have dearly liked to buy the place between the four of us; but short of money, and full of doubt about our own futures we delayed the decision for another day. Then, quite unexpectedly, the matter was taken out of our hands.

In March a temporarily wealthy young Dutchman (an acquaintance of the Muyzenburgs) came to see the property and decided, then and there, to buy it. This romantic young man (he was known to everyone as Bart) was a man who made lightening decisions, and by the end of the week he became the new owner of El Hombre Nuevo — the New Man indeed. Happy for Chela's sake, we were also aware of the changes this would bring. Bart did not intend to live permanently in Buena Vista, but rather, he said, for a year or two. He wanted Pieter and Els to stay on as managers and he intended enlarging the main house. Until this was done he would occupy the cottage where Miriam and I lived. But generous to a fault, Bart subsequently shared the cottage with us, postponing our eviction until the end of September when we were scheduled to move to the Estancia.

With Operation Raleigh's decision to replace Bolivia on their itinerary Reg Hardy made the final payment on the Estancia. So I was not worried by the loss of our home, and indeed thought it might be best if we moved to the Estancia before Operation Raleigh arrived. We could then devote our time to preparing the place for their use. It had so many rooms we wouldn't have got in each other's way. We would also be on hand to assist them with their daily activities. The idea was made possible by the good news that, between them, Reg and Raleigh were going to buy an ex-British Mission Land Rover; having failed (again) to get one donated. Since the Estancia San Rafael de Amboró was twenty miles from Buena Vista, the nearest source of supplies, it was an essential item.

Reg had cancelled his pre-Raleigh visit to Bolivia because of trouble with his health. He now agreed with my proposal to come to England; primarily to brief Operation Raleigh, but also to discuss the future direction the Amboró project should take.

The timing suited me well as I had not seen my family for some years, and leery of the Bolivian banking system, Chela Muyzenburg had agreed to pay part of my air fare if I brought the cash from the sale of her quinta with me. With Reg having already offered to pay half of my ticket, I was set to go.

Before doing so there were a number of things I had to sort out. The most important was for me to consolidate my position as Acting Warden for the Park, and to place Pieter on a formal footing with the CDF so that he could take over during my absence. As chance would have it, the CDF called me for a meeting with the Director to discuss our plans for Amboró; especially as they had received a number of complaints from the colonists living there.

It seemed the ideal opportunity for Pieter to meet the CDF and for us to petition them with a long list of items which we required. So we set off for Santa Cruz loaded down with two tons of cooking bananas which Pete wanted to sell in Santa Cruz. Riding along in their cage on top were the last ten macaws from the Rehabilitation Project, the mutilated ones we were returning to the CDF.

Referring back to my notes I am struck by the extreme optimism with which Pete and myself went to the meeting. We had divided our petition into two sections: the first, a list of fundamental necessities which would have been reasonable requests anywhere except Bolivia; the second, a list of queries, mostly legal ones, concerning the protection of the Park. When we arrived at the CDF we had to wait some time before we were shown into the Director's office. He, Nestor Ruiz, seemed to be in the middle of some crisis or other; and after glancing quickly through our petition he asked us to discuss the matter with the Chief of Wildlife, Jorge Aguirre, with whom he had a short discussion on the phone. The requests, we were told, would have to be placed before the CDF's Board of Directors, and could not be resolved immediately. We did, however, come away with an official carnet for Peter and letters of attorney for both of us. According to mine I had already been promoted. I was now Director of Amboró National Park. Pieter had been made Park Administrator. No mention was made of salaries, but with our written credentials our positions were formalised, and I felt I could leave for London with a good deal of satisfaction.

****

After spending a week with my family and finishing the transfer of Chela's money to her bank in Holland, I was ready to make the long trip to Reg's farmhouse near Abergavenny in Wales.

The house was situated on the lee slope of a hill and looked down upon an open valley where alder trees crouched over a brook (Reg's trout stream) and ended in a reedy pond at the bottom. The building, an old farmhouse of unpretentious proportions, had been renovated by Reg in a comfortable but not sumptuous manner. What immediately caught my eye, and singled out the owner as somebody special, were two Llamas Reg had brought over from Bolivia.

We didn't get down to business on the first day, but spent the afternoon idling around the grounds, feeding the poultry, making friends with the dogs, and admiring the Llamas and horses. In the evening we spent two pleasant hours reviewing a series of short videos Reg had commissioned about PRODENA's activities and the need for conservation in Bolivia. I thought they were very good.

By the time Reg's wife Laura arrived we were sleepy, full as we were with smoked salmon and samples from Reg's wine cellar. I cannot speak for Reg, but I went to bed a bit disappointed by our failure to entirely expunge the rancour between us. He was a kind host and a good one, and I tried to act the grateful guest; but there was a falseness, or forcedness that affected our chat as we felt the strain on us both our pretences created.

After an early breakfast: eggs, toast and marmalade for Reg, coffee for me, we made our way up to his study and settled down to business. The first items we cleared out of the way was the accounts and the final report on the Parrot Rehabilitation Project. As you will remember, the accounts had been a bone of contention for quite some time. Reg wanted everything down to the last penny accounted for and drawn up in a business-like manner. I had little experience of accounting and considered my amateur efforts sufficient. As there could be no evidence for most of the costs (food from the market place, wages for the daily labourers, travel allowances for CDF staff and incentives to expedite the carpentry work), the exercise struck me as futile. With receipts for only a few items (petrol, hardware and some of the medicines) Reg was having a hard time keeping his temper.

With this out of the way, we discussed the overall financial situation about which Reg was unable to sound hopeful. Now it was my turn to be disgruntled. As you will recall, among my own contacts the word was that Reg was more of a hindrance than a help to the Amboró Project. That some considered him too amateurish there was no doubt; but a major obstacle was that potential donors didn't want to pass funds destined for Amboró through PRODENA. They suspected that Reg intended to drain off some of the money for the Society itself. Reg wanted a major conservation organisation to finance Amboró through PRODENA, his Trojan Horse with which he hoped to break into the big time; and of course control me, Amboró, and, as I have already told you, the conservation movement in Bolivia. He said our hopes for funding still lay with the World Wildlife Fund; a statement, you will understand, I had to accept with a good deal of suppressed laughter. Neither of us had heard any more about their proposed visit, nor had they acknowledged receipt of my report sent them by courier back in November.

With these unsatisfactory aspects cleared from the agenda Reg, who was getting cantankerous again, turned to the matter of Operation Raleigh, to him the most important item. Given that Raleigh was a major source of difficulty between us, this was not a good omen. I had to make a very determined effort not to appear negative because Reg had me over a barrel, he had still to pay me for my air ticket. If he blew his top, I would go away empty-handed. I had already decided to keep my own points of view that I knew he wouldn't like for our joint meeting with Raleigh in London. So I sat listening to Reg with a good dose of tedium.

I nearly blew it when I handed him my CDF carnet and copies of their letters confirming my accreditation as Director of the Park and Pieter's as Administrator. In theory Reg should have been pleased the CDF had been willing to place this responsibility on us, but in the light of my knowledge that Reg intended to supplant me with someone of his own choosing, I handed him the papers with hardly concealed smugness. Having turned the tables on him, it was now Reg who had to hide his true feelings in order to keep the peace. I could see he was chagrined, and for a second I thought he wasn't going to do it.

Nevertheless, he did, just. Soon after this we recognised enough was enough, and with a common awareness that further discussion might lead to open hostility, we went down to lunch. As soon as it was polite to do so I took off for Leamington Spa to spend a few days with my mother; Reg's cheque safely tucked away in my pocket. On the way I remembered I had forgotten to mention my desire, nay, need, to move with Miriam to the Estancia.

The following week I made my way to Flood Street in Chelsea. How appropriate that Operation Raleigh's office should be located there, for stormy waters were to be the hallmark of their visit. After hasty introductions to some of the key figures supervising the Bolivian phase (as they called it), and after Reg arrived, we sat down to start the meeting. After ten minutes of introductory briefing my turn came to speak. I knew what I had to say would be unpopular, but my message to the meeting had been discussed with Pieter and Els and the CDF in Santa Cruz. I tried to keep it brief; the key point I wanted to make concerned Raleigh's timing for their visit, scheduled for 4th July to the 24th September. It was unlikely they would agree to an alteration at this late stage; but on behalf of all of us in Santa Cruz and at Buena Vista I asked them to consider the following options in the order given:

a. Postpone their visit for twelve months.

b. Postpone their visit until September/October.

c. Not come at all (because in the order of importance):

Considering that:

a. On the 14th July Bolivia was holding a General Election, an event that involved a considerable amount of 'confusion'.

b. There was trouble brewing amongst the colonists in the Park, and the Elections were likely to exacerbate an already delicate situation; not that we thought any harm would befall the Venturers.

c. The CDF-Santa Cruz had hardly been consulted over Operation Raleigh's visit to the Park (a failure we had pointed out to their Bolivian Committee) and we considered that an extra month or two would be needed to correct the situation. As far as we knew I was the only person who had informed them of Raleigh's visit but, as I was not part of the Committee I had done this unofficially.

d. July to August was not a good time to come; the weather would be cold, windy and unpredictable; even heavy rainstorms could be expected. Given that this was the Bolivian winter the scientific component of Op. Raleigh's objectives would be more difficult to satisfy, and for those who were to camp out in tents life could be miserable. (Earlier I learnt that for some peculiar reason Reg was not willing to let them sleep inside the house.).

e. The Estancia San Rafael de Amboró might not be ready to receive them. The house urgently needed a new roof among other things.

A change of timetable was not accepted, albeit for the record they had noted our comments and would follow up on our preoccupations.

Reg, who I knew was quietly smouldering, declared his intention of making his delayed pre-Op visit during May, when he would look into the possible danger to the Venturers and the disenchantment of the CDF. Also, he said, the full Bolivian Committee was expected to be in Santa Cruz shortly and would remain there for the duration of Raleigh's visit. An announcement, judging by the pompous manner in which it was delivered, his eyes fastened on mine, that was meant to dispel any misgivings by its weighty importance. And another personal warning to me, "Watch it!"

After another hour the meeting broke up and I was left to discuss many practical and technical points with Operation Raleigh's London-based staff. Whereas I couldn't help but admire their enthusiasm, that had a vicarious quality about it, it was evident they were not prepared technically; or maybe I should say (since they didn't pretend otherwise) it was unlikely they would fulfil the expectations generated by Reg's proselytising. But I could see trouble on its way, for it seems to be a characteristic of the least informed to expect everybody else to perform miracles, and among some of the nescient xenophobes awaiting Raleigh's arrival in Bolivia trouble is what they wanted.

Worried by this I called on my entomologist friends and ex-colleagues at the British Museum of Natural History and the London School of Tropical Medicine. I also went to Kew Gardens to talk to a man with whom I was collaborating on a preliminary study of Bolivia's mushrooms.

To tell you the truth I was a bit nonplussed by the derisive comments made by one and all when I told them I was looking for useful projects that the Raleigh Venturers could undertake. There was no end of things the young people could do; but there was no point in them dedicating time and energy to studies that had no finite goal.

For example, a collection of Amboró's Woodlice may be an interesting occupation, but if there wasn't anybody in the World studying New World Woodlice to give the specimens a scientific name, the Venturers final report would not be able to say more than:

Collection of wood lice (45 unnamed exx.). That wouldn't be very impressive.

On the other hand, a collection of biting flies from Amboró would interest the School of Tropical Medicine, and given their medical importance there were sure to be specialists to identify them. Moreover, this project would have a practical application because it would provide Bolivia's medical research effort with valuable data on insect vectors and information pertinent to the development of tourism in the Park, and if done well enough would benefit the campesinos living there.

But all my efforts were to be in vain, for as it turned out, the trouble I saw coming because of Reg's effete manoeuvring mush-roomed into a body much more venomous than anything I had eaten on behalf of my studies for Kew. None of my entomological projects were implemented because, in spite of everything, I had been led to believe Raleigh's policy, and I wouldn't have disagreed, was to leave the Venturers free to follow their own inclinations.

Turning away from Operation Raleigh for the moment, I should mention the famous, or I should say, infamous, competition. It had always been my wish to make a nature film in Amboró for children back in Britain. I thought I might transmit my own delight of nature to my young audience by recording the experiences of two little girls as they explored the jungle together (an English one who had never been to a tropical forest and a Camba one to be her guide).

As chance would have it a friend of mine was involved in a series of nature films for the BBC's Channel 4, and she thought they could be persuaded to take up my idea. At this point I had to leave the development of the proposal to my friend and Reg Hardy to sort out as I had to return to Bolivia the following day. One of the problems was finding the right English girl for the film, and in true "Willy Wonka" fashion the solution to this was tied to the idea of holding a competition.

I had completely forgotten about all this until sixteen months later when some students from Durham University who were staying at the Estancia (also without my knowledge) came to Buena Vista to tell me there were two English people (an eighteen-year-old youth and his father) wandering around the Estancia in a state of nervous exhaustion. Apparently they were winners of a competition organised by Mr Hardy which had appeared in the TV Times. They had, the Durham students said, been dumped on them one day by someone claiming to represent Mr Hardy (and without bedding, food, or anybody to look after them). And were now in a very sorry state wondering how they were going to get back to England. But no sooner was I told this than Reg's Landrover reappeared and took them away.

****

On the 5th of May 1985, I settled back into my British Airway's seat prior to take off. Craning my neck to look at the Amazon jungle through the potty-like window was one thing, doing the same to peer at Heathrow's tarmac would have been absurd. Inveterate window-person though I am (much to the annoyance of the cabin staff with their uncompromising passion for lowering the blind when I want it up), this time I retreated behind the pages of The Sunday Times that happened to carry an article about me. The article, Tropical Rescue, referred to my work in Bolivia. It was the sort of article librarians would file away under the heading: "Local boy makes good". I had given the interview with the understanding it would help us raise funds for Amboró by including the address of PRODENA-UK. Later I was informed this was contrary to Sunday Times policy.

To get back home. This was my only real desire after five weeks in England, a land that had once been my home; but now an estranged place devoid to me of any real interest or excitement. A land inhabited by people leading a safe grey existence, rather than the black and white of the life I had become accustomed to in Bolivia. I was missing Miriam, our Dutch friends, and the frail bird calls from the enigmatic woodlands behind my home. And Yes! trouble the black of the black and white.

"Bienvenidos, Bienvenidos," the Immigration official welcomed each passenger on our arrival at Santa Cruz; but only the Latins responded. The foreigners, failing to understand his words, continued to stare sullenly through him. "What a way to start," I thought.

TROUBLE. Again. The CDF in Santa Cruz had made staff changes. Same Director, new Wildlife Chief. "Here we go again", I said to myself, "new people to befriend, old projects to defend." You see the Bolivian Civil Service doesn't offer the same reassuring stability it does in Britain, or even the compromise to stagnant bureaucracy which it does in the United States. In Bolivia elections, whether national or local, signal an irresistible compulsion to replace all Government employees, from directors of institutions to night

watchmen with new blood. The upper echelon, (Chief Administrators, Local Government Officials, and others) are appointed
according to their political affiliations, rather than their qualifications.
In their turn each new appointee will give jobs to minor party
workers, friends, and of course family but, in a country starved of
competent experts the choices are very limited. Economists might
take over the responsibility for Education, Vets the Public Health
Service, or closer to home almost anybody could become Chief of
Wildlife and National Parks. The lack of continuity made life very
difficult for us since nearly every new CDF Director started by
making radical changes to the administration, usually with the
expressed purpose of cutting costs, sometimes with old scores to
settle. For us, clinging to the rock face of uncertainty, every change
was a threat to our efforts.

During the six years (1983-89) I was affiliated to the CDF it was
necessary for me to justify the interests of Amboró National Park and
my own position to a long line of new Directors. None of them were
biologists or ecologists, and certainly not conservationists. Half of
them lacked training in Forestry, that should have been a basic
prerequisite. The rest were either agriculturalists or, yes, an
economist.

Here is a list of these worthy gentlemen:

CDF Director's Name	Training	Reason for leaving
1983: Oscar Llado Pereira	Economist	Accused of corruption
1984: Esteban Cardona Montenegro	Forester	Underqualified
1985: Nestor Ruiz Ibañez	Forester	Accused of corruption
1986: Roger Bazan Roca	Forester	Accused of corruption
1987: Jose Serrate Perez	Agronomist	Incorruptibility'?
1982: Alberto Basquez	Veterinarian	Corruption
1988: Esteban Cardona (acting)	Forester	Incorruptibility'?
1989: Walter Landivar Gil	Agronomist	Corruption

During his diatribe at Don Miguel's Restaurant Gene Harris
accused two of them of taking bribes from the Animal Exporters:
"Oscar Llado: gets paid/box animals/no diff. Alberto Basquez CDF
vet is another Onishi man."

In our case we had a new Chief of Wildlife, Arquitecto Victor Hugo Paz; not that we had anything against him for he was an uncompromising hard worker, and though an architect by training, he had been Noel's part time assistant at the zoo.

So this time we were lucky: I knew Victor Hugo. To know somebody is to smooth the way; once again my personal contacts made over the years were proving to be vital to Amboró. This cannot be understated, without friends and through their influence I don't believe Amboró would have gotten off the ground. Stillborn, it would have faded away like any one of the Paper Parks and nobody would have noticed, nobody would have cared.

To illustrate this point, I should go back to the beginning of my story, back to 1980 when I began to take an interest in the National Park network. I had heard there were Parks in Bolivia, but no one knew where they were. There were no books on the subject and later Noel Kempff told me no compiled information either, not even in the offices of the CDF. "The best guide," he said, is a baedeker called "Bolivia Magica". When I consulted the 1979 edition I found five small pages devoted to protected areas. Here is what it said about the Department of Santa Cruz's nature reserve in the mountains of Amboró :

> RESERVA NACIONAL TCNL. GERMAN BUSCH
> Ubicación: Comprendida en los Dptos. de La Paz y Pando.
> [Location: In the Departments of La Paz and Pando.]

They had got the name right; but the location wrong as this was the "Amazon National Reserve" five hundred miles to the west of Amboró. "Forgivable mistakes," would you say? I didn't tell you that according to the publisher (Los Amigos del Libro), this was the third edition.

On the 10th of May, El Mundo published a report on the restructuring of the CDF's Department of Wildlife and National Parks under its new Chief. (Victor Hugo wasn't wasting any time.) The article stated that among the first initiatives was a new project for the management of Amboró National Park that would be implemented after the people in the area had been consulted. It also said the CDF would convene a committee to work on new wildlife regulations, and that this committee would include Prof. Noel Kempff Mercado, Mr Robin Clarke and the Brekelmans.

Pieter as a salesman, and Els as a veterinary technician, were going up in the world.

Victor Hugo was going to be a good friend; but nobody achieves anything without making a good number of enemies as well and I, as a gringo, made more than my fair share. Sooner or later I knew I would fall, and fall I did. But not yet.

And talking about falling from grace, and friends and enemies, I was soon to be reminded how easy it was, especially in Bolivia, to stumble over one of the many rocks with which the way ahead was strewn. "No llores para mi Argentina ..."

I received a letter, here it is:

UNITED STATES DISTRICT COURT
SOUTHERN DISTRICT OF FLORIDA
PROBATION OFFICE
MAY 20, 1985
RE: POLACZEWSKI, JAROSLAW JOHN
a/k/a: Thomas Michael Jensen
Dear Sir:

The above-captioned defendant is currently under a pre-sentence investigation for a federal crime that occurred on November 29, 1984. Mr Polaczewski was arrested at Miami International Airport for possession of cocaine base contained in wine bottles. Presently, a thorough investigation is being conducted in order to assist the Judge in deciding the appropriate sentence for commission of the crime. Sentencing is scheduled on June 11, 1985, before the Honourable U. S. District Court's, Judge King.

At the time of Mr Polaczewski's arrest, his briefcase contained a budget packet being sent to the World Wildlife Fund. On the paperwork contained in the package your name was listed several times. A conversation was held with Mr David Mack of the World Wildlife Fund indicating your organisation has received several grants for establishment of the above-referenced [in my address] park.

At this time, I am trying to verify your association with Mr Polaczewski. Please advise how you came to know this individual, the duration of your relationship and the exact dealings Mr Polaczewski had in reference to the establishment of the park...

Any additional comments you would like to make in reference to this defendant, would be additionally appreciated.

Your assistance in this matter is greatly appreciated.

Very truly yours,
Elizabeth Wilson, U. S. Probation Officer

"Any additional comments you would like to make..." Ms Wilson might not have appreciated any of the additional comments I could think of making. There was little consolation in knowing that Jerry would be paying for his outrage with a long stretch behind bars. I certainly hoped so.

My own position was not so clear cut. I would be tried in my absence by a court of law we all carry around with us, our personal prejudices. How my peers would judge me for my blunder could prove fatal to my hopes and could compromise the future of the Park.

****

About this time, I came up with an idea that may have been sparked off by the thought of Jerry's impending incarceration. I decided to lock up the inner parts of the Park, make it safe from further destruction by cutting a path forty miles long through the lowland forest. It would run between the area occupied by the campesino farmers and the foot of the mountains. I spent weeks poring over maps, and eventually came up with the "Red Line" ("La Linea Roja"), a frontier beyond which we would endeavour to implement the basic precepts for a National Park. At the same time, it would be important to emphasise that the Red Line did not represent a new Park boundary. It was there to demarcate an 'Inner Reserve' that we would consider sacred, and anybody crossing the Red Line would be vigorously prosecuted.

In its original form the position of the Red Line was drawn with due consideration to the needs of the Park: the protection of the mountains and remaining lowland forest; the inclusion of biologically interesting areas, such as ox-bow lakes, and the conservation of areas particularly susceptible to erosion. We realised that the Red Line would only be effective if it was respected, and without a real commitment from the CDF, or Park Guards to defend it, the Red Line would be a bit like the "Emperor's New Clothes", susceptible to ridicule. The provisional Red Line was drawn with this in mind. We cut off more land for the Inner Reserve than we could hope to keep, as we knew fully well we would have to accommodate the demands of the more powerful campesino syndicates; demands we would convert into concessions in the hope that this bluff would make them more amenable to the plan as a whole. It was a game of poker with very high stakes.

Another problem would be finding the money to cut the path. Forty miles of pathway through a South American jungle is not something which can be accomplished without tools, paint, notice boards and the manpower to cut it. As a trial Pieter and I decided to mark the position of the Red Line at those points where it crossed the six or seven footpaths which entered the Inner Reserve. We made half a dozen notice boards which stated "Entry beyond this point is PROHIBITED. Amboró N.P. CDF-SANTA CRUZ." We nailed them to the largest trees and painted bright orange bands around the trunks of all the others to either side of the track. By doing so we had fired our first salvo in a long and protracted war of attrition. By tearing them down a few days, later the colonists had answered us.

Within a month I served notice to quit on the few campesinos cut off within the Inner Reserve, including one dedicated to the cultivation of coca (an illegal activity within the Department of Santa Cruz). I also arrested a dozen hunters, some fishermen, and one or two would-be squatters who had crossed the line.

Many of my longed for bird-watching trips to the Park were brought to a premature conclusion by the discovery of these reprobates that had to be escorted back to the Police Station in Buena Vista. Because of the distances involved this was a tedious business that often took all day; and sometimes groups of angry colonists tried to prevent me from doing so. Later, when I had become wiser, I found it easier to confiscate their things, or at least the removable parts of their weapons. Personally, I thought I was doing most of them a favour, for some of the rifles and shotguns were lethal, held together with bits of string and wire, they were more dangerous to their owners than they were to their prey.

As none of them could afford to lose their possessions, they sooner or later turned up at the police station to reclaim them. At first the police were somewhat dismayed by my show of authority; but they soon got into the swing of it and summarily fined the offenders a few dollars before returning their confiscated guns and axes; but never the traps or nets. This relatively mild form of punishment worked well for minor offences, as it was more of a nuisance to the miscreants than it was a real penalty. For truculent offenders the punishment was a night or two in the cell; an horrific mud shed, open to the weather. Damp, dark and small, it was our secret weapon. The mosquitos loved it. Its floor area (about 4x4 feet) precluded the provision of a bed, and the dirty mud floor was not an attractive place to lie down.

The word soon got around that it was better to go hunting somewhere else rather than face "The Mad Gringo of Amboró, who jumps out of the bushes and puts a pistol to your head," to quote the radio and the newspapers. This was quite untrue. I have never carried a pistol or any other sort of weapon, but I let the rumour gain ground for obvious reasons.

A few hunters in a South American tropical forest are a nuisance; but they're not the same threat to the wildlife as they are in an East African Game Park. The dense vegetation, the relatively small size of the animals, their predominantly nocturnal habits and, apart from Monkeys and Peccaries, their non-gregarious nature are all factors operating in the the animals' favour.

The Federación de Campesinos tried to put a stop to my policing of the Park by asking the Military to deal with the 'Mad Gringo of Amboró.' When a military commission came to Buena Vista to arrest me I was able to produce my CDF warrant to satisfy them. After we showed them around the office, that as I have told you looked very official, they left with promises to respond to any request we should make of them in our efforts to protect the Park. Then back they would come, having forgotten all about their last visit, to arrest me again. Later, it was necessary for me to publicly denounce the Military for using the Park as an unofficial training ground. After my declarations, that I made on TV, and that appeared in the press, they went off to train somewhere else. This was a good move, for the campesinos baulked at the requisition of their villages for war games and they credited me with having rid them of a major nuisance.

For some time, I had known the Military were guilty of poaching and fishing the rivers in the Park with dynamite. Sooner or later, I knew I would have to do something about them, for their rumbustious behaviour was unacceptable in a National Park and detrimental to tourism. The final straw came in 1986 after they returned from a training exercise in Amboró carrying an Indian's lance, bow and arrows. When I asked their commander where he had obtained them, he told me a lady in the Park had given them to him. On enquiry this lady, Don Arnaldo's wife, told me they were lying, and quite casually said they must have come across an autóctono and killed him. Such a guileless statement, coming from a long-time resident of the Park, was food for thought indeed.

Albeit this was the first evidence to support my theory that indigenous people were to be found in Amboró, that of itself would be an important purpose of the Park; but at the time I was more concerned by the threat this confirmation would mean to us personally. After all, we were often to be found studying birds in the middle of nowhere and it seemed wise to consider the possibility of reprisals by people being murdered by the army. We would be sitting ducks.

Worried by the thought, I contacted the experts. The two anthropologists who examined them stated that the weapons were genuine and bore marks to suggest they were in use. They were in very good condition, the fibre of the bowstring and the Curassow feathers of the arrows had not been eaten by moths, the bow and the reeds used for the arrow shafts still flexible, the glue was fresh, etc. However, they were unable to identify the tribe to which they belonged. The fletching, the Chonta-wood bow and the string were very similar to those used by the Yukis (a group of natives living a little to the north of Amboró): but significant differences existed in the binding, the glue and the fibre used for the bowstring. To this day we are no wiser; but it would seem that somewhere in the middle of Amboró an unknown tribe of indigenous people still survive.

This could be possible, for the mountains of Amboró are very nearly impenetrable and apart from one or two gold hunters and the state-run oil company, nobody had ever explored the centre of the Park. We were never able to get the money to do so, and convinced that the area should forever remain a haven for the animals and the natives (if there were any), we didn't even try.

Free of the army, and a degree of control over the hunters, we gradually turned our attention to one of the gravest threats to the Park: the Loggers, the Oil Companies, the influential Camba landowners (not to be confused with the small traditional farmers) and the syndicated colonists.

The problems associated with logging are serious and complex. More threatening than their obvious ravages, timber companies make roads into new territory. Down those roads come the colonists who completely clear the land, once and for all destroying the home of the forest's wildlife. Forest may regenerate; but many key species of trees and herbaceous shrubs can never recover in agricultural land.

Then one must look at the nature of the logging operation itself. In a tropical forest one is not dealing with a solid stand of valuable trees. No, the apple of the logger's eye might be as infrequent as a public telephone box in a deaf and dumb school, so imagine the resulting carnage that a logger's bulldozer will leave in its wake to get to that one tree. Believe me, it's a sad sight.

Irresponsible timber companies, that are in the majority, hire hunters to provide fresh meat for their workers. Since persistent hunting over many years in any single area is extremely detrimental to the survival of the wildlife this is a serious threat to the sustainability of forest ecosystems.

Apart from these rather obvious reasons for disliking commercial logging as it is practised in most tropical countries, there are many indirect ways, as yet poorly catalogued, by which wildlife is exterminated and the recuperation of forests disrupted. Let me illustrate this more insidious spoilage by reference to a couple of related examples.

Loggers remove essential fruit-bearing trees and other plants from the forest. Though we still know too little about the separate components of topical forest ecosystems, we do know that different trees fruit at different times to provide a year-round supply of food for the insects, birds and mammals. As timber companies select the three or four highly commercial species, and may remove all of them from a forest, they break this cycle of provision. The result can be an acute loss of wildlife.

My second example may appear very similar, but is in fact quite distinct. A species of hummingbird may survive by obtaining nectar from a vine found on three or four trees in its territory. If one of those trees is removed that pair of birds may not have sufficient sustenance to survive. Maybe, as the only pollinators of that vine, the remaining vines will cease to produce viable fruit and will eventually die out as well. Their disappearance may produce a chain reaction affecting the insects, the reptiles and birds which feed on those insects, and in turn the mammals as well. There is an urgent need to study the interdependence of plants and animals so we can plan biologically sustainable systems of forest management.

Now that Amboró National Park has much larger efforts to control the loggers, (largely influential people with the power of their mafia-like associations and Government's export drive behind them) it is an on-going war. It would be to strain your patience to detail the very comprehensive laws at our disposal to keep the timber companies out of the Park. However, allow me to refer to one of them.

This law categorically states that in the case of illegal logging the logs and all the equipment belonging to the timber company may be confiscated and sold at public auction, and that fifteen percent of the proceeds go to the person or persons reporting the infraction. Suffice it to say that to my knowledge this law has never been implemented.

Certainly, of the cases of logging in Amboró National Park that we reported to the CDF, cases fully documented by statements, photographs, and sometimes videos, no such action was ever taken. In 1986, Señor Cronenbold, the owner of one of Bolivia's largest logging companies, felt free to laugh in my face when we caught his company en flagrante removing mahogany from Amboró National Park. He knew the CDF were no more serious about protecting their National Parks than they had been to collect the duty on cut wood, protect the wildlife, or manage the Forest Reserves. He knew the worst punishment he could expect would come from, paradoxically, the corruption of the law itself. That the CDF officials would use the latent threat in this law to extract a larger than usual bribe, and to him this would be no more than the loss of a bit of pocket-money. Mr Cronenbold, I take my hat off to you.

So, as far as the logging companies were concerned, our only tactic was to generate as much publicity as possible against them. Thanks to the cooperation of a handful of sympathetic reporters we were given space in the newspapers to do this. The logging in Amboró was not completely stopped but cognizant of the risk to their reputations, like the hunters, the majority preferred to go elsewhere.

Where we were fairly successful in shaming the loggers out of Amboró, the same could not be said for our campaign against the colonist squatters. Public opinion understood the logging dispute because of the clear-cut issues involved, people stealing trees from protected areas. Moreover, many journalists, especially those working in the Third World, tend to have socialist leanings and disapprove of plutocrats, as the loggers were judged to be. Removing families, and poor ones at that, from land they claimed was a different ball game entirely. The colonists were not on their own: the Church, the Organisation of Civil Liberties, Freedom of Rights, and the left-wing Political Parties, to name only a few, were quick to jump to their defence. The only course available to us was to implement the Red Line and plan to resettle those who wanted to be resettled.

As for the rest, those who insisted on their right to remain in the Park, we tried to reach an agreement with them, we wouldn't interfere with them as long as they respected the wildlife, the riverside vegetation, and of course the Red Line. This wasn't an easy task given the background of social unrest which affected Bolivia at the time. Operation Raleigh's imminent arrival, the National Elections, our discord with Reg Hardy, and a largely apathetic CDF were complicating factors which sapped us of much of our energy and time to deal with these issues.

To give you an idea of the way in which these conflicts affected our daily lives at this time, here are copies of two monthly reports we prepared for PRODENA:

Monthly Report No 3: 1st April-15th May 1985
Recent developments in the Park.

As we reported last time, people with interests in the Park have been coming in small numbers to our office in Buena Vista with their enquiries. Now we can report that this trickle of interest is turning into a flood, not only enquiries, but appeals for help, reports of illegal forest clearance, hunting and fishing. What has happened is this: people are dividing into two separate factions, those who support us and wish to see the Park become a reality, and those who have vested interests of their own and wish to obstruct our efforts at all costs. Our supporters are generally drawn from the poorer people and for that reason are certainly more numerous. However, our detractors have the power and the desire to precipitate a showdown and are now actively cranking-up to demonstrate to the people our ineffectiveness. Just one example will suffice to demonstrate our position. Señor X claims title to 10,000 hectares of land in the Park. The land falls inside our proposed area of total protection. He is now actively felling and extracting the commercial trees and leasing his land to new colonists. At the same time, he is intimidating the local population by rampaging about the place with his guns and his ne'er-do-wells.

The locals have come to us with appeals to sort this out. We have talked with Señor X without result except to receive from him threats and intimidation. Also, he intends to make fools of us in order to destroy our grass-roots support; he will rapidly do so if we don't get the means to prevent him. We believe there is only one way to handle such people (and demonstrate to the public at large that we are an effective force behind the Park), and that is to handle this affair personally.

We are now being tested, and we must respond. We don't believe the army or the police will help us because they don't understand the issues involved. Clearly, we need to do several things in the right manner, and order. We need to go to the trouble spot and see for ourselves what illegal clearance is taking place, and exactly who is responsible. Then we have to gather evidence from the locals. And finally we must catch Señor X in the process of removing timber, arrest the driver, and impound the vehicle. This will certainly stop and make him think and we will have won the first round against such (fortunately few) people. The arrested man will be handed over to the Police in Buena Vista, and the impounded vehicle to the CDF in Santa Cruz. After that, matters will have to take their legal course, and it will remain to be seen whether or not the CDF and the law will support us in our efforts.

Now all this is very well, but I ask, what does the Board of Directors think we can do in our present situation? To make the initial investigation we need to monitor the area (we have no vehicle) and spend a great deal of time (which we don't have) investigating the facts. Peter and Robin, two gringos without support, what can we do? The answer is straightforward: very little. Without real help, our credibility, and the respect we need, will be destroyed and we will lose the control of the Park (that is so nearly within our grasp). Or, we can firmly take control of the Park by prompt and effective action. We require very little to do this; two uniformed Park Guards armed with official credentials (that the CDF have promised to provide), a good 12-bore shotgun, and a vehicle.

So please, now is the time to help us. We will make the time and we will resolve the problems. The Park is within our grasp, we need a helping hand. Now is the time to demonstrate to the people we are what we claim to be, a tough force to be reckoned with should anybody wish to take us on.

MONTHLY REPORT No 4: 15th May-30th June 1985
Politics in the Park.

Our last report mentioned the disturbing developments taking place on the 'political' front. During the last six weeks these have grown to the extent that we, now, think it too late to outmanoeuvre the troublemakers. However, the actual reasons behind the trouble, and the persons responsible, are better understood.

There are two main groups actively involved in an anti-Park movement: a few wealthy Cambas (living in Buena Vista) and two syndicates on the upper River Saguayo (Sindicato Jerusalem and Sindicato Germán Busch). The two syndicates are anti-Park because they fall within the proposed Inner Reserve (all the land lying beyond the Red Line); they are illegal in that they arrived in the Park last September/October, after the Park was legally declared. They don't have titles to their land, but they are actively seeking them. The Cambas are anti-Park because some of them claim to hold titles to substantial areas (up to 25,000 hectares, both inside and outside the Inner Reserve), and they wish to extract the timber, and then lease off their land to colonists. Naturally, the two groups have mutual interests, and during the last three weeks they have launched a propaganda campaign calling for the Park to be nullified; a deputation has been to La Paz to state their case.

The events leading up to this situation were as follows: on the 25th May a meeting was held in the Park Office in Buena Vista attended by representatives of the CDF, CORDECRUZ (Santa Cruz's Development Corporation) and about forty farmers, members and representatives of a number of syndicates. As far as we know, nobody from either Jerusalem or Germán Busch were present. The meeting was opened by Robin who gave an illustrated lecture on the socio-ecological reasons behind Amboró National Park's establishment, the way it was planned, and how these plans would affect the campesinos living there. Arq. Victor Hugo Paz then gave a talk along similar lines, followed by Ing. Carlos Alvarez (representing CORDECRUZ) who put forward his organisation's interest in the Park. Questions and declarations from the campesinos followed, and a preliminary motion presented to the gathering for a vote. It was more or less unanimously agreed that the Inner Reserve would be respected, but the area occupied by the colonists should form the basis for further negotiation: specially with respect to the cut timber lying in the Park that the campesinos wanted permission to sell.

On the 1st of June, a second meeting (announced on Radio Santa Cruz) was held at the Park office in Buena Vista. It was attended by a large number of colonists, their representatives and the CDF. Victor Hugo opened the meeting by presenting the motion accepted at the last session to the gathering. However, it was soon obvious that there was considerable discord among the campesinos and his proposal to have the motion approved and signed by those present was emphatically rejected.

Representatives of Jerusalem and Germán Busch (with a total of 100 families) made long speeches opposing the Park, and flatly stated their antipathy to the gold-hunting gringos in charge. Many of the campesinos (but not all) appeared to share their view, and it seemed that many of them (including those outside the Inner Reserve) believed they would be forced to leave their homes. The meeting broke up in discord. Clearly, the people opposed to the Park were, for the first time, taking the Park seriously and had come to the meeting to win support for their point of view. A desultory motion was made to have another meeting the following Saturday, June 8th (but nothing came of this).

At 7.30 a.m. on Saturday, June 15th, representatives of the CDF and CORDECRUZ arrived at the quinta to inform us that a meeting had been called and would commence within the hour. On arriving in the plaza, it was immediately obvious this was to be a jumbo-sized meeting; there were hundreds of colonists present, divided into separate meetings scattered around the plaza. By ten o'clock the crowd had doubled and, somewhat alarmingly, three trucks arrived full of angry people (about a hundred) standing in the back; these carried placards calling for the dissolution of the Park: "RECHASAMOS EL PARQUE AMBORÓ. 'RECHASAMOS A LOS YANQUIS IMPERIALISTAS". Out from this crowd stepped a dozen colonists with a placard all to ourselves: "RECHASAMOS A LOS GRINGOS", that featured a picture of a gringo (Robin) hanging from a tree; worse, they had brought the rope. This ugly mob stood in the road outside the Park office demanding we come out.

Throughout the last three weeks we had been dealing with the chief representatives of the campesino syndicates, one of whom was also the local representative for the MNR (Movimiento Nacional Revolucionario); he stated that he was in accord with the Park. The main mass of people now moved down to the MNR office where they held their own lengthy meeting. Eventually, they emerged, and we all went down to the Alcaldia where the inner courtyard offered sufficient space for the entire crowd. Here the meeting was opened by the spokesmen for the Sindicatos Jerusalem and Germán Busch. They set the tone for the rest of the speeches to follow. They called for the Supreme Decree declaring the Park to be set aside. If that were not possible, the boundaries of the Park should be changed so that all the lowland sector remained outside the Park; and that on no account were they going to allow gringos to enter the Park because, as everybody present knew, they were stealing the gold and precious stones, taking photographs (i.e., spying); and intended enslaving the Bolivians in much the same way as they had done to others the World over: "VIVA VIVA BOLIVIA! VIVA VIVA! VIVA LOS BOLIVIANOS! VIVA VIVA!", the crowd chanted loudly together. It was quite plain that the issue of the Park had taken on a political tone. Arq. Victor Hugo and Ing. Carlos Alvarez made rather feeble speeches of short duration (not surprisingly considering the circumstances) to the affect that nobody would be thrown out of the Park. A moment of comedy followed in which one young lady from the village was heard to shout: "VIVA LOS GRINGOS! VIVA!" The meeting ended without mishap.

There have been no further official meetings with campesino representatives since that time. Victor Hugo warned us to stay out of the Park, and not to open the Office, because rumours were circulating that some of the colonists intended killing one of us so that the rest would go away, and the problem would be over. Victor Hugo and campesino political representatives appeared at different times on TV putting forward their corresponding points of view. We kept the office open and paid two visits to the Estancia, now owned by Reg. A few campesinos drifted into the office to state their support for the Park.

On the 22nd of June the colonists apparently held a further meeting of their own. Pieter and Robin were surrounded by this boisterous group of, not unfriendly farmers. They wanted to know more about our activities and the aims of the Park, how we could think of throwing them off their lands, etc, etc. We explained what the Park was for, and what we were doing. This meeting went off without incident. Five minutes later we had another impromptu meeting with some of the campesino leaders (mainly from the Yapacaní side of the Park, but including Felipe Rodriguez who represented Germán Busch). We asked them why they were intent on disrupting Government plans for the Park, and the Law, why the colonists were lying about the gringos, and why they made threats against them, and we suggested that we should get together to form an official committee to resolve all the differences. The discussion proceeded in a reasonable manner until two drunks turned up, demanding to know: Why we had been stealing gold which truly belonged to them? Pieter and Robin prudently retired.

All things considered, if it were not for the threat to kill a gringo, and the potential disturbance to Operation Raleigh's plans, we think the last few weeks have been very positive. The Park has become the focus of attention because it is being taken seriously. The strength of the opposition is now known, and their arguments against the Park have been publicly stated.

The opposition is clearly well organised, no doubt more so because of the element of electioneering for National Elections on the 14th July. It is clear that a number of campesinos, including some of their spokesmen, are not against the Park. At least the issues have been crystallised, and when the sediment of political fervour has settled down, we should be in a position to hold positive discussions with individual syndicates and, VERY IMPORTANTLY, cut the pathway (the Red Line) to delimit the Inner Reserve. The opposition to the gringos is mainly political and we believe the manifest aggression will be of short duration. Some campesinos obviously think that without the gringos the Park will fold, leaving them to carry on as before.

Whatever, we are all in agreement that the National Elections will have an important bearing on the issue. It was Banzer's Government which issued the 1973 decree to protect the area of Amboró; one might suppose that should he win this time [he lost], he will make sure the Park is strengthened. Certainly, the Buena Vista representatives of his party, the ADN (Acción Democratica Nacionalista) have adopted the Park, and their support for it, as one of their major issues.

Reg Hardy arrived with Ralph Peterson, and two Army Captains to investigate the rumour that the colonists were going to kill a gringo, the first one that stepped inside the Park, and in what way the recent developments were likely to affect Operation Raleigh, which is due to arrive shortly.

After I had read this latest monthly report to them, and some administrative matters had been dealt with, Reg, Ralph, and Pieter went to the Estancia to view the repairs made to the roof that had been completely renovated and the new kitchen. I stayed in Buena Vista to get away from Reg for a moment. He had become progressively sullen during the reading of the report. The few comments he made suggested he held me personally responsible for the hostile attitude of the colonists. Once again Reg was looking for a scapegoat.

At the end of the day, we all went to Santa Cruz for the night because we had an early morning meeting with General Gary Prado and Victor Hugo Paz.

The General was the military commander for the Department of Santa Cruz. He had made a name for himself by being accredited with the capture of Ché Guevara in 1967. Later he was the victim of an assassination attempt which left him paralysed in both legs. When we got to his office, he was sitting in his wheelchair chewing over a brief about the threat to the gringos. I could tell he had some bad news for us.

While we awaited the arrival of Victor Hugo (imagine keeping Monty waiting!) we were served tiny cups of sweet black coffee. Reg ran through the events to date emphasising, as he always did, the importance of Operation Raleigh's visit to Bolivia and the Park. The General took this all in with a quizzical look in his eye, the man was no fool.

He was also a politician, for when Victor finally arrived, the General invited him to present his thoughts to the meeting. Victor did not waste time on needless bush beating; he was of the strongest opinion that Raleigh should not be allowed into Amboró, or for that matter anywhere else in Bolivia, because of the emergency situation pending the National Elections. Reg sat in stony silence. With a noncommittal expression the General voiced his tacit agreement with Victor's remarks that he said were much the same as those made by his own people investigating the matter.

Reg was hardly able to contain himself, and throwing his hands into the air interrupted the General with something like: "For the love of God. Why, as a member of the Raleigh committee, didn't you voice your opinion before?" This shocking breach of diplomacy was received by the General with the statesmanlike disdain it deserved, silence! To break the awkward hiatus, I requested the meeting be closed, and the situation be reviewed on the 15th July, the day after the Elections. Reg was too angry to demur.

Nevertheless, when we got outside Reg insisted Pieter and I accompany him to Kempff Mercado's office at the zoo. Noel must have had an inkling things weren't well, for the moment we stepped into his office, and had completed the ritual handshakes, he spent an inordinate time fidgeting with the papers in the lowest drawers of his desk. But Reg, in no mood to pull punches and without waiting for Noel to sit upright, launched into a diatribe against Victor Hugo, who, as you may remember, was Noel's previous assistant and friend. Reg swearing he would get him fired from his job as Wildlife Chief and repeating his searing comment about the Bolivian members of the Raleigh committee, which included Noel, was embarrassing to the rest of us. Then losing all control, he threatened Noel, Bolivia, and anyone else, glaring at me, saying that if there was any disruption to

Operation Raleigh's plans he would make quite sure there wouldn't be another penny for further conservation work in Bolivia. After this patently ridiculous outburst, Reg got up and stomped out. "At least," I thought, "I am in good company." Reg now had four scapegoats: the General, Noel, Victor, and myself. Pieter, his face the colour of a beetroot, could only keep repeating, half to himself, half to me: "This is a bad busy-ness. A very bad busy-ness."

Back in Buena Vista we spent two frustrating weeks awaiting events. Ralph Peterson eventually rang up to say that Operation Raleigh had arrived and had gone off to stay on a private farm to the east of Santa Cruz. They planned to move to the Estancia as soon as the elections were over. What he didn't tell us was that the Military members of Raleigh's supervisory staff had gone into Santa Cruz dressed in their British Army uniforms, and having been taken for mercenaries, were arrested by the police.

It was not difficult to feel sorry for the 'Venturers', mixed up as they were in a typical Latin American scenario. A potpourri existed of xenophobia, jealousy, damaged pride, mismanagement and distrust; and the situation with the colonists that, with Raleigh's overriding need to proceed with caution, was viewed as life-threatening. From our seat in the stalls in Buena Vista we thought a lot of fuss was being made about nothing. The Estancia was private property owned by Reg and we saw no reason why the Raleigh members shouldn't move there without delay. Victor Hugo's opposition stemmed from a failure in protocol, he should have been fully consulted about Raleigh's intention to work in the Park, and given the campesino threat, he now had some vindication.

Reg had made a major blunder in seeking the support of the General. He was not in a position to override Victor (even had he wanted to) or risk the political storm (now that Reg had put him on the spot) should one of the Venturers get hurt.

The chief obstacle was Reg. Fuelled by his penchant for bullying (which he rationalised by his conviction he was fighting for a just cause) he did everything he shouldn't have done and failed to do anything that he should have done. We thought the solution to the problem was obvious: the campesino leaders should be called in for consultation, and if necessary, a timely rebuke; Noel should be left to handle Victor; Pieter should be allowed to manage Raleigh's day-to-day affairs; and Charlie King (Raleigh's expedition leader) and I should get on with the work in the Park.

In fact, we went ahead, without involving Reg, and did all of these things. The campesino leaders were somewhat amazed we took the threats to the gringos so seriously, and though they wouldn't give any guarantees, suggested we ignore them. Victor accepted this information with some reservation but was appeased by my promise to represent the interests of the CDF and the Park: he didn't seem to quite believe this had always been my guiding principle.

He also wanted guarantees that both the Land Rover and the Estancia would be handed over to us, as stipulated by Reg on numerous occasions. He also expected to be kept fully informed by the Raleigh committee. With that the threatening situation disappeared, temporarily that is.

On the 8th July 1985 Reg, Ralph, and two members of Raleigh's organising committee arrived in Buena Vista. It was quickly agreed the Venturers could move to the Estancia as it was now ready for occupation.

Then Reg, in a fatuous show of force, dropped his bombshell. The decision to hand the Land Rover over to us, and my move to the Estancia would be delayed, pending a reassessment of the situation in September, after Raleigh had left Bolivia. Obviously we were to be held to ransom pending the success of the Raleigh mission, and the future of the Park was to be jeopardised by the bungling of Reg and his committee. Naturally we didn't take kindly to the threat of a vendetta should things go wrong.

More personally, Reg's ominous declaration left me no choice. I had to find somewhere for Miriam and me to live. We had promised Bart we would move out of the cottage at the Quinta. Since there was nowhere else we could go, my only alternative was to buy a plot of land and build our own house. Moreover, it would have to be done before the rains made construction work difficult.

Charlie King was not happy with the situation either. Though he had paid for half of the Land rover out of Raleigh's funds, he didn't think it wise to press Reg on his decision for the time being. As far as the Estancia was concerned, he could do nothing; but it did put a different light on Raleigh's plan to renovate the house prior to its formal hand over to the Park. He couldn't justify financing the work if the Estancia was to remain PRODENA's property.

As far as Victor Hugo was concerned, Reg's threat had put me in a difficult situation. I took the only alternative available to me. I went to see him to discuss the situation. He was adamant: Operation

Raleigh would not be allowed into the Park until Reg fulfilled his promises. In the meantime, I was to leave the matter in his hands. Had I not been honest with him at this stage I would have been out on my ear, and without my presence I didn't think Amboró would survive. Once again, because of Operation Raleigh's visit I found myself "Entre La Espada y La Pared" the equivalent in English would be "Between a Rock and a Hard Place".

After the four of us (Pieter and Els, Miriam and myself) had discussed the situation we decided to distance ourselves from Reg and his bullying and get rid of him as soon as possible. Until then we would do our best to help Raleigh without compromising our relationship with the CDF.

These were, then, the disturbing developments taking place in Buena Vista and the Park prior to Raleigh's arrival. As you will recall Operation Raleigh should have been represented throughout the period of its stay by its Bolivian committee and no one else; but here is a brief analysis of the committee's status at the time:

a) Reg Hardy: suffering with back problems, was only able to be in Bolivia for part of the time.

b) Alan Hutchinson: was tied up by business in the States but appeared for two weeks during August.

c) Ralph Peterson: had gone to the States at the end of July because of a problem with one of his legs.

d) Noel Kempff Mercado: was too busy to do much, and anyway, almost certainly considered his nomination to the committee as an honorary position rather than an active one.

e) General Gary Prado: probably held the same point of view (it is the Bolivian way of seeing things) and would have been far too occupied with his military duties, especially with the imminent National Elections, to have thought otherwise.

Operation Raleigh was in trouble.

****

Any possibility of my help had been neutralised: first by Reg, then by Victor Hugo. Fate also took a hand. After the meeting on the 8th of July, Reg handed me ten one-hundred-dollar bills; $500 for my monthly allowance, $500 to pay the builder for the repairs at the Estancia. Pieter, Els, Miriam, and I were once again solvent. After the others had left, we went down to Blanca's for a modest celebratory drink; the $1,000 safely in its wallet in the pocket of my shirt. By the time we got back to the quinta it was dark and I jumped down from the back of the pickup to open the gate. Full of beer and food, we were glad to go straight to bed.

The first thing I did in the morning was to discover my wallet was missing. As an immediate search of the house and grounds failed to uncover it, we all sat down to discuss it's disappearance. We came to the conclusion it must have fallen out of my pocket when I jumped down to open the gate, and being dark, had probably lain there all night. The first person to pass by in the morning probably found it. Our only hope of recovering it was to keep an early watch on the road to see who this might be. By this tactic we identified four possible suspects: a group of three women on their way to sell yuca in Buena Vista were invariably the first, a man on horseback the second. They were all vaguely known to us. If none of them, the money was lost, as a steady stream of people followed soon afterwards.

Leaving the house on the fourth day of my vigil I discovered my wallet in the middle of the driveway. Somebody had had the decency to return the wallet, but not the money. The important thing was its return, suggesting we were dealing with a guileless person, someone unaware of the clue this provided. This led us to believe our hopes lay with the three women.

I decided to tackle the man first. He was educated and sure of himself, and as an ex-soldier, proud. We had something in common and I thought the best ploy was the straightforward one. Had he found my wallet? No, he hadn't. He was brief, but convincing.

The three women presented much more of a problem. They were uneducated campesinos, timid and diffident by habit. My talk with them left me in doubt, I simply couldn't tell. The only promising plan was to let it be known there would be a reward for the recovery of the money. Chances were, whoever had found my wallet had never seen a one-hundred dollar note and would be imprudent enough to show them to a friend, somebody who would know its value. Then they might make the mistake of changing the notes in Buena Vista where $100 dollars was small fortune. Furthermore, so much money might lead to jealousy. If we were right, we thought that the offer of a reward could elicit the information we needed. Miriam and I spread the word.

Nothing happened for a month. With the loss of the money, I saw the option to build our own home slipping away. Obsessed with its recovery, I fell to brooding, a condition which made it easy for me to comply with Victor's order to leave Raleigh in his hands.

Then, about eight o'clock one night (unusually late for visitors) someone tapped softly on the window shutter. Three bashful campesino men stood before me when I opened the door. "We've come," they said, "to claim the reward. Doña Rosa, one of the three women, showed someone in our village a lot of American money. Now you can go and retrieve what you've lost."

But knowing who had the money, and actually getting it back, were two different things. Involving the police was out of the question. I didn't want to get the woman into trouble; and the police could not be relied upon to return the money to me, more than likely they would split it between them. The only way was for me to personally confront the woman, shame her, scare her into returning the money voluntarily. The trouble was, as a campesino woman, shame might not enter into it. She would probably believe in the simple "finders-keepers" morality. And who could blame her?

Luckily for me Doña Rosa was a decent sort, and by ten o'clock the following morning the business was over. She soon admitted her daughter had found my wallet; but was now afraid because her daughter had spent a good part of it. Her daughter, she said, was not always well behaved, and as she handed me the six hundred dollars left over begged me to forgive them.

Indeed, I did. After I paid the reward, I was left with $500 to pay for the repairs to Reg's Estancia. By using every penny we had, I could still build a small house, but I would have to do it myself, and I would have to wait. I still had to earn my allowance if allowed to.

****

On the 20th July, Victor Hugo arrived in Buena Vista with furniture for the Park office. He had just finished telling me the good news that the CDF were going to rent a new one and take over the responsibility for its running costs and, even better, start paying Miriam and myself a nominal salary. Victor was just about to go into the details when in walked twenty-five members of Operation Raleigh. Victor's face contorted with rage, and slow on the uptake my own with the realisation of what had happened, a hunted look.

We knew Operation Raleigh was due to arrive soon because Ralph Peterson had left a telephoned message with Miriam, but my assumption that Victor also knew was mistaken. Neither of them had bothered to tell us when they were coming (people rarely did) and only the devil could have arranged for both Victor and Raleigh to arrive at the same time. With the office made small by the crowd of boisterous gringos, I had to take Victor to the café next door to calm him down. He was furious with me. When I reminded him that he had expressly forbidden me to liaise with Raleigh, an instruction I had followed to the letter, and told him my only mistake had been my failure to check with him (after Ralph's message), he quietened down a little. I then pointed out he was powerless to prevent Raleigh from going down to a private Estancia, but of course, if he insisted, he could refuse them entry to the Park. The conversation was made more difficult by the frequent intrusion of the Venturers, who unaware of the difficult situation, were trying to be friendly. They wanted to get going since they had arrived sweaty, dirty, and sunburnt after six hours travelling the dusty roads in the back of an open truck.

By the time Victor calmed down, the Venturers had grown impatient. But tired, now, they were getting angry, as most of them had learnt there was a bureaucratic problem. Keeping the peace was like conjuring doves from thin air, a slight of hand that depended heavily upon Victor's negligible English and the Venturers' equally poor grasp of Spanish. When at last the stalemate was broken, and the Venturers free to leave, one of them discovered his camera had been stolen, a minor crisis that threatened to plunge us all back into the jaws of defeat by delaying their immediate departure.

Five days later Victor returned, and Pieter and I accompanied him to the Estancia where we were to interview the Raleigh members individually. When he first saw the Venturers, dirty and tired from their trip to Buena Vista, Victor's irrational prejudice changed; before he referred to them as a bunch of Boy Scouts, now they became a gang of Hippies. In order to slake his thirst for retaliation he insisted each Venturer be made to explain in what way he or she was qualified to help the Park. This patently ridiculous provision, with its potential for badgering, turned out to be mental torture worthy of the Spanish Inquisition. With great relief we finished the last interview just as the second wave of Venturers turned up. Now we had to start all over again. Several hours later, everybody shaking with anger and frustration, we got away at last. Operation Raleigh was now free to enter the Park at their own discretion.

A few days later, when they were ready to do so, the weather turned hostile. A "sur" (a strong south wind) carried a stream of cold air below the warm clouds above and it rained without stopping for 36 hours. Flood Street indeed. Apart from washing out the track between Buena Vista and the Estancia (that between us all we eventually repaired) the rain put a severe dent in Raleigh's itinerary. It was not until the middle of August that they really got going. Together with Charlie King, and two or three young Bolivian participants, I made reconnoitres into the Park to palpate the current attitude of the campesinos. Apart from a few snide comments about gold hunters, and vaguely cynical enquiries about the role of the gringos in Amboró, we heard nothing of undue concern.

It would be tedious for you to have to read an account of Raleigh's time spent in the Park. Apart from some verbal abuse none of the Venturers had any trouble with the campesinos and serious accidents were avoided by the application of a little common sense and close supervision. Nobody broke limbs, drowned, or as far as I know fell seriously ill. The nurses set up a makeshift clinic to dispense first-aid, and cure minor ailments among the Venturers and the campesino population. With the help of experts from the Missouri Botanical Gardens, the University of New Mexico, and an ornithologist friend of mine from Peru, Raleigh did accomplish some of their scientific objectives. Their report included a creditable list of the plants, mammals, and birds in the Park.

Six weeks later they left as they had come, the cause of dispute and ill will. One not-so-young wag carved 'FUCK REG HARDY' on Reg's travel chest that, as Charlie King wrote me later, "caused a lot of unnecessary hassle." And they failed to return a valuable mountaineering rope Pieter had lent them and someone stole the stores and minor equipment which had been set aside for the Park. This came as no great surprise as I had been shown an Operation Raleigh memorandum stating that theft among the Venturers was one of the major problems expedition leaders had to deal with.

And to get me off his chest, a few days later Reg informed me he had decided to cancel my contract with PRODENA, place the Estancia under the care of a sereno and withhold the provision of the Land rover until the following year, by which time he thought everything would be clearer. To me it was already clear. I told him I didn't think it mattered any longer since we had already taken the decision to have nothing more to do with him. I won't repeat what he said to that.

****

With the departure of Raleigh, and Reg for now, I was able to dedicate all my time building our new home. I was negotiating the purchase of a three-acre plot of land only a stone's throw from the perch where I had sat on my first visit to Buena Vista. A lot of things had changed since then; but the green valley of the River Surutú with its mountainous backdrop would always remain the same. I have spent hundreds of pleasant hours gazing at the scene with its backdrop of ever-changing cloud formations; whether dark with the flurry of low rain clouds, or white with the early morning mist, or burnt yellow by the summer sun, it has always remained beautiful. On the other three sides of our new house forest hemmed us in, and as I have said before, friendly trees rustled in the wind to whisper sylvan secrets through our bedroom window.

During my first bird-watching rambles I had been struck by the beauty of the hilltop, and its potential value as a site for a lodge in the future. I was already planning for the day when the Park would have to get itself organised for the tourist trade. We got the green light when Miriam's tentative approach to the owners proved positive, they were willing to sell and by September I was ready to turn builder. From then on, every day was the same. Coffee and three cigarettes outside our tent at five-thirty whilst I planned the day's activities. At six the arrival of my helper. Furious unabated digging, sand shovelling, cement mixing, brick laying, sandwich eating and beer drinking until, too exhausted to lift another finger, supper at seven-thirty, wash, and bed.

On the 12th of December, Miriam and I had somewhere to live. We moved all of our possessions out of the tent, a large double bed which had taken up the whole of the tent's floor space was the first to go together with an assortment of boxes stored beneath it. An hour later everything had been moved into the unfinished house. The day we moved in we had the first real rain, that up to then had been notably absent.

Over the next few months, I worked on the house, and when it wasn't raining turned my attentions to the garden. Some garden! Apart from a few yards around the house we were hemmed in by ten-foot-high weeds. Among these was our own brand of stinging nettle that the Cambas call "pica-pica", with large fan-shaped leaves. I'd been told they could cause the death of children but thought it to be another tall story until one very hot day I chopped through the stem of one. As it fell its leaves slapped me across the face and within half an hour, I was immobilised by a pervading weakness which left me all but unconscious. I soon recovered, but the experience was a salutary one. I dare not think what the outcome would be if one of the tree-sized ones had fallen on me! By the time I had cleared the land immediately around the house to a respectable distance and the wild grasses had covered it with a rough lawn, the place began to look like a theatre. The low-lying auditorium to the front; the hilltop and the house perched on its grassy slope, the brightly lit stage with its backcloth and wings of tall trees, and on the stage the silent actors: bushes of "bulls' balls", "dogs' balls", "cats' balls", "goat's foot", and many palms and Cecropia trees.

I was happy. Where else in the world could a foreigner fix himself up with a job, a $500 resident's visa and own land in his name? Thanking fate that I'd stumbled upon Bolivia, I returned to my task with renewed enthusiasm.

But, as I have said more than once, trouble was never far away. Now Els was sick, really sick!

For some time, Pieter and Els had been talking about returning to Holland. With the withdrawal of my PRODENA beefsteak, and my CDF salary of $60 a month, I couldn't help them out any longer. They were also a little tired of Bartholomew and his idiosyncrasies. Unfortunately, there was no way of making a living, except in Santa Cruz, an option they emphatically rejected.

Then up pops Reg again, and renews his offer to Pieter to manage the Estancia in return for a small stipend, "to keep you alive," as he said. "Of course, this means that you can't have anything to do with Robin" he added. Naturally, I was sure Pete would turn him down. But no, "busy-ness is busy-ness, Robin"; but he did tell Reg his private relationships were his own affair.

Even so, I felt a wee bit betrayed. Torn between my love for my friends and my chagrin at their understandable disloyalty, I threw myself once again into a bout of house finishing.

Then, quite suddenly, Pieter and Els were gone. Her illness turned out to be a particularly virulent form of hepatitis and she wasn't responding to treatment. Their vague talk of leaving now became a race to save her life. Even today I sometimes get messages from them to say they pine to come back but, not able to leave their vehicle spare parts business unattended they haven't done so yet.

This reminds me; before I could sign my contract to come to Bolivia, the British Government insisted I have a hepatitis vaccination. I demurred. I reasoned that after nearly thirty years spent in Africa and other parts of the world, I was one of many people with natural resistance to the disease. They said that if I didn't have the vaccination, they couldn't offer me the job. A standoff. I proposed I get myself tested for hepatitis resistance. They said that was expensive, two or three times the cost of a vaccination. Impasse. I proposed I pay for the test if I was wrong, they pay if I was right. They agreed. I went to the government doctor and he carried out the test. It was positive. "What's more, you were quite right to insist on the test because; once a person has been vaccinated their natural resistance is destroyed".

The last time I saw Reg was in the plaza in Buena Vista. He was sitting alone on a kerbstone outside the Notary's office. He looked mortified, like someone struggling to overcome a temper tantrum in the face of nescient disrespect. No doubt Don Mario the Notario had failed to keep his appointment. Reg could have crossed the road to sit on a bench in the shade of the plaza, but apparently determined to milk the situation of its last drops of humiliation, he stayed where he was, his sweaty red face scarified by the rays of the midday sun.

He was still there an hour later. Feeling concerned for him I approached him with the offer of a chair in the office and a cool glass of tamarind juice. My! he was pissed off. For my trouble I received an uninterrupted series of four-letter expletives, that even managed to capture the attention of the unconcerned village folk. Bye-bye, Reg.

So that is how things turned out. Far away from the madding crowd, sitting here in the middle of the jungle, at times (but not always) desperate for a bit of company, for a friendly gringo voice, I lost them all. First Operation Raleigh, fifty or more people with whom I might have spent a couple of enjoyable months; then Pieter and Els and, finally, Reg. We had been forced apart by a confused mix of politics, economic difficulties and personal vanity, so very typical of the Third World's situation. The sadness of it all.

Author's note: Reginald Hardy died in May 1989. He was posthumously awarded "The Condor of the Andes" by the Bolivian Government for his contribution towards the conservation of the country's natural resources, and his part establishing Amboró Nacional Park.

An apt honour for a man who had an Imperial manner. Before he died
he called himself Lord something or other, a title he bought.

7. Grape Shot

Bolivia's first protected area, "La Reserva Fiscal Cerro Tapilla", was established in 1940. By 1986 nine National Parks, six Reserves, one Biosphere reserve and three Wildlife Sanctuaries had been legally declared. Of these only three were considered active. The rest, the "Paper Parks", legally constituted but politically abandoned, represented Government's lip service to the idea and remained little more than areas marked on a map. No one seemed to mind. Nobody noticed.

By 1986, with the curmudgeonly support of the CDF, they published a map of the Park showing the position of the Red Line. With that we had gone some way towards consolidating Amboró. I was determined it would not be added to the list of casualties. Moreover, I started talking about Amboró as being Bolivia's showcase, its model Park, but how could we achieve this? We had a serious problem with the colonists who were determined to undermine our efforts, albeit publicly they issued statements to say they were not against Amboró and conservation in general. They insisted on their right to stay where they were, and they wanted freedom to use or abuse their own lands without interference from the Park authority. As a rider, they added that the Park should be administered by Bolivians and the gringos should be thrown out.

To me, this rider was tantamount to losing the Park. They knew, at this stage of its development, without the gringos the Park would suffer a rapid decline and be forgotten. The colonists would get their way over the privacy of their own lands and bite by bite they would gnaw away at the Inner Reserve until it, too, was forgotten.

What happened was a compromise. I was demoted to Scientific Advisor and an interim Park Director was appointed. Together with the CDF we started serious negotiations with the Colonist's own elected representatives. I was allowed to attend these meetings but I was not allowed to act as spokesman for the Park. Since I had to take a back seat my tactic was to prepare copious notes with which to brief the CDF representatives before each meeting. At first, I was only

247

rarely allowed to speak and the Colonists never spoke directly to me. However, it gradually became clear to everybody that the negotiations could not advance without my direct participation, and the campesino representatives were not in tune with the desires of their own people, often concerning minor issues that only a few locals and I were aware of. The CDF officials, largely ignorant of Internationally accepted norms for National Parks in general and the problems in Amboró, were not really interested in the proceedings and often failed to turn up for the meetings. I calculated that since its establishment two years ago the CDF staff had spent about thirty man-hours in the Park, whereas I had spent at least eight hundred.

With this undesirable scenario impeding negotiations we did eventually manage to produce a somewhat unsatisfactory draft resolution:

1. For the protection of the Park.

1.1. The Campesinos accept the legality and integrity of the Park with its boundaries as declared.

1.2. The Campesinos respect the Inner Reserve as marked by the Red Line.

1.3. The vegetation protecting rivers and streams would be protected.

1.4. Logging would be prohibited, and further forest clearance only allowed by special permission.

1.5. The killing or capture of wild animals and birds, other than those considered plant pests (including rats, armadillos, and some grain eating birds), or those that persistently killed chickens (such as the Roadside Hawk), but not the Jaguarundi would not be allowed.

1.6. Fishing with dynamite would be prohibited and fishing with nets only allowed at certain times of the year.

1.7. Some Sites of Special Scientific Interest in Campesino held land would be declared wildlife sanctuaries, and the Campesinos would respect their protected status.

1.8. The Campesinos would be held responsible for their livestock, and by the provision of adequate fencing would ensure that these did not stray into the Inner Reserve or graze on vegetation protecting the watershed.

1.9. The campesinos would assist in the implementation of these basic rules. To make the agreement effective the CDF would supply one Chief Guard and four wildlife guards, and the Campesinos 28 honorary guards.

2. For the well-being of the Campesinos.

2.1. The rights of the Campesinos, as owners of their land, would be respected.

2.2. The authorities would resettle those Campesinos who did not want to continue living in the Park.

2.3. The authorities would help the Campesinos with technical and financial support to improve their incomes through the development of agroforestry systems.

2.4. The authorities would establish transitable roads and tracks within the Park, and if possible, a permanent river crossing.

2.5. The authorities would provide schools, clinics, and cabins for tourism that the Campesinos would be taught to manage.

2.6. The extraction of precious stones, gold and other minerals would be studied, but would include the right of Campesinos to take oil from existing seepages.

Apart from the obvious contradictions and shortcomings of this provisional agreement there remained many contentious issues that would have to be resolved at a later date. The main weaknesses were twofold: the Campesinos' right to do what they liked on their own land, and the lack of money to finance their demands. Later, a more serious weakness became apparent that was to plague all our efforts to establish a satisfactory accord. As soon as we finished and signed one draft, the Campesinos would complain of being sold down the river and demand that a fresh agreement be negotiated by new representatives.

This unsatisfactory state of affairs continued for some years and, indeed, to this day a firm agreement has not been reached. Meanwhile the frustration of the Colonos expressed itself in a series of negative acts, some quite outrageous.

****

The first sign of trouble came during our first (1986) census of the people and the land they owned in the Park. One of the preconditions made at the start of negotiations was that the colonists should provide a list of their members and plans of their cooperative holdings. Together with those of a dozen or so private landowners the plans enabled us to make a reasonably accurate map of the campesino settlements, that in its turn allowed us to establish the exact position of the Red Line. Many of those responsible for submitting these documents refused to do so, as they questioned our right to undertake the census in the first place. Only after the CDF published their legal obligation to do so in the newspapers and publicly broadcast it on the radio did we receive a marked response. Even then, it took another twelve months to complete the census and only after we announced that their failure to register would imply loss of official recognition. The results showed that eight hundred and twelve families divided between 39 Syndicates and Cooperatives and 14 private holdings claimed to own land in the Park.

Three years later we received funding from US-AID (through a programme called PL-480) to undertake a more detailed census that included statistical data on each family. From the outset this census was plagued by problems, least of which was a call by campesino politicians for a campaign of civil disobedience that fortunately was largely ignored. The census provided data on family size and age, land use and cropping systems, and year of tenancy:

Total campesino population:	4,642
Number of families:	908, an increase of 100 since the 1986 census.
Average family size:	5.1 persons
Total land occupied:	40,000 hectares (88,000 acres)
Total area of pasture:	4,000 (10% of total)
Total area of arable land:	4,000 (10% of total)
Total area of scrubland:	20,000 (50% of total)
Total area of forest:	12,000 (30% of total)
Year of tenancy:	
Pre-1973	272 (30%)
1973-1984	391 (43%)
Post 1984	245 (27%)

Whereas the Politicos had largely failed to obtain support for a boycott of the census, they were more successful in manipulating the Colonists' response to the last question on the forms. To me, this item was the most important since it dealt with the issue of relocating the people in the Park to new land. The Politicos wanted the Colonists to state their objection to being moved out of the Park, or in their own words: "Don't hand over your lands to a bunch of gringos who will then exploit the gold and export the parrots."

At the time the following one-liner became popular and was sung to me on several occasions: "Soy un árbol y nadie va a sacarme" ("I am a tree and nobody is going to uproot me.")

However, the politicos were a bit late getting their campaign going and it was not until the census was half-complete that their efforts to influence the outcome were reflected by the completed forms coming back to our office. The responses to the last question: If it were possible to relocate you outside the Park, what would you want? were as follows:

	Month 1	% of total	Month 2	% of total
a. To be given new land:	117	29.5	32	6.6
b. A cash compensation:	47	11.8	4	7.0
a+b, Willing to be resettled:	161	41.3	66	13.6
c. Unwilling to be resettled:	223	56.2	401	82.8

****

The next trouble was the repeat of an earlier one, the placement of official Park notice boards. Experience had shown us we needed to put these beside the house of somebody willing to look after them, otherwise they were soon vandalised or stolen.

Ignoring our advice, the CDF posted large notice boards at the five main entrances to the Park. Within a month four of these had disappeared. Later, I learnt that the planks from one of them had been incorporated into the altar of the new church at El Carmen; the settlement where I had considered establishing the Parrot Rehabilitation Project.

It was not until 1990 that these notice boards were replaced with new ones under the care of responsible campesinos, where they remain to this day.

I cannot set down all the troubles we had with the Campesinos over the years but here are some more examples reflecting their attitudes.

****

On the 14th January 1989, Miriam left Buena Vista to guide a group of teenagers (12-16 year-olds) on a weekend excursion to the northern corner of the Park. Since it is her story, I translate what happened in her own words:

"At 7 a.m. on Saturday 14th January, on a clear sunny day, fourteen young people (comprising 6 girls and 8 youths) left in two vehicles to take a trip to the area of the Rivers Yapacaní and Colorado. We were enthusiastic, as it was to be the first visit to the Park for most of the participants.

"By 9 a.m. we arrived at a point where the road was so bad we had to leave the vehicles and continue on foot. After several hours we arrived at the River Yapacaní, opposite the point where it is joined by the River Colorado. We were very excited to see the natural beauty of the place. We ate lunch and then crossed the River Yapacaní to find a camp site on the bank of the River Colorado where we could pass the night. We had hardly finished eating our supper when a violent storm blew up and we had to retire to our few tents and scraps of plastic sheet that some of the others had brought along for their protection. It rained all night.

"In the morning it was still raining, and we decided to return to Buena Vista without further delay. We made our way back to the River Yapacaní only to find that it had become a raging torrent. There was no way we could have made it across, so we went back to our camp site beside the Colorado River.

"On Sunday and Monday, 15-16th of January, it rained without stopping. We were in a sorry state, covered by insect bites, soaked to the skin, and very hungry. We had been trying to feed ourselves on Chonta palm nuts and some wild fruits growing in the forest.

"By Tuesday the 17th we were desperate to return home. We decided that we couldn't wait any longer for the weather to improve and we returned to the River Yapacaní. We went in search of lianas to make a rope for the crossing and by two in the afternoon we had supplied ourselves with one fifty metres long. After great difficulty, and completely exhausted, we finally managed to get everybody across the river.

"With our spirits raised, we set off back up the track to cover the eighteen kilometres to the one vehicle waiting for us. By 7 p.m. we arrived wet, filthy, hungry, and very tired. We decided to load the jeep with our baggage and that we would follow on foot. We were still twenty kilometres from the nearest main road and we knew we would have to find food on the way.

"Almost immediately we caught up with the jeep, that had been unable to proceed because of some small trees blocking the road. We thought they may have been blown down by the storm and we set to work to clear them out of the way. Slightly further on, we caught up with the jeep again. This time it had been stopped by a series of very large rocks in the middle of the track. They were too big for us to move. Disconsolate we sat down to discuss our predicament. Then, quite suddenly, we were surrounded by about twenty very drunk Campesinos armed with machetes. They were very aggressive and told us that we were trespassing on their property. They also said they were not going to tolerate this type of activity any longer and that they had decided to close the road to strangers. Prudently, we decided that there was nothing we could do, and we went in search of other, more friendly, Campesinos to help us move the rocks, or, at least, to sell us something to eat. But as it turned out nobody was willing to help us, and we had to spend a very miserable night camped in the muddy clearing of a logger's camp called 'Campamento Oliva'.

"On Wednesday the 18th in the morning, too hungry and tired to move, we just sat looking at each other to share the pain. We knew that our only hope lay in being rescued, but we also knew that the rain had made the road almost impassable. At 10.30 a.m., Father Paul and Robin somehow managed to get through, and found us. How happy we were to see them, and the food they had brought us! After we explained what had happened to us it was decided to take all those who could fit into the vehicle back to town. We had to leave four of the boys behind with the rest of the food. We were so happy to get out safe and sound, and to be able to give the good news to the parents of the kids in our group."

Miriam ended her report by asking for disciplinary measures to be taken against the campesinos of the area and thanking the three rescue groups that came looking for them.

On Friday, January 19th, a CDF vehicle accompanied by two policemen, returned to Campamento Oliva to recover the jeep that had been left behind. No legal action was taken against the campesinos.

****

Three park guards returning from their three-day patrol of the headwaters of the River Semayo were surprised by a friendly campesino who dashed out of the bushes at the river's edge to warn them that around the next bend a dozen armed campesinos were waiting to ambush them.

The terrain is perfect country for an ambush; the narrow river is hemmed in by small cliffs partially covered by dense bushes and tufts of tall grass. With some difficulty the guards were able to scale the cliffs and pass around the would-be assassins.

No attempt was made to investigate the affair, nor identify the colonists involved.

****

CEPROINCT, a Bolivian organisation dedicated to the formation of future scientists and biologist was considering the adoption of Amboró National Park as their training ground. Since their first expedition, with more than one hundred young people, had gone off relatively smoothly, they decided to finance the construction of a large thatched shelter as a base for their future visits. They built it just inside the Inner Reserve on a bank overlooking the River Isama.

The colonist who claimed jurisdiction over the land (though he said nothing at the time) took exception to the shelter, and together with some of his friends burnt the place to the ground the night before the next contingent of student scientists were due to arrive.

Although the culprits were identified no legal action was taken against them. And CEPROINCT never came back.

****

In 1989 the problems with the colonists came to a head when a deputation visited the President's office in La Paz to complain about their treatment at the hands of the Park's administrators.

Their complaints were set down in this misleading statement:

"YOUR EXCELLENCY, CONSTITUTIONAL PRESIDENT OF BOLIVIA, the Central Campesino Federation has come to inform you about the extremely arbitrary acts, acts of an alarming nature, that have been committed against peaceful and legal settlers. Mr President we represent the interests of more than thirty campesino communities, with a population of 10,000 families, who have lived (with legal title to their lands) for 25 years in an area that now comprises Amboró National Park.

Given the present situation, we have found it necessary to come before you in order to publicly denounce the intimidation, the intolerable abuses and the threats made against us by the staff of the CDF and the Director of Amboró National Park, who, with their malicious and bullying ways, ignore our legitimate rights to live in the area [and] whose wicked acts have deprived our communities of agricultural credit, medical posts, access roads, and most important of all, salaries for teachers to instruct our children; paradoxically when education is the chief obligation and responsibility of the State.

Actually, the arrogance and the systematic intimidation to which we are subjected is mainly caused by the DIRECTOR OF AMBORÓ NATIONAL PARK, AND AN ENGLISH SUBJECT— SCIENTIFIC ADVISOR TO THE BRITISH MISSION.

We have arrived at a point in time when we cannot tolerate the situation any longer.

"Your Excellency, it cannot be possible that we Bolivian farmers (notwithstanding our poverty, and the sacrifices we have made to contribute to the economic development of the zone) are prevented from working our own lands; and, furthermore, we have suffered efforts to remove us by force and violence.

"It is necessary to point out that we have never been responsible for the destruction of Amboró National Park. We, in our humble state, know that the Park plays an important role in the conservation of the ecology and the protection of the wildlife of the area. Actually, we were among the first to protect these natural resources. We should also point out that our settlements were legally established 25-30 years ago, and that our titles were conferred on us in May 1962 — titles that we obtained only after many years of effort, and a great deal of sacrifice."

[Here the letter continues with one short and one long paragraph summarising the Law.]

"We ask that those responsible for the distribution of credit for agriculture, educational facilities, sanitary installations, and other basics needed for our rural development, attend our petitions favourably; which (thanks to the wickedness of the Director of Amboró National Park and the British Advisor) have been rejected in the past; causing us, our families, and our children grave and irreparable damage, both moral and material.

"We ask for justice".

Signed: Zacarias Quispe Ch. [Representative for the Campesino Syndicates]. The Confederation of Agricultural Workers. The National Workers Union [and others].

Received: Presidential Office. 1130 hours, 27th October 1989, La Paz.

****

On a lighter note, a local theatrical group called "Despertar" ("To Awake") presented a play entitled "Un Joichi y un Metichi en el Amboró", which translates as "A Fucking Nuisance and a Meddler in Amboró". The 'Joichi', called Rocke Fehler, [presumably J. D. Rockefeller] was supposed to be me (your own "Bulldog Drum skull"), and the 'Metichi' the Acting Director of the Park.

Unfortunately, I didn't see the play when it was first shown in Santa Cruz, or later when the company toured the country.

As Noel Kempff would have said, 'At least they're taking notice of you, Robin.' And truly, I was a little flattered by the attention even though the play represented an indictment of our efforts for Amboró.

****

Every year I visited the same river, the Semayo, to see if a pair of Crested Eagles had returned to occupy their nesting site high up in the fork of a dead tree. These magnificent raptors are difficult to separate in the field from their close cousins the Harpy Eagle, that have also been recorded in the Park. Since both species are registered as 'endangered' in the IUCN Red Data Book they are a considerable natural asset, and of great attraction to visiting ornithologists and tourists alike.

One night, watching the local news on television, my attention was caught by a report about a brave young lad who had killed a huge eagle in Amboró National Park. He was accompanied by his father who held the spread-eagled bird for the appreciation of the viewers. When interviewed his father praised the action of his son, "who has rid our village of this chicken-stealing menace". I was annoyed by the impudence of the Campesino for appearing on television to publicise his son's brutal and illegal act. Moreover, it was extremely unlikely that a self-respecting Crested Eagle would be interested in campesino' chickens. They like monkeys and sloths.

I was also annoyed by the solicitous attention of the interviewer, an employee of the University's TV channel, who said nothing about the moral bankruptcy of people prepared to kill precious animals within a National Park.

I recognised the campesino as one of the most troublesome colonists in the area of the River Semayo, and to pay him off for his cockiness, I insisted that a park guard be sent to the Semayo to confiscate his shotgun. This was done, but on the payment of a small fine the police returned it to him.

Fortunately there are still Crested Eagles nesting in the area of the River Semayo, until, that is, the same boy and his father decide to return with the same gun.

****

For some years I was in the habit of visiting a small Curichi ("Oxbow lake") where I would pass many happy hours watching and photographing the animal and bird life. The place was an impenetrable thicket of tall weeds, patujú ("Heliconia") and the thorny Tacuara bamboo. To get from one part of the curichi to the other involved crawling on all fours under vegetation that stung the skin and tore at one's clothing. To take my photographs, I first had to wade up to the armpits in stagnant water, running the gauntlet of submerged tree stumps and limbs that moved with the disconcerting motion of overfed saurians or their uneaten prey. It was marshy, verminous, and wretched, and to be quite frank, a bit of a hell-hole.

For the wildlife it was a Paradise, home to a surprising variety of birds and a family of not-so-shy Capybaras. More than a score of Hoatzins had adopted the curichi as their home and breeding site. Greater Anis gathered in close circles to bubble and wheeze in lengthy debate, and once I saw a solitary sun grebe, the first record for Amboró National Park. Fascinating as these and other birds were, pride of place had to be given to twelve breeding pairs of Boat-billed Herons: the only colony registered in Bolivia.

Only fifteen minutes' drive from Buena Vista, it was one of the few places where I could give visitors a taste of the Park at the drop of a hat. And in spite of the uncomfortable conditions, they were, to a man, dumb struck by the bizarre assemblage of animals and birds to be found there.

It was quite obvious this worthless tangle of twenty acres deserved special protection and I took steps to make this possible. The problem was that the land lay just outside the Park boundary and was owned by two aggressive Cholos, two "Uruks of Mordor".

I offered to buy them some much better land. 'No.'

I offered to buy the curichi for twice its worth. 'No.'

I doubled that offer. 'No.'

In addition I offered them a small stipend to look after the place. "No. Vaya a la mierda." ("Fuck off").

Desperate, (for they were gradually clearing the place, and one of them had built his decrepit thatched hut too close to the Boat-bills) I took the case to the CDF. I took the case to the CDF every month for two years. I wanted them to declare the "curichi" a "Site of Special Scientific Interest" and apply the law which allowed for the compulsory purchase of such places. Eventually the CDF sent up one of their disinterested lawyers to look into the matter. He wasn't able to reason with the owners any better than I.

Uruk 1: "No. We don't want to sell any land to the gringo."

CDF: "This land will belong to the State, not to the gringo."

Uruk 2: "No. Leave us alone."

Then I heard from one of the park guards that the campesinos had cut down and burnt off all the vegetation surrounding the curichi and killed the birds. When I got there, so it was; a grisly message from two of the "Nature Haters".

Nothing was done about it.

****

After we received information that colonists of the Syndicato Colquiri were hunting in the Inner Reserve, the acting director of the Park sent a note to their secretary to inform him that the CDF would take appropriate action if the hunting continued.

In a letter addressed to Amboró National Park the syndicate replied as follows: "On [this day] the 19th July, the Colquiri Syndicate wishes to make it known that our community does not accept the dispatch of [official] directives, or alarm-raising notes. Contrary to this, the next time that such wicked acts are perpetrated in our community, we cannot be held responsible for whatever accident might befall [your employees]. You have been warned: Don't interfere with our Colquiri community."

This last, their idea of a community, that it was their community, that nobody had any right to even set foot on their lands, was often the reason why the Government, and us as their representatives, had such difficulty in our attempts to come to terms with the colonists. Even today, in their altiplano home, this severe xenophobia leads to inter-community wars that end with the death of some of their own people.

We were not the only persons to have experienced the dark side of the altiplano peasant's character. This is what Peter Matthiessen, in his book "The Cloud Forest", had to say about them: " it is hard to make out whether they are stupid as they seem or just perverse."

Well! Not exactly politically correct, but...

****

Here I feel I should bring the account of our problems with the colonists to an end with a discussion on the rights and the wrongs of hunting in the South American context in our modern day and age. Before doing so, I had better say that I do not intend discussing the control of crop-damaging wild animals, a subject closer to pest control than hunting per se, even though it's becoming one of the most contentious issues for today's working conservationist. A whole book could be written on the legal and moral implications.

Many conservationists claim that whatever was acceptable in the past is no longer so, and given the crisis threatening the fauna of a world strapped by ecological problems, hunting in any form should be banned. This opinion has been answered by those who point out that subsistence hunting, including the right to sell surplus game, is a vital component contributing to the survival of indigenous peoples, and by sportsmen who consider hunting a "God-given right" that every man should be free to indulge in.

So, what is the truth? Are the fruits of hunting a shameful harvest or are they fair game? I believe the following broad statements may help narrow the debate to manageable proportions:

True: As all hunting scares wildlife away, it directly impinges on every person's inalienable right to live among animals in their natural state; and can damage the commercial interests of those involved in ecotourism.

True: The consequences of hunting in an area will vary with the conditions found there, but irrational exploitation will always be destructive to wildlife.

True: The destruction of habitat, especially forest, is far more damaging to the survival of species than is hunting.

Probably true: Unless wildlife is valued as an important economic resource, people will continue to destroy natural habitat to make way for other land uses.

True: Hunting is not compatible with the aims of protected areas because these are areas set aside for natural organisms, whether they be plants or animals, to live a free and unmolested life.

True: Nor is hunting compatible with the aim of a forest reserves' long term sustainability because many animals and birds are important pollinators and dispersers of seeds and without them many types of trees cannot regenerate. [Incidentally, at the time I was writing this the Senate was debating revision of the Forestry Law. I hope, at least, they take measures to stop the logging companies employing hunters to feed their workers.]

Nearly True: A few hunters working the periphery of a large forest reserve or wilderness area will not seriously affect the population levels of their prey.

More True: Hunters, whether many or few, working isolated woodlands can quickly exterminate all the game animals living there.

True: Hunting rare or endangered species can irreversibly damage their population levels and cause their extinction.

True: Selective hunting, where the target is limited to one or two species (such as Cats, Caimans and Peccaries for the skin and hide trade), has a long-term detrimental effect on these animals from which they may never recover.

Probably true: Some gregarious species (Monkeys, Peccaries) are particularly prone to hunting, whether it be for meat, or specimens for medical research.

True: As stated for gregarious species: the capture of large parrots like Macaws for the pet trade, legally or otherwise, can never be justified.

Probably true: Some large birds like Curassows are vulnerable anywhere because as relatively easy targets, they frequently fall victim to hunters of larger game when returning to home empty handed.

True: Hunting (or for that matter fishing) using traditional methods: bow and arrow, lance, rod and line are honourable pursuits (if one must kill), and (maybe true) may even benefit animal or fish populations by selecting against the least fit among them.

True: Laws to control hunting are ineffective because they are rarely based on a technical assessment of the actual situation and because; they are difficult to implement. If the latter is true, then education is the only effective tool available to us.

Legislators recognise the elements of truth contained within these statements and have devised laws to mitigate the worst effects of hunting, even though they know that until the population as a whole acknowledges the right of animals to live out their lives as useful members of our Biosphere (and not just treat them as objects), and understand why this is important to all of us, these laws will remain little more than lip service to the idea. In countries like Bolivia, with many poor as well as poorly educated people, we must also seek ways to overcome poverty if we wish to eliminate the problem.

So, yes, as educated and concerned people we should set an example: "thou shall not kill", capture or buy wild animals; steal their eggs or young; wear their skins or use wild animal products; eat wild game; allow your children access to guns or catapults, or fail to educate them in the arguments for and against the protection of wildlife, and so on.

But what about the rights of ethnic populations and poor Campesinos? Should they be subject to the laws of the land like everybody else, or should they be free to disregard those they consider incompatible with their traditions? Shouldn't we say: "Tolerance should be extended to those who wish to follow their tradition but, as far as hunting is concerned, guns are not part of that tradition." As soon as anyone chooses to carry a firearm (and here we could add manufactured traps, nets, or weapons of any sort) they should be subject to the laws. [In Brazil they don't bother with semantics, they have a special 'Law of Diminished Responsibility' for Native People, Mentally Disabled Persons, and Children.]

But the laws, or at least their implementation, are equivocal. On application to a hunting and fishing club, one becomes privileged on payment of a membership fee, a privilege (as everybody knows) not extended to the poor Campesino. Here, then, we have identified one of the many paradoxes: "The man who says he's a sportsman is allowed to hunt, the man who says he's hungry isn't."

In truth, the hunting laws are full of such dilemmas and represent many biased and arbitrary opinions. To illustrate this, we can extend this argument further and in two separate directions. For example, the arrows of most indigenous tribes are made with the feathers of large protected species of birds, like Curassows and Harpy Eagles. Should they be allowed to make genuine arrows for the tourist trade? International agreements, that Bolivia is party to, say not. So, "No" must be the answer.

Folklore and Hunting: A variety of animals and birds are slaughtered or maimed to provide decoration for ceremonial costumes. The feathers adorning a single headdress, like those of the Beni's "Macheteros", contain the tail feathers of several Macaws. The altiplano Indians use hundreds of "Suri" (Lesser Rhea) feathers for the same purpose. Should people be allowed to kill and mutilate hundreds of these protected birds each year in pursuit of what is folklore to some, and a barbarous custom to others?

Pet Lovers and Hunting: As every traveller will tell you the indigenous peoples of South America demonstrate their love of animals, or at least this is the myth fostered by stereotyped images of Native American families sharing their hammocks with innumerable pets. But, to want pets is not to love animals, for true love can only come from respect, and to respect an animal is to leave it alone. Moreover, an indigenous person's pet could be a pet until he gets hungry enough, then it's into the pot.

As anybody who has visited South America, China, or the Far East will tell you, there are many peasants wandering the city streets and market places offering the sale of protected species of wild birds and animals. And the authorities do very little to stop this completely illegal activity. But who is to blame? The pet-loving public of course! If you don't buy a pet, nobody will offer to sell you one. But if you buy one, the benevolence of a government which does not like to rob you of your freedoms prevents it from taking action against those who rob animals of theirs.

So, finally, what are we to do? We have seen that hunting and attempts to regulate it affects everybody in one way or the other. But legislation is not just a question of pandering to the sympathies of nature lovers, it can mean life or death for the people who depend upon hunting for their subsistence. Wise legislation based on scientific study could provide an equitable solution. Until there is money for these studies all we can do is give it our best shot, but it'll probably be "a shot in the dark".

****

And whilst we are talking about scientific studies and hunting, we shouldn't forget the professional biologists, not those studying hunting, but those hunting to study. I was one for many years, with 200,000 or so dead beetles in my collections. But scientists may also be in a position to make a plea of diminished responsibility. My own plea was that the science of coleopterology could not advance without specimens; the same argument used by several prestigious North American universities to justify killing Bolivia's birds in large numbers. And I single out the ornithologists because the mammologists are not as bad, concentrating, as they do, on rodents, bats and other smaller animals, rather than Monkeys, Tapirs and Jaguars, which is to say the ethics of the matter are somewhat mitigated by the size of the subjects they study.

Whereas I've seen hundreds of drawers full of skinned birds, I've never seen one full of tapir skins. Mind you, this doesn't mean there aren't any.

The ornithologists maintain that many of Bolivia's birds cannot be reliably identified in the field, and the number killed for museum specimens was inconsequential in comparison to natural rates of predation and habitat loss. I said they exaggerated the difficulties of identifying birds in the field, and where there are real problems (the smaller Flycatchers for instance) we wouldn't make headway unless we spent time developing field techniques to resolve them.

In other words, I thought they should spend less time catching and skinning birds and more time watching them. They countered by saying they didn't have time to do this; I answered in true Hardy fashion, "then find some".

They pointed out that their expeditions were expensive and the committees that granted them funds wanted to see hard evidence of their labours. I said this was a problem of working from a foreign country and the correct procedure was for them to move down to Bolivia if they wanted to study Bolivia's birds. They replied that this was impossible [they didn't want to], but they were training Bolivia's own ornithologists. I said I didn't like the way they were doing this because they were teaching the Bolivians to do the same thing, that studying birds meant killing them, an activity already favoured by Bolivian naturalists with their field-shy attitudes. And now, the bit really between my teeth: "Why should people, barred from killing birds in their own country, do so in Bolivia?"

Then there were more specific points: they needed gene samples to study kin relationships between different groups of birds; one always needed reference specimens as a back-up to field obser-vations; and new taxa (species, subspecies) could not be described without specimens. I maintained that genetic material could be obtained from the museum specimens they already had (only a droplet of blood is needed), and that if they required permanent records, they could net the birds, photograph them, and then release them.

I also understood the special case of new taxa, but out of the many thousands of birds they killed, less than one in ten thousand were likely to be new to science. And, I added, as far as new species were concerned, it would not be a terrible sin to leave them for some Bolivian scientist to discover in the future, as I had done for three or four potential new ones in Amboró. And the hurry was not to catalogue all the birds, the hurry was to save them.

And while on the subject of new species or endangered ones, I maintained it was all very well for them, as experts, but how did they expect budding Bolivian ornithologists to differentiate between highly endangered species and very common ones, an easy mistake to make, until it was too late? What would they say if an overenthusiastic student presented them with a heap of endangered species, or worse, a drawer full of Horned Curassows!

And whilst we were on about it, why didn't they put all the money used for research and collecting into new parks and reserves?

Then I was told they were making snide comments about my work: 'We think Clarke's bird list for Amboró is impressive. What a pity there aren't any specimens to back up his work." That, we could have done, but we didn't allow any collecting in Amboró. Anyway, how could we tell resident campesinos that they weren't allowed to hunt birds but, gringos could?

Then another comment: "Doesn't he realise he has probably made mistakes? Really the man's not a scientist, just a Naturalist!"

We thought it better to risk making the occasional mistake than spend our lives killing birds, an activity completely anathema to our own personal philosophies and the aims of a National Park. But to be called a "Naturalist" was a great honour.

Piqued, I would then go on about how time wasted preparing bird specimens would be better spent studying their needs so that we could learn how to save them from extinction, and you couldn't do that whilst you had your head down ripping their skins off.

Here I should add that it used to be common practice for field ornithologists to make a stew from the bodies of birds they collected. I'd even heard one joker talk about the number of Red Data Book species he had eaten!

None of these arguments got me anywhere, and they only made me more enemies; and I already had enough of those. The thing is, it's not just a question of ethics, it's more to do with setting an example. Before Louisiana State University went to the Noel Kempff

National Park they prepared a list of birds for the area, but this wasn't a list of birds found in the Park, it was a list of birds they wanted to kill! Now, the indigenous peoples of the Huanchaca area that includes the Park were nearly as hostile to the Park as the colonists were in Amboró; they said it was wrong to filch their traditional hunting grounds just to provide amusement for a bunch of gringos. Then the bird collectors from Louisiana arrive, and the locals hear nothing but gun shots from dawn till dusk.

Not sound public relations to say the least.

****

Now that I have dealt with the problems of both hunters and Colonists in Amboró, at least for a while, I must return once more to the chronology of my story. Unfortunately, more bad news was on its way, but first, a bit of good news.

On the 27th of June, 1986, Victor Paz Estenssoro, Constitutional President of the Republic of Bolivia, signed Supreme Decree No 21774: ARTICLE 1 'Se declara la veda indefinida y con caracter general para la captura, acoso, acondicionamiento de animales silvestres y sus productos derivados como ser: cueros, pieles y otros a partir de la fecha.

In English, my approximate translation: "I declare the ban for an indefinite period, and for other reasons include the sale of thick skins, skins in general from now on"

At last, after a five-year wait, we had won! No longer could any-body legally capture, harass, or encage wild animals and birds, or deal in their secondary products.

The only exceptions were the poor old caimans, people still want their crocodile-skin shoes and handbags, and wild pigs, why anybody should want their skins, 48,000 of them every year, I have no idea. The fact is the demise of these two animals in Bolivia is a sad tale I've not even hinted at until now, nor will there be space to correct the oversight.

But all the other birds and animals, well, they were free! What a joyous moment this was for us. Once again, I could look a monkey straight in the face. But as you will see, the privilege was going to cost us more than we bargained for.

****

Miriam and I often went to a cantina in the village to play "sapo", a game in which the players try to throw lead counters into the open

mouth of a cast-iron toad (whereby the game gets its name). On this
occasion we interrupted our game to watch an interview with Noel
Kempff on the television. After fifteen minutes of stilted conversation
(the habits of various birds and animals and the problems facing
wildlife in general) Noel began to talk about his up-and-coming
expedition to Huanchaca and, of more significance to us: the efforts
being made to establish Amboró National Park.

****

Amboró is progressing with the help of some English people, a
Mr Reg Hardy, and PRODENA.' He did not mention me. Noel still
had not understood that I didn't work for PRODENA, or, rather, he
was forgetful of the fact and he often talked about the British Mission
when referring to my work in Amboró. Whilst he thought I was still
receiving an allowance from PRODENA, or a fat salary from the
British Government, he would not be aware of the financial
constraints preventing us from implementing the Amboró project in
the way we would want; and would do little to help resolve my, by
now, dire financial situation.

Moreover, it was galling to see the credit for my work given to
somebody else and to realise that after all these years Noel still knew
next to nothing about the Amboró project. If I was Reg Hardy's
Cinderella, I was also Noel Kempff's Belesario. As you can imagine,
I was tired of it and I decided to go down to Santa Cruz and have it
out with him once and for all.

With this decision made we resumed our game, but with Noel's
words still rankling in my mind Miriam gained an easy victory. Sapo
is a game that requires a relaxed laid-back temperament, reasons why
the locals were so good at it.

****

It seemed as if I had only just got back from Santa Cruz and my
talk with Noel, in fact it had been a few weeks when, lying in bed one
evening the radio mentioned him. "Profesor Noel Kempff Mercado,
illustrious son of Santa Cruz de la Sierra, was killed by unknown
persons in Huanchaca".

The report went on to say, that even after several days, very little
was known about the circumstances of his death, but it was believed
that Noel, the pilot of the plane in which he had been travelling, and
his guide had been gunned down on an airstrip when they landed. It

stated that a Spanish biologist who had accompanied Noel had been rescued. And it advised listeners to stay tuned for further information.

Over the following weeks the story began to emerge, and I set down here, together with a bit of past history, what was believed to have happened:

11th May 1977: US citizen, Bruce Lindenberg, buys part of a land concession from Ariel Coca Aguirre (ret. air force Coronel) to extract timber from the mesa of Huanchaca. Priority is given to the construction of a new airstrip as the existing one is considered past repair. [From here on I will refer to this new airstrip as the Huanchaca airstrip.]

It soon becomes obvious that Lindenberg's true intentions are very different from logging or tourism. Apparently, with the financial help of Bahamas Islands' disgraced Prime Minister, Sir Lynden Pindling, he spends large sums of money importing equipment and drums of chemicals used for converting crude cocaine paste to pure crystals. By the end of the year, Lindenberg finds himself in a US prison for drug trafficking operations, but not those in Bolivia. Here ends the first attempt to establish a drug operation on the top of Huanchaca.

28th June 1979: With Supreme Decree 16646 the Bolivian Government declares Huanchaca a National Park.

1980: The British Geological Mission, working under the auspices of Britain's Ministry of Overseas Development, restore Lindenberg's Huanchaca airstrip.

February-June 1985: A distribution centre for the equipment and chemicals needed for purifying cocaine is established a few hundred yards from the Huanchaca airstrip. Later, witnesses testify that this Drug Centre, as I shall refer to it from now on, was owned by Jorgé Roca, others that the new entrepreneurs were thought to be Colombian or Brazilian citizens. The truth: may be both.

December 1985: Noel Kempff asks José Cabot Nieves (a well-known Spanish zoologist and conservationist) if he would be interested in forming a team of Spanish-Bolivian biologists to study Huanchaca National Park.

July 1986: After Spanish biologists finalise their project proposal to study the Park, Noel Kempff makes arrangements to set up a scientific base camp on the Rio Paraguá at the Moira Logging Company's sawmill, 750 km north-east of Santa Cruz and 60 km west of the Huanchaca mesa.

3rd & 19th August 1986: After a number of denuncias made by Coronel Coca, he, the US Drug Enforcement Agency and some Bolivian officials overfly the Huanchaca Drug Centre. According to Coca's testimony the plane, an Air Commodore 1000 piloted by James W. Farnan (DEA), was equipped with a television screen on which the drug encampment was clearly seen.

Thursday 14th August 1986: Four biologists from Spain's Doñana Biological Station and two taxidermists arrive in Santa Cruz. They are joined by three more Spanish biologists two days later.

Tuesday 19th August 1986: Having been delayed by bad weather, the Spanish-based biologists and their Bolivian assistants leave Santa Cruz in two jeeps and a truck. They spend the night at the old reduction of Concepción, still a day short of the Moira base camp.

Wednesday 20th August 1986: After admiring the beautiful Jesuit church of Concepción the motorised party leaves for Moira where they arrive at 5.30 p.m. A little late, having run out of petrol 50km short of their destination, Noel Kempff arrives chauffeured by Ruben Poma [my good friend who first took me to the Estancia San Rafael de Amboró].

21st August-2nd September 1986: The scientific team establish an advanced camp 60km to the east of Moira and a few kilometres short of Huanchaca's cliff-face. They conduct faunal surveys in the lowland forests of the area and dedicate much effort searching for a way up the five-hundred-metre-high cliff guarding Conan Doyle's Lost World. The CDF overfly the mesa of Huanchaca and report the presence of an airstrip 60 kms to the east. Whether or not this one is the Huanchaca airstrip is not clear. They don't report anything unusual, no oil drums or camp. Noel abandons the original plan to scale the cliffs and decides to contract-hire a plane (owned by the Bishop of San Ignacio) to fly the scientific party there on the 2nd of September.

Saturday 30 August 1986: Owners of the Moira sawmill advise the Kempff party that the Bolivian narcotics police are planning to talk with the scientific team. Presumably about the wisdom of pursuing their studies in an area known to be occupied by drug dealers. For some reason Noel gets very angry at this news. But, in the event, the police don't show up.

Noel sends a progress report to the newspapers that El Mundo publishes on the 5th September. After a brief summary of their progress the article states the following: 'Next Tuesday [this would be the 2nd September] we plan to fly in the aeroplane [how else?] up to the high part of the mesa, with all the team, in order to start the study of the fauna and flora found there.'

Sunday 31st August: Curiously, Noel aware of the growing consternation among the Spanish biologists, all of them anxious to get to the summit, admits that there may be only enough money to finance a couple of flights. [How, then, did he plan to take all the team, 16 people and their equipment, to the Huanchaca airstrip for thirty days and back? Even if we suppose that a "couple of flights" meant there and back twice, the chartered single-engined Cessna can only carry three or four passengers and their personal equipment, not loads of scientific equipment, formalin, alcohol, tents, tarpaulins, bedding, cooking gear, radios, water and whatever fuel they might need. In fact, a couple of trips would only permit four people to get to the Huanchaca airstrip, stay for a day or two before being picked-up again.]

Monday 1st September: To the relief of the expedition's members, they are told there will now be enough money to finance five flights to the summit of Huanchaca. [This is more realistic, but I still don't see how this will be enough for all the team to go unless they pull people and equipment out on each return flight. I labour this point because it has considerable bearing on what happend later.]

Tuesday 2nd September: The expedition abandons their camp near Los Fierros and return to the Moira sawmill to await the arrival of their Cessna at 9 a.m. By midday they are sufficiently worried by the non-arrival of the plane to radio the Bishop's staff in San Ignacio. Worried before, they now become bewildered as well, since no one at the Vicariate has the slightest idea about plans to fly to Huanchaca.

Wednesday 3rd September: [Isaiah 40:31 "They that wait upon the Lord shall renew their strength; they shall mount up on wings as eagles".] Not so; their wings don't come, nor any message from the Lord.

Thursday 4th September: At 9 a.m. the Cessna's pilot passes a message to say the flight had been postponed because smoke from grassland fires grounded the aircraft. In the afternoon he sends another message to say he would send another message later on in the day. [Isaiah 30:7 "Their strength is to sit still."]

A few hours later a plane lands at the Moira airstrip. With high hopes two members of the expedition rush off to the airstrip only to discover that the new arrivals are a Malaria Control team.

At 7.30 p.m. their Cessna pilot radios to say that he will arrive on the morrow at 9 a.m.

Friday 5th September 1986: 9 a.m. The eagle has landed. [Matthew 14:31 "O thou of little faith, wherefore didst thou doubt?"]

No doubt, the frustrating three-day wait, with all their things packed to go and nothing to do, and the inevitable suspicion that there wouldn't be enough flights now precipitated some fierce arguments among the Spanish expedition's members over, who should leave on the first one. The original plan was that the first flight would be an exploratory one to locate the airstrip; the pilot taking only one of the Spanish biologists and Moira's administrator, Papi Moreno with him. Moreno had overflown Huanchaca before and had seen an airstrip, so he would have been a good choice to go, especially now, when the pilot admitted he had never flown over the mesa. Others thought that the first flight should be made up of the youngest and fittest team members since it was likely the landing strip would need attention.

Whatever, all these arguments came to an end when Noel decided to change all the plans, scrub the idea of an exploratory flight [maybe a good thing given there weren't enough planned for] and go himself, together with Franklin Parada, his guide. Just before leaving, the pilot decides to remove the back seats to make enough room for the three Spanish team members, their equipment and his large plastic drum of aviation fuel. Seeing the drum of fuel go in, the team decide to chuck out the battery for the expedition's auxiliary UHF radio that had been bundled aboard at the last minute. [You're confused? So am I. So were they!]

Noel, his guide, one of the Spanish biologists, Vicente Castelló (who was on board because he was also their radio operator), and finally the pilot of the Bishop's plane, Juan Cochamanidis, leave the Moira airstrip at 10.40 a.m. to fly to the summit of Huanchaca. The plane is scheduled to return between 1 and 2 p.m. to pick up other members of the Spanish team.

After thirty minutes the plane carrying Kempff and his three colleagues begins its descent towards a well-maintained airstrip they spot from the air. On approach, just a few hundred yards to the north of the airstrip, they see an encampment of several large tents and a considerable stack of oil drums. Suspecting something very fishy, Vicente Castelló urges them to abort the landing, but is brusquely overridden by Noel and Franklin Parada. [Apparently the pilot makes no comment.]

Cochamanidis then makes a single pass over the runway but not the camp to check its condition before making a perfect landing. No sooner has the pilot cut the engine than they all get out and hurry up a track going off in the direction of the drug encampment. They notice that the track is well used, with many fresh footprints and heavily rutted by tractor wheels and other vehicles. After a few hundred metres Noel and Vicente decide to turn back, leaving the pilot and guide to continue.

Shortly after returning to the shade offered by the wing of the plane, Kempff perched on one the of the wheels, Castelló standing beside him, they see their two companions running desperately towards them whilst being shot at by two heavily armed individuals giving chase.

Then, once they are all together by the plane, and in spite of Castelló's hurried explanation of who they are and what they are doing there, the two assassins raise their guns, and at point blank range shoot the guide Franklin Parada. Noel jumps up from his perch on the wheel but has only time to shout out, 'No! Don't do that!' before he was shot as well.

Meanwhile, the Spaniard, Vicente Castelló, sees Cochamanidis running for his life away from the plane and immediately takes after him down the airstrip. With bullets whistling past them, Castelló, younger and fitter than the somewhat portly Cochamanidis, soon overtakes the pilot, who now follows the biologist's lead as he leaves the open ground and rushes down one of many paths entering the forest from the runway.

Castelló, now on his own, sprints down the path for a considerable distance before turning off it to enter a dense thicket. Terrified of making the slightest sound, the young man throws himself to the ground and covers himself with leaves as best he can. Maybe only now the real horror of the situation takes hold, as the two assassins spend the rest of the day and part of the night searching for him. Sometimes close, sometimes further away, once firing their guns into the bushes, the young Spaniard lying frozen with terror listens to the sounds of the manhunt. Then, just when he thinks the killers have given up, he hears the tuc-swink of a machete and the shuffle of footsteps very close to him. Vicente, jungle-wise, keeps perfectly still; oblivious to the mosquitos, the ticks and the sharp ant bites, and ignoring the sweat pouring down his face, resisting the heat and the thirst which tortures his body and mind, he lies as quiet as a mouse.

****

Back at the Moira airstrip the rest of the members of the Hispano-Bolivian Zoological Expedition await the return of Cochamanidis with, at first, mounting impatience, and then with the ticking of the clock, a paralysing anguish that insinuates itself into the thoughts of each and every one of them. Maria Corvillo, biologist spouse of Vicente Catelló is bravely trying to hold back the tears that with each passing moment spring more readily to her eyes.

[Amazing as it would seem, the aircraft provided by the Bishop of San Ignacio, came with a damaged radio, one without UHF, the only band available to the group waiting at the Moira airstrip. To add a touch of grim irony, the expedition's inoperable radio, now gave rise to some optimism. 'Maybe,' some said, 'the advanced party was still busy cleaning the landing strip.']

At 7.30 p.m. the waiting group returns to the Moira sawmill where the administrator notifies the Santa Cruz air traffic control that all contact with the Bishop's plane has been lost.

Saturday 6th September Just before dawn Vicente Castelló abandons his hideaway in the forest and makes his way back to the path. After slaking his thirst from a small puddle, he comes across the body of Cochamanidis which has several gunshot wounds in the back. Fighting the instinct to turn about and flee down the jungle path, now that he realises he is probably the only survivor of the advanced party, he forces himself to creep forward to the edge of the airstrip. Keeping just within the border of the forest Vicente makes his way to a point opposite the, now, burnt remains of the aircraft.

Satisfied that he can no longer help any of his comrades, Castelló makes his way to the far end of the airstrip where, in relative safety, he is able to command a good view of the area while he awaits events.

****

Having received no response from his radio contact with air traffic control the previous night, the sawmill's manager now radios his boss at Moira's Santa Cruz office, and by 10.30 a.m. they're told that rescue aircraft are on their way. At 1 p.m. a twin-engined plane belonging to the armed forces and a private light aircraft land at the Moira airstrip. Both planes soon take off and the remaining biologists return to the sawmill to monitor the rescue mission's radio traffic.

At 2.10 p.m. the military pilot, Captain Añez, comes on the air to ask the Moira administrator to give him the details concerning a twin-engined plane, also burnt, on one of the mesa's airstrips some time in 1985. As soon as Añez realises the description does not fit the burnt-out single-engined Cessna he has just seen, he bursts out 'There's been an accident! There's been an accident! [Kempff's] burnt-out plane is in the middle of the airstrip. We will try landing. There are survivors. There are survivors.' Captain Añez has just witnessed the two [some say three] gunmen running to the burnt remains of the Kempff party's aircraft.

Vicente Castelló has waited more than eight hours hidden at the end of the Huanchaca airstrip. Frightened, lonely, hungry and thirsty, attacked by biting flies and irritated by the population of ticks which have colonised his skin, he suddenly becomes aware of an approaching aircraft. This is a decisive moment for him. Is it foe or is it friend? He really doesn't have much option.

****

Captain Anez's excited confirmation that there are survivors is almost instantly followed by his shout, 'There's a man coming out of the forest!' Vicente has just made his life's most important decision.

Fortunately for him, the senior Spanish biologist, Francisco Braza, sitting behind Captain Añez, also sees and recognises his teammate. Armed with this extra piece of knowledge he quickly realises something is seriously amiss; why would his friend and colleague, Vicente Castelló, have separated himself from the other survivors by as much distance as possible? In mid-landing approach he turns to Añez and persuades him to keep away from the burnt Cessna, and instead land where Castelló is desperately waving to them. Captain Añez, not slow on the uptake, surprises everybody, his passengers, the assassins waiting in ambush below, and probably Vicente Castelló as well, by making an astonishingly acute turn that brings his plane within inches of the incinerated Cessna, but which enables him to land and roll directly along to the waiting Spaniard. With the help of Braza's sure arm Castelló does not delay boarding the military bimotor; and screaming out the gist of the attack, ('What a disaster! They've killed everybody! They're pichicateros! I didn't want to land! Let's get out of here, they've got machine guns!') convinces Añez to take off with all possible speed.

****

Back at Moira, the waiting listeners don't pick up Vicente's screams but do hear Captain Añez's sudden shout, 'Don't touch it!

Don't touch it!' as his co-pilot makes to give more flap to the desperately low altitude of the plane's foreshortened take off. [If he'd increased the lift the plane would have stalled and down would have come, Añez and all.]

After what seems an interminable delay, but which is but a heartbeat, they hear the military captain announce the recovery of one survivor. Among the small crowd of listeners only Maria, Vicente's wife, could verbalise the inescapable question: 'Who is it? WHO IS IT?'

Afterwards, they said even the roar of the plane's engines couldn't muffle the answer, albeit a bit insensible, it comes through loud and clear: 'The Spanish biologist. We're coming back.'

The workers at the Moira sawmill get wind of Kempff's murder and they all run away. [Later, it transpires, the airstrip near Moira had often been used during the transit of chemicals and equipment to the Drug Centre, presumably with the help of the workers at the sawmill.]
****

The news of the murder of the Kempff party reaches the media in Santa Cruz. [But not Buena Vista, cut off from the newspapers; through our own aversion to the local radio stations, Miriam and I don't hear about the incident until several days later.]

Sunday, 7th September 1986: The President of the Unión Juvenil Cruceñista, Marcelo del Rio, together with other civilians and members of the armed forces arrive at Moira. In the afternoon they fly over the mesa of Huanchaca and see four armed men guarding the airstrip. Prudently, they return to Moira. During the night they report hearing four light planes leaving the Huanchaca airstrip.

Monday 8th September 1986: As soon as it is light, at 6 a.m., del Rio and his civilian companions take off to fly to the Huanchaca airstrip. They are closely followed by soldiers of the armed forces, who aren't responsible for antinarcotic operations, and four hours later by UMOPAR who are. On landing, del Rio and his party immediately find the burnt-out plane with Noel's body inside and then that of Cochamanidis still lying in the forest. They are unable to find any trace of Franklin Parada.

About half a mile to the north they come across the pichicata storehouse. They are astonished at the size of the operation and report the following: 12 very large tents and 100 beds. 1 wooden house, that they believe belonged to the boss of the operation. 4 tents containing laboratory equipment and solvents for the purification, drying and packaging of crystallised cocaine. 1 store house with 3 month's supply of food; tractors and trailers, motorcycles, a helicopter landing pad, generators, water pumps, televisions, radios, and 1,700 forty-gallon drums, 700 of them containing ether, the rest acetone and other cheaper solvents like sulphuric acid. This was not a Pichicatero's Supermarket, it was his Hypermarket! Everything a drug processor could possibly need was here, it was the Drug Centre.

Del Rio ends his press interview criticising UMOPAR: "It took them more than three days to arrive and only an hour to loot everything of value".

Tuesday 9th September 1986: The remains of Noel Kempff arrive in Santa Cruz. The body of Cochamanidis arrives in his hometown of San Ignacio.

Thursday 18th September 1986: The body of the guide, Franklin Parada, is found and taken to Santa Cruz.

Friday 19th September 1986: General Hugo Banzer, ex-President of Bolivia (1972-1979), is asked by the press what he thinks about the Huanchaca affair. El Mundo quotes him as saying: 'It is very sad, this accident…'

****

The death of Don Noel and his two companions, as I have said, turned into one of the biggest political scandals to shake Bolivia in modern times. The municipal authorities in Santa Cruz immediately declared him 'A Martyr to Science' and announced a period of thirty days mourning for the loss of their illustrious son. 'What will happen to the zoo?' 'Will the Botanical Gardens have to be abandoned?' 'Who will teach us about the birds and animals, and who will protect them like Noel did, with so much dedication, and so much bravery, to stop them being exploited and smuggled out of the country?' 'How can we ever find a replacement for him?' All these questions were included among the cries of anguish that filled the newspaper columns.

The fact that Noel Kempff was killed by Pichicateros precipitated an essential change in the way Bolivians viewed them. Fundamentally freebooters, most Cruceñans were prepared to ignore the illicit nature of the coca-cocaine trade. A great many took to the business with the enthusiasm of a gold rush, that it was, and given the relatively simple technology involved, there was opportunity for everybody. No one was too lowly or too ignorant to profit from this new money-making miracle. An English newspaper, The Guardian, described Santa Cruz as 'The town that got high on cocaine.' Ignored by President Hugo Banzer, a Cruceñan himself, the trade flourished, but the real boom, the national boom, had to await the presidency of Garcia Meza. "With him", one US official declared, "the drug mafia of Santa Cruz has bought itself a Government".

The death of Kempff changed all that. The most important drug barons, Roberto Suarez ('The Cocaine King') and Jorge Roca Suarez ('Strawhead'), and others were no longer regarded as local heroes pitting their wits in a relentless war against the interests of the United States; they were seen for what they were, killers.

Now there were calls for their capture and vilification, and many, many people turned away from the business in disgust. The people demanded justice for Noel Kempff Mercado and looked expectantly to Government to produce those responsible for his assassination.

The political opposition and the public, alike, demanded to know how it was that the security forces, and those specifically charged with the war against the pichicateros (The Minister of the Interior, UMOPAR, the Fiscal, the Sheriff, the police and the US Drug Enforcement Agency had taken so long to get to the scene of the crime? How was it that the first to arrive there were civilian volunteers? How was it that UMOPAR arrived three days after it was known that the Kempff party had been attacked and three of their members left dead? And how was it, when many people knew about the Drug Centre, Kempff did not? As El Deber (8th October) reported: 'In the opinion of one of his friends [Ruben Poma] the real killers were those who knew about the existence of the cocaine drug centre.

The Minister of the Interior Fernando Barthelemy Martinez's explanations for the unfortunate delay: unforeseen difficulties, orders and counter orders, were swept aside by an avalanche of public opinion and accusations that he received money by freeing many suspect drug dealers prematurely. [Even so, as a measure of the official connivance in the affair, it was not until 1992 that Barthelemy lost his seat in the Senate. With him went dozens of other officials, both civilian and military.]

People were screaming for the affair to be resolved. Four separate commissions were convened to investigate the triple assassination and the subsequent cover-up. The commissioners, themselves, were accused of corruption, as being made up of people desperate to hide the truth under layers of paperwork, smoke screens of false accusations and dubious theories. The whole affair had been politicised, one political party accusing another, until it was quite obvious to everyone that none of these official investigations were going to come to anything. The public knew what they had always known: the politicians could not be trusted; they were going to close ranks to protect one another.

Of the Congressional Committee's findings, the General Secretary of the National Union of Workers declared (El Mundo, 7th November), 'It is an insult to the intelligence of the Nation, a National disgrace.'

Then one man, who represented a very minor political party, stepped forward and promised the people he was not like the rest; he had, he said, got to the bottom of the scandal and would release his information in a television programme the following night. The public held its breath. This man, Edmundo Salazar Terceros, 39-year-old father of three children, and supplicant governmental deputy for Santa Cruz, had forced his way on to the Congressional Committee. Then, on the 10th November 1986, in the centre of the city, a few hours short of his planned revelations, Edmundo Salazar was machine-gunned as he entered his house, and died. Along with the loss of his life, went the loss of his documented evidence.

Now I must return to my more mundane world and let my new neighbour, Don Ezekiel have a good grouse, not the bird!

****

Now I've got to know 'im, that Don Robin isn't really so bad. And he knows a good piece of land when 'e sees it, that's how we became neighbours, I sold him a bit. I know I used to think of 'im as that twit from the British Mission but it's true what he says about all that ecological stuff, and it does all make you think.

For us, we campesinos, living 'ere amongst the disappearing forests of our beloved America, it's natural that our thoughts should often turn to this sad thing. What we 'ave seen is the alteration of our land from one completely covered by trees to great big weed-choked clearings, some of which rich folk have bought on the cheap and turned into grazing lands. We don't like what we see because it's a dramatic alteration of what we're used to, and change can mean uncertainty. Unconsciously, we feel that we're making a terrible mistake, that...somehow, all those trees that 'ave taken hundreds of years to grow are being wasted, and for what? We're not insensitive to the bond, the liking that humans 'ave for trees, and we, too find the open fields monotonous to the eye. We know that much of the wildlife's gone, a sense of loss that the last animals and birds, with their sad calls, remind us of it. Even amongst us, so used to them, the gobbling of monkeys, or the grating of Guaracas, is reason to pause for a moment to listen to them, smile, and, should we be in the company of a friend, comment upon them. We like 'em for they remind us that we're not all bound up with the cares of everyday life.

'What a pity it's all going,' we say, while we clear away a bit more forest to make the yuca field a bit bigger, or extend the grazing for our few cows. But there ain't no alternative; yuca means something to eat, pasture means milk and a bit of money in our pockets. Everybody knows that; why, even the value of my land depends upon the amount of space I've cleared. I know if I 'ave to sell-up, my cleared land will be worth much more than my bit of forest. Reasonable you see, because clearing land is hard work, and worked land must be worth more. Get it?

But don't get me wrong, now. There's advantages to clearing the land. Fewer snakes, fewer mosquitos, and less problem with damp. I don't get so many rats in the house, and it must be a good few years since we last had carachupas in the roof. It's cooler too. As you see I've left a couple of shade trees outside where I can sling my 'ammock and swing in the breeze.

I've still got a bit of forest on the bad land down there, and that's good too because it gives me a bit of firewood, and the pigs like it.

It's a pity though. It's not really small folks like us that's spoiling things, it's the rich folks; like that Ramiro, for instance. What a shame. 'Ave you seen what he's done? Cleared that lovely bit of forest behind his house; must be at least an 'undred hectares. And what for? For growing rice. 'E must be rich, aye?

Wish I was rich. My kids want to be rich. They don't want to stay 'ere; they're not interested in getting their 'ands dirty. They want to be computer technicians. When the day comes, and I'm too old to carry on, I s'pose I'll 'ave to sell my beloved 'ome and move to the city to be with 'em. Got to 'ave somebody to look after me and my missus in our old age. Thank God we 'ad enough of 'em; so many died, you see. But we're all right. We'll survive. At least, I 'ope we will. Bit worrying though. All these problems of the world and the ecology.

Don Robin's told me a lot. I lie in my 'ammock and go through all he's said ..., don't understand it all. I can't really worry about ozone, or atomic bombs, or pollution, or any of those problems I can't see. But I do worry about what's mine, and not only because I don't like change, but because I am told that the earth is getting 'otter because of what we're doing here, chopping down all the forest. Makes sense. Trees, shade, cool, you know. I wouldn't want it to get hotter, it's unbearably hot already. They say it's all got something to do with c-o-two.

Which reminds me, I must take advantage of the dry weather and burn off the field, and plaster those bloody fire ant nests with that stuff Don Robin gave me. They're a curse, and there seems to be no way of stopping the little blighters. Mind you they throw a lot of chemicals around out 'ere. They say you can't eat the blessed tomatoes, they're too poisonous. Wouldn't surprise me. There's no government control nowadays. You only 'ave to look at what they're doing to the forests. It's all illegal you know, at least much of it. But what can we do?

They say we 'ave to save the forests, but, do we? They say that in the USA and Europe they've planted millions of acres of pine woods. They say that in Russia there's a forest even bigger than the Amazon. Aren't all those trees enough? Do they really need ours as well? And if they do, what can we…I'm sorry, just going round in circles, aren't I?

So let's say they're right, we must save our trees to save ourselves, and we 'ave to figure out a way to do it. Sacrifice. There's a word we 'ear a lot. We'll 'ave to make sacrifices they say. I'm not sure what they mean by it. Sacrifice what? What I 'ave is nothing to sacrifice because I don't 'ave anything anyway. It's okay for them, they're rich.

I think I'll go over and talk to Don Robin, see what he's got to say. He knows all the words and could argue the hind leg off a donkey. If he's in a good mood maybe he'll invite me to a beer. He's always drinking the stuff. He's got money, not much though, you can never tell with these gringos. They're all rich.

Anyway, he's singing, he must be in a good mood.

There was an old woman who swallowed a fly,
To kill the fly she swallowed a spider,
To kill the spider she swallowed a bird,
To kill the bird she swallowed a cat,
To kill the cat she swallowed a dog,

To kill the dog she swallowed a man,
To kill the man she swallowed the oil,
To kill the oil she swallowed the argument,
To kill the argument she swallowed us all,
To kill us all she swallowed up greed,
To kill the greed is quite impossible,
as you know, know, know,
as you know, ho, ho.

So you want to hear a bit of common sense. Okay, Don Ezekiel, sit down. As it so happens, I've just been jotting notes down for that book I'm writing; got to say something about it. So here it is, straight from the dreamer on the logs: I have spent hundreds of hours thinking about the World's problems, more especially those concerned with the deteriorating condition of the environment, as we affect it here, with the uncontrolled clearing of our forests. I often read that the Third World can't afford to wait for the developed countries to take all the initiative, and we, too, must make sacrifices. This is fair comment, but I would have to free my mind of its cluttering prejudices first, and then consider where we, South Americans, could afford to make those sacrifices.

Okay? So let's review our prejudices, and then put forward a possible solution.

It is true that the problems of the environment, whether they be pollution, the loss of the ozone layer, or the greenhouse effect, know no international boundaries. Whatever the fate awaiting us, awaits us all.

It is also true we must forget the past, that the developed countries created their wealth at the cost of their own natural resources. It will not help us to copy them, for we must learn from other people's mistakes. The real problem is the developed countries don't learn from their mistakes. They're still going down the same old road, still gobbling up the oil, still over-exploiting the oceans, still demanding tropical hardwoods, etc., etc. Theirs seems to be a policy of 'Do as we say, not as we do'.

They say the economies of the richest countries will see a tenfold increase, Yes! a tenfold increase during the next forty years, insatiable monsters that will outstrip the world's last resources long before they are satisfied. Are we also asked to forget that very little of this economic growth is going to trickle down to us here? We already know the USA, with only 7% of the world's population, causes 60%

of the damage to the web of life, and that Europe and Japan take care of the rest. We know, to speak metaphorically, that if you take a cake to a child's birthday party, the American kid's slice will be fifty times larger than your kid's slice.

They say the world's population is five billion, twice as high as it was fifty years ago. They say it will double again in the next fifty. They say the greatest increase will take place among the world's poor, and that unless something is done half the people in this world are going to die of starvation. They want us to stop having kids. They want us to stop clearing the forest, but we do so, to feed the kids we have. Makes sense, dunnit?

They say if we clear the forest, the pollution in the atmosphere is going to get so bad that the earth will warm up, the sea level will rise, the water will become undrinkable, and the trees, themselves, will die. They want us to keep our trees to clear up the mess they make and are going to make whilst they make even more money. Fair?, No! What seems to be the trouble?

Okay, so forget the present too. We are used to having nothing, but nothing is better than being dead. So let's stop bleating and discuss this thing sensibly. Let's try to come up with a plan to save our forests that will satisfy the needs of the global environment and will be acceptable and fair to everyone, including us.

So what are we going to do? What are we going to do about the vanishing rainforests? All those forests covering the lowlands and mountains of the Third World, and what about their wildlife? To my way of thinking rain forests suffer two distinct processes, over-exploitation, and destruction.

Let's discuss exploitation first. The exploitation of forests is undertaken by loggers. There are two very different types of loggers, large corporate businesses (financed by foreigners) and smaller locally based logging companies.

The corporate conglomerates use their economic and political power base to obtain very large concessions (often running into millions of acres). With their huge machines and large work forces they exploit the forests in an unacceptable way. All too frequently they are not interested in managing their concessions to allow for long-term extraction. They are after the valuable hardwoods (for the lucrative part of the timber trade), and the soft woods for the production of processed wood products and pulp for paper. Theirs is a policy of get in there, get rich, and get out. This is often done with the intention of completely removing the forest (as part of the original contract) so as to provide cleared land for further speculation (sugar cane, or soya bean production, or more commonly beef production), or for colonisation projects. In other words, none of these people are interested in sustainable practices.

Large corporations should not be allowed to sell timber since the world's requirements could be supplied by local family businesses. They should restrict themselves to the production of pulp and wood-based secondary products, for which the supply of softwoods could come from agricultural clearing by implementing a system along the lines used for recycling any other waste products. Subsidiary businesses could be set up to cut, collect, and transport the logs to the local paper mills and wood processing plants. If this means a roll of lavatory paper is going to cost more or be ecologically smaller (not so wide) so be it, but at least more of the money is going to stay where it is most needed, in the hands of the local people.

The smaller city-based and rural logging companies work much smaller concessions (measured in thousands rather than millions of acres) or depend upon buying trees from the private sector (both large and small landowners). These family-owned companies (those with concessions, at least) have a vested interest in sustainable methods because they wish to pass their businesses on to their children. The problem for them is that they are rarely the owners of the land they work, and when they do obtain concessions these are often for a few years only. Whatever they do to maintain the value of their concessions (replanting trees, selective cutting, sparing the mature seed-producing trees, and other important aspects) may not benefit them directly.

The smaller logging companies could be made to manage their concessions more wisely by ceding them large areas with long-term leases of fifty years or more, but always and when an acceptable system of management exists. Government, or preferably independent bodies (like NGO's, of which there are many) could monitor their activities, and register the logs to prove that they have come from a recognised sustainable forestry project. Similar schemes are already planned for introduction in the near future.

The answer to the problem of the loggers, be they on a large or small scale, resides in the application of national laws to enforce sound management practices, and where this fails, international pressure. Since their businesses fall within the established commercial framework (which lends itself to legislation and control) sustainable practices can be enforced, or, better still, encouraged through the application of grants and subsidies.

The process responsible for the destruction of rain forests is a much more intractable problem involving the never-ending search for more agricultural land, whether this be for large-scale or small-scale arable farming. Neither group has an interest in trees, other than the immediate gains to be made by selling valuable hard woods felled during routine forest clearance. The softwoods are considered an unavoidable nuisance that the wealthy landowners bulldoze aside, and that the poorer farmers leave to rot, or worse, burn in situ. The recycling of the latter by intermediates (as mentioned above), or by the farmers themselves, could make a real contribution towards our efforts to conserve the forests which remain.

Large-scale arable farming is totally destructive because it has to be mechanised, and mechanisation requires tree-free, stump-free land from which forest can never regenerate. This practice is favoured by wealthy farmers since it allows the rapid conversion of the land to pasture when crop yields fall below economic thresholds, or when the bottom drops out of the market. But large scale farming, like logging, is part of the established commercial framework and can be controlled by legislation, grants, quotas, and subsidies. Again, money could solve the problem.

But what are we going to do about the campesinos and their subsistence farming? Individual campesinos don't threaten the future of the rain forests, but together they represent a destructive force larger than all other vested interests put together. Because they are considered too small to form part of the established economic framework it is very difficult to control their activities. Having said this, it does not mean that ways to bring them into the system cannot be found, and in the long term I believe this will become the principal factor in any initiative to deal with the campesino problem.

South America has about seventy-five million people working in the agricultural sector, most of them depend upon subsistence farming to supply all or part of their needs. To supply the demand, it is common practice for South American countries to hand out between twenty and two hundred acres to each and every campesino family. Very often the land ceded to them is selected for political motives (settling the frontiers, vote rigging (as in the case of the

Colla-Camba situation) rather than its suitability for farming. About 1,000 million acres are already in the hands of these poor peasants, and the demand for more land will greatly increase amongst them because they are expected to account for most of the world's population increase. The expected shortfall of land to satisfy the burgeoning rural population will be exacerbated by families seeking resettlement as the land they already own becomes worked out.

Neither is the plight of the campesino farmer likely to be improved by technical innovations. So far nobody has developed an economically viable solution to the problem of subsistence farming on Latin America's poor soils. Many research projects investigating the problem, like the British Mission in Santa Cruz have come to realise the difficulty, and not only have been unable to come up with anything new, but have also had to recommend a return to the traditional practices adopted by the first people to arrive in South America thousands of years ago: slash and burn in the lowlands, land terracing and irrigation in the highlands.

Even so, I would maintain that the problem of the campesino has not and is not being faced up to because it seems to be beyond our economic will to do so, and for political reasons. For the moment the policy appears to be one of denying the campesinos real suffrage by keeping them poor. Most South American governments are terrified of coming to terms with, what they call 'the Indians', who through lack of education, and a history of exploitation are generally resistant to schemes that require their voluntary participation. Thankfully, they are also deeply suspicious of those who try to exploit them, like Peru's Sendero Luminoso.

And whilst we are talking about voluntary participation, and just to get it out of the way, we should discuss the "let's-pay-them-all-to-do-nothing" philosophy that is gaining ground amongst the fluffy-headed Neoliberals, including myself.

Whereas the Latin character may lend itself to being paid to do nothing, any attempt to manipulate their right to do as they like (which includes having lots of children) will result in a war more literal than the use of words. The South American does not like to be manipulated, and will not be, unless he sees a marked improvement in his standard of living and some way to survive his old age with dignity. So paying them all to do nothing is going to be a very expensive business.

How expensive?

If we consider some of the poorer countries in S. America only (Colombia, Ecuador, Peru, Bolivia and Brazil, where the problems are most acute we find that the average income is $1,110 per annum, with the Bolivians the poorest ($400), and the Brazilians the least poor ($1,610). If we are going to buy the cooperation of the sixty million or so subsistence farmers in these five countries by paying them an acceptable minimum income, let's say a modest $1,200 each (to include a no more than two kids clause), our total cheque would be $80 billion a year. We might need about $20 billion more to implement the scheme.

The figure of $100 billion annually might convince adherents of the philosophy that it is totally unrealistic. After all, total AID relief to South America is only in the order of $1 billion at the present time.

But this does not mean we should completely forget such ideas, because today's problems are urgent for us, tomorrow's critical for our children. And whatever scheme be proposed, it has to be said that the fewer of them (children) the better, whether they be rich man's or poor man's.

What we need is a scheme to save the Neotropical forests that is non-coercive and self-propelling because it will benefit those making the effort to make it work, and for which funding can be gradually phased out (or diverted to other needs) as it takes hold. The scheme would also have to ensure that the money gets to the people themselves, and not swallowed up by inept bureaucracy and corruption, faithful bed mates to any South American initiative.

Since we are talking about saving forests it is necessary to make several important points about them. There are two ways of defining forest: large wilderness areas of virgin or semi-virgin forest (which are normally state owned) and privately owned rural lands which are partly forested, or at least have the potential (through regrowth) of returning to forest. For the sake of our planet, we must be pragmatic and accept that we cannot only include the state-owned wilderness forests within our scheme because in most South American countries such forests are too few to provide the global service we require (i.e., enough vegetative cover to lock up enough carbon dioxide to ameliorate the greenhouse effect), and unless we help the campesinos they will sooner or later be forced to invade all state owned forest in their search for land. Indeed, I would go further, and say of the two the need to persuade the campesino landholder to look after his forest, and extend it by allowing regrowth is by far the most important, because through regrowth carbon dioxide is absorbed from the air and the carbon sunk into the plant's tissues. Also, we should not forget the importance of privately owned forests (that can be very extensive) in this context.

Instead of agriculture, we could base campesino settlement on low-tech methods of sustainable total forest use. Low-tech forestry is ideally suited to campesino needs and philosophy, and sustainable forestry geared to stability and long-term needs. Sustainable total forestry must include the harvesting of secondary forest products ((pharmaceuticals, gums, resins, fibres, fruits, nuts, etc) to make it totally viable. As most of these minor products can only be gleaned in small quantities, only the poor campesino, motivated by small daily exigencies (fags for dad, shoes for mum, school fees for the kids), is going to bother with them. There is a good market for secondary products, but no traders to complete the link between campesino and merchant. Secondary products aside, campesinos would still need to log their lands for their bread-and-butter income, that at current values could be supplied by three or four good-sized logs each year. Using low-tech equipment, such as portable sawmills (which are ecologically acceptable because they waste less wood than traditional sawmills) by cutting planks in situ, and bullock carts to haul them to the nearest collection point, most of the casual damage inflicted on forest ecosystems by the coming and going of heavy machinery would be avoided. With the campesinos organised into syndicates and cooperatives (as many of them already are), essential components such as equipment pools, logging quotas, logging densities, and other technical requisites could be established to keep costs down, and provide what we are looking for, sustainability for man and nature both.

A scheme to cover all of South America's forests could be financed along the lines of the debt-for-nature-swop idea, that in its simplest form consists of cancelling X millions of dollars of a country's national debt in exchange for strengthening their conservation effort and establishing new protected areas. My proposal is a logical expansion of this idea whereby countries will be encouraged to conserve their forests by paying them a fixed amount of money (or by giving them an interest-free loan to the value of this amount) for the total area of forest they maintain.

At the technological level the assessment of each country's worth (in terms of forest cover) is a relatively simple procedure. One computer linked to any of the satellites which already monitor the world's surface could supply all the data required. Under the scheme each participating country might receive, say, $1 million for every thousand square kilometres of forest they maintain. Most of this money would finance the campesino sustainable total-forest-use scheme, and it would be paid directly to the campesino cooperatives. In other words, we would pay people to offer a service, not governments.

Under our scheme, then, local cooperatives will receive the money, and they will truly benefit when their efforts become efficient. The more forest each community sets aside, the more money it will receive. Some of the money would be withheld to pay project consultants (mainly rural development and forestry experts, conservationists and ecologists), and to provide such things as managers for the cooperatives, improved subsistence farming (for which there will be a need), and environmentally friendly cottage industries (together with master carvers, carpenters, weavers, and artists to pass on their skills), and research aimed at improving forest management and marketing unusual products. The generation of income from their own forests would provide sufficient incentive to ensure the success of the scheme, and a point will be reached at which the campesinos can't free themselves of AID dependency and the scheme can be run down. That there will be many glitches to overcome, and that it will take many years to arrive at this point there is no doubt.

Bolivia has about 500,000 square kilometres of forest that, using my rough figures would earn them $500 million a year to share out amongst the campesino cooperatives. At present about one million campesinos would benefit from the scheme, each receiving the equivalent of $500 a year for their participation. Not a bad incentive for a family whose income is less than this, and whose members would still be free to augment their collective earnings with traditional farming. The money for implementing and monitoring the scheme, and for additional research into forest harvesting methods, could be made available from existing AID budgets, and by switching the terms of reference for many on-going agricultural research projects. For Bolivia, at least, an attractive proposition that would be well within its practical capability, and consonant with the recent emergence of politically aware campesinos. Now it would be the campesinos themselves who would put pressure on the government to maintain their forests, because their livelihood would depend on it.

Colombia, Ecuador and Peru, together, would require a further $1 billion annually to implement the scheme; not unreasonable considering the benefits to the world as a whole. The country which presents the biggest problem is Brazil. With 5 million square kilometres of forest, and its huge relatively well-paid population, about $5 billion would have to be found to put the scheme into practice there.

Even so, if our total of $7 billion per annum to pay for a scheme that could save the forests of the Amazon Basin was three or four times more, it would still be a lot less than the $100 billion the United States Department of Defence intends cutting from their 1994 military budget And please take notice, cutting its budget, not financing it, that is much more!

Do I hear you say, 'Visionary hyperbole'? Well, quite frankly, it is, but as many an expert will tell you it's the lack of vision that's the main obstacle to a healthy World Climate.

And if vision has no place in today's world, what is the alternative? Should we keep going down the same old road when the collective evidence shows that if the road is leading us anywhere it is leading us to destruction?

All these ideas could be implemented in practice. The visionary element is the part dealing with the politics of the situation. Wealthy countries won't come up with the money, not even for pilot schemes, and receiving countries won't look at it because they wouldn't want the economic power behind the scheme to fall into the hands of their poor people. And yet it is the knowledge that they are forgotten, expendable as far as World Trade and politics are concerned, that sows the seeds of discontent among the poor. "Money talks, bullshit walks", and the campesinos know this and resent it. They tolerate the abuses of their own national governments, because, for them, it's a familiar scene with its back cloth of inefficiency and corruption. They view international concern about the clearing of forests as just another gringo plot to disinherit them of the little they've got. I would maintain that this attitude is one of the main factors mitigating against our ability to resolve the problems. The scheme proposed here would put the onus on to the campesino, make him a participant in world affairs, give him a sense of purpose to life, give him back his dignity.

Lastly, it should be stated that minor attempts have been made to implement similar ideas without success. Of course, there will be failures to begin with as we are dealing with elements we know nothing about. And we don't know anything about it because we are stuck with a bureaucracy deeply rooted in tradition: helping the Third World means improving their food production, and of course this means jobs for the boys, my indispensables, the agronomists. And of course, money will be squandered through inefficiency and corruption, that's part of life here. But these are not reasons to abandon the only hope we have of coming to terms with the problem. But, as I inferred at the beginning, there are just too many congenital nit pickers for proposals like this to be taken seriously, and nobody does.

'I do.'

'Thank you, Gunga.'

Author's note. Many of these ideas, written 30 years ago, now form the basis of Carbon Sequestration schemes

8. Return To The Park

The heavy rain of yesterday has left the surface of the path with a slick layer of mud. We are forced to follow the narrow trail pressed flat by the feet of the Campesinos, who having kept to the slightly higher ground, have left a wiggly line down the winding track. Our slow progress was further delayed at the many swampy depressions where the footprints of both humans and horses give way to deep holes gouged in the mire. "Why is it", I say to myself, "that the people who use this path every day have never made the slightest effort to improve it?" Day in and day out, dozens of villagers living in Amboró National Park are quite content to slip and slide their way to town and back. Only occasionally do we come across a spot where somebody has momentarily stopped to cut a few "Patujú" leaves with which to cover the mud, or thrown down a branch to offer an uncertain footing.

My musing is suddenly distracted by a loud sluumph and Miriam's preferred expletive, "fuck!". I turn to see her splayed out on the mud. Then, as all South Americans do after minor disasters, she gives way to a spell of unbridled laughter. I put my rucksack down, reluctantly, because the trail we have been following is not wide enough to prevent it being fouled with muck. Miriam is stuck; and as we quickly discover when I pull her out, missing one shoe. The problem is that in her stumble she has covered quite a few yards and now has no idea which of the many knee-deep holes has custody of it. Half an hour later we find the shoe and withdraw to the nearest puddle of stagnant water to wash ourselves as best we can.

I stand looking down at her, consumed as she is every few minutes by a fit of giggles. I'm not sure whether to join her or to give way to my growing impatience. I have just decided to predispose myself to the latter when I become aware of a growing clamour behind me. Too late; I'm lifted up and slammed against a tree by a cow, where I am ignominiously held, my feet hanging ineffectually a few inches from the ground. Miriam, now lost in renewed paroxysms of laughter, can't help me. Then alarmed by my uncharacteristic lack of indignation, my quiet acceptance of this latest humiliation, her laughter is suddenly cut off. She springs into action and belabours the cow around the head with a rotten branch. Helpless, I try to watch the proceedings over my shoulder, but my rucksack blocks even this germ of participation. Oblivious to Miriam's efforts, and having established a certain degree of superiority over me, the cow and its calf back up, and I drop back to my feet.

"Are you hurt?"

"No, not really. Luckily the damned animal batted me in the rucksack."

Free from anguish, Miriam gives way to a particularly undisciplined burst of laughter, that for the sake of my wounded pride I try to emulate. The cow, resentful of our levity, moos once and runs off down the track followed by her calf. My attempt at good humour having failed, I vent my spleen on the young lad whose warning shouts came too late to save me from such prosaic degradation. If I had been pinioned by a truculent Tapir my image would have grown by the incident. But a cow! My spleen continues to rumble in unison with the departing thunder somewhere to the north of us. Only when I remind myself that we are on our way to the Park, and the best trips are those that start badly, does my rancour slink off to the shadows to await another opportunity. Miriam does her best to encourage its return; contemptuous of my verbal abuse of the child she answers my questions with perfunctory sullenness and a lengthening of her stride that I find difficult to match.

My mood takes an unconscious upturn when we pass a Cattle Egret sitting on the thatch of an abandoned hut. I am sharply reminded of a curious incident nearly forty years ago in the jungles of Sarawak, that encapsulates my unappeased desire to move all these colonists out of the Park.

****

We had been exploring the headwaters of the River Baram, somewhere in the vicinity of what is now Gunung Mulu National Park. Going from longhouse to longhouse we received a cordial welcome from the friendly River Dyaks, who stuffed us with food (rice roasted in bamboo, sweet yams, manioc, and chicken) and got us drunk on "Sake". They placed the prettiest of the village's maidens to attend our every need. I kept mine busy preparing cigarettes with their feral tobacco rolled up in bamboo leaves. And filled with unquenchable curiosity they watched our every move, including the attendance of the whole village should one of us slink off to the bushes for a crap. All of the longhouses we visited had apparently been full of happy, healthy people. Then we came to another, dirty, dilapidated, its roof partly fallen-in, where the natives seemed to be under some sort of blight. The food was poor, there was no tobacco, and the whole place emanated a profound sadness.

"Why", we asked, "are you living in such misery?" They said they had fallen on hard times owing to a natural calamity. Earthquake? Typhoon? Disease? I wondered what they could mean.

About a year ago they had moved out of their present longhouse and gone to live in a beautiful new one about ten miles away. There they had laid out fresh fields of rice, planted manioc, peanuts, tobacco, and fruit trees. The whole village had been happy and prosperous. Then one morning a white bird had perched on the roof. To the village shaman this was a terrible omen which forced them to abandon their new home without a moment's delay. We had found them back in their old place; short of food or comfort and wracked with illness. The headman doubted the people had the energy to build another longhouse because most of the strong young ones had gone to join relations some thirty miles downstream.

****

I don't hurry after Miriam to tell her what it was I had recalled. But I do conjure up from nothing the perverse satisfaction that there are a lot of white egrets in Amboró, and maybe those Dyaks knew what they were talking about. Being a Camba, Miriam would approve the sentiment. I am stopped from further reverie by recalling a quote I had just come across from the Bible: "I will never again curse the ground because of humanity" (Genesis 8:21).

And: "Dear Wormwood, we want a man hag-ridden by visions of an imminent heaven or hell upon earth, ready to break the Enemy's commands in the present, if by so doing we make him think he can attain the one or avert the other, dependent for his faith on the success or failure of schemes whose end he will not live to see. Your affectionate uncle, SCREWTAPE".

I feel free, however, to curse the muddy path and the shabby campesino farms to either side; a constant reminder to me of the Bolivian Government's failure to safeguard their protected areas with the vigilance they deserve. It was seven years since Noel Kempff had been murdered; and as long as the Government continued to turn a blind eye to the problems of campesino squatters and illegal logging it wouldn't be long before someone else was assassinated. Chico Mendez died because those who saw profit in his demise felt free to kill him.

As Chico said: "Stop! That's enough! That cry in our throats as the new year begins. Enough of death threats and killing for profit and personal gain; destruction and misery; the violence of development projects and ecological holocausts. Enough of the government's abdication of responsibility and connivance with all this death and destruction. Let's all get together and put a stop to it! Let's build an alternative to all that!"

Noel died for much the same reasons. People with vested interests saw him as a threat to their livelihood, as a man who would sooner or later persuade government that the country's National Parks and wildlife should be sacrosanct and its forest reserves only available to those willing to harvest them in a sustainable manner.

But let us not stop there, because the problem is a worldwide one. As long as people want mahogany doors and pet parrots there will always be someone willing to supply, and woe betide those who get in their way. Educate the people and the bureaucracy by making them understand the forests for what they are, a valuable but vulnerable natural resource, and we will go a long way down the road towards solving the problem. Tropical hardwood doors and furniture are not addictive, and pet parrots not habituating, but in the long run the desire for them may be more destructive to our health than alcohol or drugs.

The old arguments parade through my mind as steadily as my feet plod towards the Inner Reserve. Only there, I know, will my spirits really lift; unencumbered by the sight of half-hacked-down forests and weedy fields, my eyes will send back happier messages.

But for how much longer will the Inner Reserve be there? There are strong rumours that government will yield to the demands of Amboró's itinerant Campesino population and give them the last of the lowland forest now protected by the Red Line. In other parts of the Park large numbers of new colonists have taken over thousands of hectares and government has failed to take any action to dislodge them. The Campesinos are not fools, sooner or later they will really get organised; and free of official censure will grab what they want.

Even now, as Miriam and I stride past their farms, we are met by hostile eyes and impertinent remarks about our right to be there and our continued pilfering of the gold that they claim belongs to them.

The loggers and the animal dealers can be dealt with. The United Nations could establish a Nature Crimes Commission. But: what are we going to do about the millions busily felling the forests for food and polluting the rivers in search of gold? As far as I can see we don't have many alternatives. We could kill them all. Or we could share our wealth with them. "Kill them. Help them. Kill them. Help them", my improper thoughts seesaw around my head in time with my step.

We throw down our rucksacks beside the Isama River. At last, after three hours of forced marching, we have left behind the last of the campesino holdings. In the old days we used to come up the river; now we use the well beaten track. In the old days the river route took five or six hours, but they were hours well spent admiring the plants and sharing a moment with the highly coloured butterflies and birds attracted to the water. In the old days only heavy rains stirred up the ooze; we stepped through the clear water and where we placed our feet we disturbed many fish. Now the water is permanently muddy and fish only found in the deepest pools. Where a cluster of rocks had been marked by the wet prints of Otters and other animals, or signposted by their rank smells. Now these same rocks are marked by abandoned pieces of soap, empty shampoo sachets, and plastic bags, and human faecal odours emanate from the adjacent shrubbery. Where the heron stood, now stands the woman, menstrual blood coursing down her thighs as she unashamedly raises her skirts to wash herself clean. Once there was a gringo and his girl, always fighting and berating the campesinos, confiscating their axes and guns; now the same two walk humbly past, filled with memories and thoughts about what could have been.

I sometimes turn to self-pity to explain my failures.

Beside the rock beside the river where we sit, my hummingbird and me, chewing over memories of better times, the rain or the river (I know not which) has left a muddy pool. The gently inclined sides are home to a variety of tiny animals, mostly beetles and collembola. With a mouthful of stale bread and a morsel of cheese halfway to my lips I notice something very unscientific happening before my eyes. I say unscientific; what I really mean is that my fellow entomologists would say I am being unscientific when I tell them what it is I am watching.

It is a bright, crimson-coloured dragonfly. Because of its colour I know it's a male, the females of this species are green. Dragonflies hunt on the wing and will follow animals and humans because their activity puts small insects to flight. But here is something different. Here the dragonfly is deliberately creating its own disturbance. He is methodically working his way round the edges of the pool, splashing the sides with water. He dives towards the surface and flicks the water with the tip of his abdomen. Every now and then he snaps up a small insect washed on to the surface of the pool. This is a new one for science: an insect using a tool (the water) to capture its prey! But will they believe me?

You think I'm kidding? What about this from a fellow bird watcher only a few days ago: "Do you think they believe you when you say you have seen Horned Curassows in the Park?"

****

Miriam wants to go. I can't tear myself away. I tell her what the dragonfly is doing but it means little to her. Yes, she is interested, but she doesn't appreciate that this is a scientific first, a real discovery. Or it would be if I were willing to kill the dragonfly, send it to the British Museum for determination (that's scientific jargon for getting it named) and write a field note about it to some entomological magazine. But I won't. It's not laziness on my part. It's something more complicated, personal. Only madmen try to share their visions with somebody else. The dragonfly represents part of my vision, part of the magic that makes my life worthwhile, like my love for Miriam. Unshareable. Some say unbelievable. And maybe they wouldn't believe me.

Well, I should have stuck to these sentiments, but eventually decided to ask a friend to put me in contact with somebody who would be interested. The obvious choice was Mr Pool, yes, that really was his name, the Worlds leading dragonfly expert, who replied:

"Dear Mr Clarke, this sounds to me very much like scoop oviposition (i.e. egg-laying behaviour)".

Yes, he believes males lay eggs!

****

Now that the dragonfly has gone, and my cigarette smoked down to the filter, I have no excuse to tarry longer. But a strange lethargy has overcome me, partly due to the heat, partly because I am aware that we still have another five hours to walk, but mostly, I think, because I am tired at heart; sick of corrupt Bolivians, sick of nescient colonists. Sick of the constant fight to save the things I love: Miriam, my lifestyle, the Park, the trees and the wildlife, and my freedom.

An all too familiar sensation now creeps over me, anger! For a moment I wonder why. Then it dawns on me, we are sitting on the Red Line. But this is not where it's supposed to be, it's been moved! It should have been half a mile back downstream. I get out the map, but Miriam is there before me: 'Yes, Robin. Nobody else cares like you do. Nobody here thinks like you do.' She comes over and puts her arm around me, kisses the sweat from my brow and looks into my eyes. I slip my arm around her waist, and there we sit, silently waiting for my bile to sink back down.

Marmalade Brown, of East African renown, always said that the white man had to go through three phases in his relationship with Africans: the first, short and sweet, in which the Africans could do no wrong; the second, sometimes permanent, in which the Africans could do no right; the third, a wiser course, in which the Africans could go and do what they bloody well liked.

Something passes between us. An unspeakable truth. I love her but am unable to feel her love for me. Maybe this is a man's problem, not his capacity to love, rather his failure to feel loved. It's a one-way relationship which will always keep us apart, isolated within our own psyches, completely alone.

I look at the trees. I can't see two the same. They are all different. Not for them the solid companionship of temperate woodlands. They also stand alone; and like many of us locked in a grim battle with their neighbours A fight for light which sooner or later must end in the death of the loser.

I examine my relationship with them. When I was a kid, I prac- tically lived in them. "Where's Robin? No, don't tell me." My mum shouting towards the general direction of the woods at the back of our house. "Come down, Robin. Your lunch is ready." Here it is very rare to find a tree to climb. The first branches are always way out of reach twenty or thirty feet up. Do I know them? No. How can one. You can't pick their leaves and taste them, or crunch them in the palm of your hand and smell them; you can't even see them, just yards of trunk disappearing into the canopy overhead. They remain impersonal, any attempt to communicate with them like staring into the ocean or looking at the stars, an unsatisfying experience. Do I like them, love them? "I love trees", I say to myself. But do I? When they are clustered together there is something in my psyche that makes me feel threatened. And not only me, I see it written on the faces of my companions whenever we are in the forest. The trees separate us from the sun, block out the view, and hem us in. But do I hear you say, monkeys live in the canopy where there is sunshine, open space and security. So, if Darwin is right, like monkeys we should feel at home among trees. True, but our closest ancestors were terrestrial savanna species, like Baboons, for them isolated trees were their friends and they escaped to thorny Acacias to avoid being eaten. And it's true, I love isolated trees, sometimes kiss them, often stroke them, always apologise to them on those occasions when I have to hurt them. I have a one-on-one relationship with trees, but lots of them repel me. In this I don't think I am alone, people like parklands, forests frighten them.

****

My silent reverie is interrupted when Miriam nudges me to draw my attention to the call of Titi Monkeys somewhere to the side of us. Their turkey-like gobbling, which continues for several minutes, is answered by other more distant ones. We count them. There are five or six groups now calling across the top of the trees, all of them back in the direction we have come. This species keeps to the lowlands, leaving the mountains to the Spider Monkeys, Capuchins and Howlers. Once, I am told, there were many more. The Cambas didn't kill them. Now they are being persecuted by the colonists who say they eat their maize. I know Squirrel Monkeys do, but Titis? Never.

We turn away and look at the mountains which we haven't seen clearly for some time. For the last three months there have been so many fires that smoke has hidden them from view. The Colonists have been burning off the forest in preparation for the planting season. Recently I flew from Sao Paulo to Santa Cruz, a distance of more than a thousand miles, without once seeing the ground for smoke. Here, too, the palls of smoke from the lowlands drift hundreds of miles, even reaching La Paz high in the altiplano; where visibility got so bad they had to close down the airport. Doctors complained about the sharp rise in the number of bronchitis and asthma victims and the newspapers were making weak appeals for the government to do something. But every year it's the same: chop chop chop, burn burn burn. The whole world is going up in smoke.

Now the thunderstorms of the last few days have put all the fires out and the smoke has thinned away to a light blue mist. We are close enough to the mountains to pick out the individual trees that together cover the slopes in a thick green carpet. Only here and there can we see bare rock faces or cinnamon coloured erosion scars like rat bites out of a birthday cake. My eyes repeatedly return to them, mesmerised by their repellent presence. I dread to think what the place will look like when, if, the colonists invade these mountains.

If the British Government has its way, that's what they will do. You think I'm lying. I can only suggest you write to the Ministry of Overseas Development and ask them about the CARE project in Amboró National Park. Ask them why Her Majesty's Government financed a multi-million-pound CARE project designed to establish migrant squatters in Amboró. Ask them if they understand what CARE means. I do, it means Care and Relief EVERYWHERE. Yes, ev'rywhere, including National Parks! Let me know what they say about this very "care-less" use of public money.

That reminds me, what can we do about the Salesian priests? They have been encouraging the locals to disregard Amboró National Park and take what land they want. They don't seem to understand the fragile nature of these hills, with its loose sandy soil and heavy rainfall, nor their importance to the stability of the regional weather patterns. Put quite simply, they don't seem to understand that man depends upon his environment. It can never be otherwise, the place in which man lives will dictate whether or not he survives. These priests, who say Man must come first, are attempting to make Gods of us all and flatter us with a doctrine whose fallacy is denied by the simplest of our ancestor's parables: "The Goose that laid the Golden Eggs".

So yes, all my thoughts end in those twin feelings of frustration and anger. "But what's the point of feeling angry?" I ask myself this same question dozens of times a day. Anger, it's always lurking just around the corner of my mind, waiting to come out and pounce on it's unsuspecting victims. Why can't I cool it? Everybody else seems to have settled for what they've got. Bolivians rarely get angry or, rather, rarely show it. They allow it to smoulder away until it can be appeased by revenge, usually in some petty form more often harmful to themselves than their intended victim. In other words "they cut off their noses to spite their faces". I sometimes get cross with the maid because she persists in some irritating deficiency in her work. I tell her off. Two days later she quits. She goes to work for somebody else who abuses her all day long and pays her half the wages I did. Then back she'll come, full of remorse, and ask for her job back. She'll correct her mistakes for a time, but harbouring a grudge it's not long before the deficiency returns. My anger is not like that, it's a subterranean furnace whose heat will never die, because the fuel that supplies it is an endless reservoir of other people's apathy, idleness, corruption and damn right dishonesty, catalysed by their latent stupidity and lack of education. Then there is greed. "Oh, foul despoiler of mankind's hopes".

Ancient civilisations included thinkers, wise men and women, among their most privileged people, the work of the muses, who thought about how to make life better, the country safer, stronger, better educated and artistic. Nowadays they don't pay anybody to muse or think, or at least they say they do, they call them "Politicians!"

A slight breeze stirs the vegetation at our backs and skitters across the river on a path of glassy water. Miriam tugs my arm. We rise, shoulder our packs and wordlessly proceed along the rocky beach. We both know that the next few hours will be difficult ones. The rain from the last storms has yet to drain from the mountains, and the river will be deep and difficult to cross in the narrow gorges up ahead. Fortunately, we are able to avoid many crossings by following a crude trail cut through the adjacent forest. This trail was made earlier in the year when a rescue force of more than a hundred people went in search of three Boy Scouts lost in the mountains. Even so, I swipe repeatedly at the encroaching vegetation with my machete. Palm leaves, Elephants' ears, stringy vines, and fallen limbs conspire to block our way. One ubiquitous thorny vine, I call it "trouser-ripper", eventually has its way and opens up an L-shaped hole in my new trousers. My face, neck and arms are covered with repellent, but the exposed skin of my thigh, intermittently bathed in the river, remains a naked invitation, a window of opportunity, for the mosquitos and horse flies waiting to exploit any advantage.

An hour later we are sitting on the rocks again, staring at the river. We have already crossed it five times, but here, where it bursts from a gorge, it's narrower, and looks dangerous. The water is muddy and hides the rocks and snags. The turbulence swamps the stillness marking the deep pools. Downstream looks even more dangerous, a long series of white crested rapids crashing between large rocks, with a few ominous whirlpools busily collecting the foam and debris the river brings down. We look at each other. 'Should we go on?' is written into our concerned expressions.

We haven't spoken for an hour, and there is no need to now. We both know. I motion Miriam to stay where she is while I move off to cut us each a stout pole. I return. We hold each other tightly around the upper arm and step into the flood. We are too experienced to hurry or panic. We proceed very slowly, each of us waits until the other has stepped forward and found a solid footing. Only then does the other one move. I have her on my upstream side, on my left, leaving my pole free to counter the water's force. To lose our footing, and go with the flow, well, we don't want that. The water rises above my waist and touches Miriam's breast. We glance at each other with increasing frequency and rising consternation. We have to edge a little downstream to permit Miriam's passage around a submerged boulder. But there's another to my right, and rather than risk the loss of my pole's comforting support, Miriam sees the necessity of climbing up and over the boulder. As she does so, throwing more weight on to me, my stake gives way, and for a second or two we struggle to keep upright. I stab at the riverbed with the stake and quickly find a new purchase, but it's not a good one. We start to go. She flashes a look of terror at me, I try to flash confidence back. Then the stake catches again, and with all my willpower I resist the current and force us upright. We seem to spend an eternity correcting our relative positions. During the ordeal Miriam has been swept to my right, and after a telepathic note from me, she very carefully makes her way to my left. We shuffle forwards like two drunks, the water tugging at our bodies. Miriam flashes another unspoken question to me. I silently tell her it's all right. A large indigo coloured hornet inspects her ear. Leave it, Miriam. Don't bat it away. Please. Fortunately, it decides the hole is too waxy to make a good nesting site and side slips away. We back up again to avoid a snag, but now we are approaching the shallows, and the manoeuvre is not difficult. Even so, here we can expect deep pools, and we continue to proceed with exaggerated care. Miriam finds one on her left, we sidle to the right. We let go of each other's arms, the breaking bond almost painful, but we're safe now. I go ahead, the water falling from my waist to my knees, the tops of smooth, rounded rocks now above water. We keep to the river for another thirty yards to gain a slight dip in the bank that offers us the opportunity to climb out. We do so, the water gushing from our clothes makes the clay bank slippery and I reach out with my stake to help Miriam up.

Oh, the relief of solid ground. We stand there a moment to look at the river again. I know she is doing the same as me. We are making mental notes that could be the difference between life and death: the speed of the current, the amount of white water, the level of the river against a brightly coloured pebble stuck in the bank. We store this information away. It will help us gauge crossings yet to come and enable us to make wise decisions on our return. I hope the rain is truly over and our mental preparation superfluous, but to treat the river with contempt is folly. Nature expects respect; pay her her due and she will see you through. Just to make sure I cut a short stake, peel a ring of bark away from the middle, and stick into the sand at the edge of the next beach, the ring flush with the water. Now we will know exactly how much deeper, or shallower, the water is on our return. Miriam smiles and looks up at the cloudless sky, that wanders overhead like a winding blue sheet. The canyon with its thick covering of vegetation would have been considered a World Wonder in biblical times. A huge, wild, "Hanging Garden of Babylon". Now it's just another piece of tropical wilderness.

We walk on. "Fear not ye who passeth here". We try not to, fear I mean. But the way is full of small reminders that death, when it comes, can be sudden. A fallen tree. The clicking of a mini-landslide. The sharp crack of a pebble that has fallen from the top of the cliffs. A tomb-sized boulder surrounded by pieces that were once its own.

I remember advising one young ornithologist who had accompanied me to the Park not to sit in the place he had chosen, on a rock just below a V-shaped hanging gully. Such places were doubly dangerous, not only for the frequency of falling rocks, but also because the sides of the gully funnels all these towards its centre. He didn't want to move. I left him to it. Then I called him over to where he could see a pair of Cliff Flycatchers I was watching. This interested him, and he did move. He had hardly reached me when there was a roar like an oncoming train and we saw a huge limestone boulder come over the lip of the cliff. It struck the rocks where moments before he had been sitting and, even from our relatively safe vantage point, we had to duck as pieces of rock splinter flew in all directions. The gorges of Amboró are dangerous, especially if one does not take heed.

The Park has hosted hundreds of visitors over many years and still no warning notices have been set up to advise people of the dangers.

Then there are many minor hazards: a particular sort of rock that seems to offer a sure footing but is always very slick, wasp nests that hang above the trails at chest height, exactly at those places where one has to stoop to pass; or others, impossible to detect, that one species makes on the back of a leaf, hiding their comb with an exact replica of the leaf's surface. And wild bees. All these can be dangerous in tight spots.

Once, in thick jungle undergrowth, my companion and I were attacked by African Bees as we made our way along the edge of a cliff. My friend unwisely stepped on a fallen palm trunk lying across our path. His foot broke through the hollow trunk and within seconds we were engulfed by the angry swarm. We only had one option, and we took it. We threw ourselves off the cliff into the bushes some twenty feet below.

Miriam and I read all the signs as a bag lady would a New York street. We skirt trouble, but unlike her, we do not touch the many objects of minor value. That blue orchid flower. The sweet guapomo fruit. That rock crystal. The hallucinogenic vine. That lump of Curupaú resin. The hummingbird nest. That mica plated pebble. The tiny poison arrow frog, its black livery adorned by two amber coloured patches. That Jippi-happa. The seed of an Oxe-eye vine. That Vanilla plant. The drowned Morpho butterfly: its lifeless blue wings offering to those who know a small fortune locked into their mystery. And would that be a glint of gold? Fools! Hurry away, for we are not here for that.

We are in the forest, grappling our way up a steep rise. Now it is our hands that proceed with utmost caution. Left one…reach out…inspect the hand hold…grip. Right one…extend…eye that inviting stem…no, its home to a most pugnacious ant…but not that one either, several fudge-green caterpillars with stinging hairs (and how they sting!) have taken up abode on it. This one…hold…wait …grip. Now I pull my body up. We proceed automatically, the decisions made with lightning speed, the movements more laborious. In five minutes we breast the hillside and slide down the other side to the river. Another deep crossing by-passed.

But I have to turn back. During our uncontrolled descent a Grey Tinamou shot up with the alarming explosion of a Grouse put to flight. I search around for a few minutes before I spot its eggs, like bright blue billiard balls between the buttress roots of an Almendra tree. Only now do I regret my camera's absence. They lie there, like the corner of some fairy queen's boudoir, seven azure orbs more beautiful than jade, damp rotten leaves for a casket.

I recall Fawcett's uninformed remarks: "the abundance of fish and game at the mouth of the river [Verde] deceived us. With care our scant food supply might have lasted for three weeks but, the peons are voracious and eight days later our supplies had gone. On October 2 [1908] one of the dogs found a Gray Tinamou's nest with four large sky-blue eggs in it. The dog received one as a reward, and the other three did very little except make us more conscious of hunger…we were starving now".

I would like to rename the bird Fawcett's Tinamou, for just as the Cliff Flycatchers saved my friend, those eggs, as Fawcett himself intimates, gave him an extra twelve years of life.

On the next bend we have to make another river crossing. It's a bit deeper than most, but free of rocks and the current isn't so strong. Which is just as well because we have to use our staffs to help sling our rucksacks over our heads. We look like two inexpert hunters returning with a kill. The staffs being too short, my pack swings into the back of my head, and Miriam's away from her, as we step forward.

A surprise awaits us on the other side. Bear prints! Since Bernie Peyton's visit to Amboró I was always on the look out for them. We had found bear prints in this same river not very much further upstream. Bernie was surprised to see them here, because nowhere else had the Spectacled Bear been recorded at such low altitude (1,200 feet), and he didn't know of any other locality where bear, tapir, and jaguar lived side by side. Normally the Spectacled Bear doesn't come much lower than 3,000 feet.

Over the years I have come to realise that much of South America's biological data has been collected from studies made in Venezuela, Colombia, and to a lesser extent in Peru. These countries have relatively well-defined geographical regions: the cold high Andes, the subtropical foothills and the hot lowland jungles. In Amboró the situation is not so clear cut because of the seasons. For example, in winter the tropical lowlands can be cold, and birds, and animals like the Spectacled Bear from the higher regions, can be found at much lower altitudes than is generally realised.

These are the clearest set of prints I have ever seen and from the look of them they are very fresh. They lead up the beach towards a

natural lean-to formed by two large slabs of rock. The Spectacled
Bear is one of the smallest species of bear (even smaller than the
Himalayan Sun Bear), it's about the size of a well grown sheep. I
won't say German Shepherd. And it's timid. With these facts to the
forefront of our minds we follow the trail into the cave, but not soon
enough. Our only reward being the noise of some animal entering the
forest we can see through the far side. We just missed it! I know there
is no point in attempting to track it, the steep hillside with its thick
vegetation would make that impossible.

We are not too disappointed; we saw its tracks, smelt it, heard it.
We three shared a moment together and that makes us feel happy. To
hell with the 'Twitchers', those bird watchers who run around the
globe trying to be members of the 1000-Club'. I had better explain as
we walk along.

****

Let's call him Steven. Steve is an amateur ornithologist who
spends one month every year visiting a different part of our planet.
He has been to Alaska and the Arctic, Europe (including Siberia),
Malaya, Thailand and New Guinea. He has done North Africa and the
National Parks in Kenya. He knows Mexico and Costa Rica well,
Venezuela and Peru, but this is his first time to southern South
America. Wherever he goes he takes along the appropriate volume of
Peter's Checklist of the Birds of the World. Every time he sees a new
bird he gets the list out that he keeps safely inside a zippered plastic
pouch and, with a ruler and a pen underlines the bird's name. In the
margin he neatly writes a number, which he informs me indicates he
has now seen 971 different sorts of birds. He does all this on the spot.
To glimpse the bird is enough. He doesn't want to watch it or know
anything about it. As soon as he has seen it he loses interest in it; and
now looks at me as if to say "right, where's the next one?" Half an
hour later he has written 972 beside the name of a bird I know he
didn't see because he was taking a leak at the time and 973 beside one
we only heard. He doesn't know anything about birds and is totally
dependent upon me to identify them. I feel like tossing out a few birds
that only occur in Brazil or Tierra del Fuego, but I don't. Why destroy
his only interest in life no matter how irritating I find him?

****

There's something odd about Miriam today. Whatever I think
about she says or does something triggered by my thought. Before,

when I was thinking about monkeys and trees, she calls my attention to the Titis. A moment ago I was thinking about snakes and she points out a small one hunting in a pool. Now, as I think about the bird that Steven didn't see, she wants to take a leak. She knows if she squats down here beside the river she will have to contend with the "Maruhuí" (sandflies of the genus Phlebotomus whose stinging bites smart for hours afterwards). With one's pants down their mass attack feels like a blast of bird shot in the bum. So I wait whilst she makes her way into the forest where the "Maruhuí" won't follow.

The "Maruhuí" are my friends. They keep a lot of people out of the Park. The fly leaves a small subcutaneous bloody spot where it bites. Later these turn black. In some particularly bad places, where the rivers are rockiest, the campesinos have become tattooed to such an even blackness they look like it is their natural skin tone.

Miriam is back. We walk on. We have no time for bird watching or scene gazing. We have to concentrate on each step now. It is rocky and slippery and one false step could mean a nasty fall or a twisted ankle. The latter would mean the end of our trip; whose purpose I now remember I haven't told you about. It's something military.

****

But first a true story. Back in 1984 the Bolivian post office issued a series of stamps depicting seven species of Macaw. The one with the highest value depicted the "Military Macaw" correctly identified by its scientific name, Ara militaris. When the stamps were issued members of "The Armed Forces Committee" made a complaint about what they thought was an ill-conceived attempt to poke fun at them.

We're going to see if the Military Macaws are still nesting on the cliffs below Cerro Amboró and, if so, how many there are. This large parrot gets its name from its scarlet forehead, recalling to mind the caps worn by the British Military Police. Somebody preparing a book about endangered parrots has asked me for information on the Bolivian subspecies. The Military Macaw is another asympatric species (like the River Dolphins I told you about). The "Type" form, Ara militaris militaris is the subspecies that lives in a narrow strip of mountains from Venezuela to northern Peru. It is separated from the other two subspecies by a considerable distance. The commonest one of these, Ara militaris mexicana, as the name implies, hails from Mexico. In Amboró we have Ara militaris boliviana, the rarest one.

We were expecting to camp below the cliffs where we have seen the birds nesting, but our slow progress and the depth of the river have combined to make that impossible now. We decide to make camp at the place where the ill-fated Boy Scouts chose to do, a raised beach adjacent to a deep water bend of the Isama River. In the old days we used to call the place "The Cinema" because the smooth rock face in front of us had a screen-like section. Now that part of the rockface has been partly overgrown. We have time to go further but it's the last good camping spot before the river deepens to pass through a narrow gorge only twenty-foot wide, a place where we might get our clothes and bedding wet. If we wait till morning and if it doesn't rain during the night the water level should drop.

A lot of 'ifs' again, I know, but Amboró is like that: the "If National Park".

After we make camp, not that there is anything to make camp with, we get a fire going to make some coffee. Miriam wants to go for a swim. I don't. I'm a cat, I don't like water much. I ask her to hurry, because I am hoping to see an otter later. Not that we will.

One of the Park's problems is some fifty miles downstream, where most of the Park's rivers flow into the River Yapacaní, a Brazilian fellow has put a dozen nets across the river and now very few fish of breeding size are reaching the Park; or anywhere else for that matter. One greedy Brazilian is allowed to deprive the local people of a much-needed source of protein, and seriously affect the ecological balance in Amboró. When I tell you his activity is illegal and that our complaints to the CDF have fallen on deaf ears, I know, by now, you will not be surprised. I just hope I am not depressing you as much as I am depressing myself. Am I? Do you think I could make you angry enough to grab paper and pen and write to our self-proclaimed first Green President of Bolivia about it? As Chico Mendez said: "Government's abdication of responsibility and connivance with all this death and destruction".

I should tell you that Señor Jaime Paz, Bolivia's first Green President, has not done all that much for nature conservation. Since he became President in 1989, saying 'there must be something for everyone,' he has given away the Isiboro-Secure National Park to the coca growers and a band of self-styled Indians bent on taking over from where the logging companies were made to stop. He has permitted the development of plans for the colonisation of the Pando, Bolivia's last green frontier, that will put an end to the Manuripi-

Heath National Park. He has turned a blind eye to the constant invasion of legally protected areas by colonists and loggers and has taken very few steps to punish corrupt members of his administration, especially those responsible for handing out illegal titles to state owned land, as in the case of Amboró. But he declared a "Truce with Nature", a package of laws designed to protect the environment which are so ambitious they are farcically impractical. And he does wear a green leaf in his lapel, that, no, wait for it, does not symbolise the desperate plight of the mahogany tree, but rather the coca leaf. Yes, Bolivia's President is green, he wants more coca bushes.

Miriam's finished her swim sooner than I expected. She says there's a lot of "sardinas" biting her, small active fish that together with "bogas" and "bagrés", no, I'm not using bad language again, are now the surviving otters' scant diet. They, the otters might be justified in saying to me "Bulldog Drumskull, did you make that promise to us just so we could die of hunger? Why didn't you let us die in peace?"

My glum reflections are reinforced by a total lack of the patter of otters' tiny feet. But, as we sit beside the river in the gathering dusk, we are rewarded by the sight of two Oilbirds that skim across the water to drink. Each time they dip their bills to the river's surface they leave their own ring of bright water. "Always look on the bright side of life. Deedum deedee".

Back at the campfire swallowing lumps of damp chicken sand-wich, we are invited to participate in Bolivia's favourite pastime, a Beauty Contest. A parade of Curucucis. This is one of the most charming exhibits of South America's night life. Curucucis are one-and-a-half-inch long Click-beetles that have two large luminous spots on their backs. Unlike fireflies these spots glow continuously. As we stand holding our wet clothes over the campfire to dry, a dozen or more of them are flying amongst the low vegetation around us. Their comings and goings making no sound, they look like a candle-lit fairy procession. Some of the lights are a pale luminescent green, others yellow. Occasionally one or the other settles on a stump with the aid of a ventral light, a powerful white landing light that is switched on to throw a beam a foot or so below the flying beetle. The temperature drops a point or two and more start to settle. Now the scene looks like Kennedy Airport, mini-aircraft landing all around us. One particularly large one approaches my face, its landing light shining directly into my eyes. "Hey, you! Stand still, laddie".

I do, mesmerised by the scene. There are smaller lights everywhere, hundreds of ordinary fireflies, and discrete trails of
phosphorescence where we have scuffed up the dead leaves. Miriam
takes my hand in hers. I know what she is thinking: 'What does all
this mean? What message is there for us, for mankind? Why, large as
we are, do we feel very small surrounded by these minor beings?'
Then, quite suddenly, a cool breeze disturbs the scene and all the
lights switch off. Now only the leaf mould and the embers from our
fire separate us from the inky darkness under the trees.

We remain where we are, still holding hands, like two pilgrims
shut out from heaven, the gates having closed on the glimpse of
beauty within. We feel denied, melancholy, even guilty. Paradise lost.
I turn to comfort her, my cheek pressed to hers washed by her tears.
As we turn to bed the rhythmic low purring of a Pygmy-Owl seems
to carry its own forlorn message to us.

I can't sleep. The cool breeze has died away and it is hot again. It
is too late for Curucucis, but individual fireflies flash for a mate. I lie
watching the moonlight seep into the treetops. The soft noises from
the river punctuated every now and again by the "pi-link" of a bush-
cricket somewhere close to me. Miriam turns in her sleep. A dry
Cercropia leaf startles the silence with its fall. I scan every inch of the
canopy overhead in the hope of seeing one star or a piece of moonlit
cliff, but the tree canopy is too complete. Somewhere down river a
Great Potoo roars, just once but, loud enough to unsettle the
uninitiated. A pebble falls from the cliff and enters the pool with a
plop. A Longhorn beetle nearly lands in the fire.

****

I drift away to other times. Now with Guillaume in Ethiopia. We
have been collecting beetles all day long and have now sat down with
a bottle of rum to play cards by candlelight. It is our evening ritual
with which we break the monotony of a collecting expedition. The
game is "Mau-mau". The stakes are the new species of beetles we
have caught. Guillaume bets a rare longhorn. Having caught nothing
of equal value I have to match it with two "Paussids" gleaned with
much pain from a colony of vicious ants. His hand I know, is good.
We play furiously for a short time. His score mounts up and seeing
through my end-ploy makes the hundred and nonchalantly sweeps my
prizes to his side. Now I have a good hand. I stake an arboreal Tiger
beetle that neither of us has ever seen before. He offers a diamond-

studded weevil. I demur; it's very nice, but not so rare. He withdraws it and offers a small plain black ground beetle instead. He smiles at the light in my eyes. I return his smile in acknowledgement of his expertise. We both know that, small and undistinguished as it looks, he may have offered me the day's prize. He lets me win it. I pick it carefully from the cloth and put it away with my small collection of unbettables. A few years later I call it after him: Harpalus rougemonti Clarke.

****

What good times those were. The age of the Naturalist, the best of the field biologists and in spite of our card games, non-competitive. Not like now, with all these so-called scientists looking to make their names. We were young, indomitable and as owners of the World breaking new ground in entomology. Acknowledged experts in our field, every few years we would scrape enough money together to return to the dusty museums of Paris, London, Brussels, all the capitals of Europe. But who was I now? To the Bolivians: nobody. To the Americans: an oddball. To some: a legend. To others: a hero. To my neighbours in Buena Vista: a man like them, as their own song says:

"A gringo has come to live among us
Who, like us is not rich
But who has a life-style
Many a millionaire would envy
Gringo be happy
For this is our gift to you.
For this is our gift to you."

Now it's Lake Victoria. I'm ten years old. Though early in the morning it's already hot. Rusty and I are floating on our inner tubes, trembling, not with cold, it's because we are a bit frightened. Scared my dad will see us and give us the hiding we deserve. Scared that we have talked ourselves into a rash new test of manhood. We paddle a little closer to them. We take comfort from the milky calm surface of the lake as it allows us to see exactly where each one is. We think there are about eight of them, really close now, closer than we have ever been before. The one we are making for sinks just below the surface, but as far as we can tell makes no effort to leave. Its movement creates an eddy that sucks us forwards. I dart a look of panic at Rusty. The Hippo sweeps its head to the side and looking directly at us gives a titanic yawn before settling down to sleep again. Rusty and I are paddling for all our worth towards the beach. "We did it! We did it!" We nearly touched the hippos!

From that day on we never pressed our luck again. The hippos became our friends, gentle, tolerant creatures allowing us to have our fun. Then we got bored and thought up something new. Nobody could do that now. Not on your life. Some of the locals spent their meagre earnings on .22 rifles and considered it sport to use the Hippos as target practice. The puny rifles didn't kill them, just wounded them. Nobody did anything to stop it. The hippos became unreliable, their soft grunts turned to bellows of rage. No longer did they come to my bedroom window at night to scratch their bristles on the mosquito netting. I was chased down the beach by a mother hippo protecting her young. An Italian had his bum bitten off. Somebody was killed. The hippos were declared a public menace and a squad of policemen (Askaris) came to put an end to them. Man had shown his superiority again. Another paradise lost. And who is there to cry for what we have done?

My watch tells me it is time to make a move. The undisturbed sand tells us that, indeed, only time and the river have passed whilst we slept. My marker at the river's edge tells us that the water level has dropped six inches during the night. A wheeling, mewling Black Hawk-Eagle tells us that from his vantage point high above the canyon he can forecast another sunny day.

The river flowing peacefully through the twenty-foot-wide passage looks passable. Free of our backpacks, that we have hidden amongst the bushes not far from the camp site, we feel our way forwards, our staffs picking out the best route. We have passed here many times before. There is no danger now, but the water is cold, and a dunking would ruin our binoculars which we have tied to the tips of our stakes. Patiently we follow the line of a submerged sand bank we know runs through the narrows. It starts on the left, swings to the right, then returning to the centre peters out, to carry on again after a space of ten or twenty yards. We know that here the water could be deep. How deep? The current isn't strong so much as insistent, so we stop for a moment to gather our breath and take the opportunity to absorb the sculptured beauty of the passage. Just above high water mark, twenty feet above our heads, the grey walls are tinted green with moss. Only here and there, where the smooth rock offers a tiny foothold, Orchids and Bromeliads grow, some in flower. Overhead the passage is roofed in green, the trees reaching out from either side to touch each other.

We push on again against the current to arrive at the half-way point. We know the sand bank is running out because we feel the bed of the river changing to mud. Miriam waits. I step gingerly forwards. The water rises to my armpits, touches my collar bone, then spills away. We can get through. I turn and help her, lifting her slightly through the deep. Our troubles are not over yet. We have another sixty yards to traverse; it's not so deep as we try to pick up the sand bank again. We find it. Everything will now depend on how close it will take us to a shelf of rock that guards the deep pool at the upper entrance. If it's only a yard or so we know we can launch ourselves across and take hold of the ledge with our elbows. If it's farther, we will have to swim for it and with our binoculars that will be difficult. We are nearly there, the soft sand angles towards the shelf, closer, closer…then it plunges away towards the bottom. I turn to look at Miriam, she shrugs. I back off a little, and she passes me her staff. I edge forward and gently lower the binoculars on to the ledge. It takes all my strength as the weight of our binoculars come into play. They don't make as soft a landing as I would have liked and are now precariously balanced on the edge of the hard wet shelf nine feet from us. Miriam comes forward and together we launch ourselves into the deep pool. We throw our elbows over the ledge, hang there for a moment and, taking care not to dislodge our binoculars, haul our water-logged selves out.

Now we are cold. It has taken us twenty minutes to pass through. We hasten towards the sunny beach up ahead. We strip our wet clothing from our bodies and spread them and ourselves out on the warm rocks. I take my cigarettes and lighter from their water-proof container and we share a well-deserved smoke together.

I look at myself. Pale, paunchy, wasted. In this setting I could easily be mistaken for Gollum. White body, little white balls, little white dick, white thighs, white bum. It almost hurts to look at Miriam. Lithe, brown, firm, fit and beautiful. One could be forgiven for mistaking her for that famous other, Bolivian, Raquel Welch.

But do I want to change? Keep fit, go jogging, give up smoking, quit drinking, attend for medicals, have my skin ultravioletted, my nose remodelled, my face lifted. No. Of course not. It's not dignified. Are we really too damn important to be a bit scruffy, smell a bit, look as old as we are? Do we really have to put ourselves on a pedestal, make gods of ourselves? Our social interpretation of the world has reduced everything, but us, to mere objects which have no value of their own except as mirrors to our own narcissism.

There is nothing wrong with recapturing your young body, but the rest of it is the New Evil that will divide the "haves" from the "have-nots" even further and might through economic and social blackmail, turn to vindication of a new repression no less sinister than Mao Zedong's marxism. The western cities are yours; you can keep them.

I am not against products that make you look better and feel better, but don't come down here telling every young girl (struggling to find a job that will pay her sixty dollars a month, if she's lucky), that she is nothing, a nobody, that she must look sophisticated, fashionable, appetising — her skin too dark, her teeth irregular, her eyes black, her tits too small, her armpits too smelly, her clothes too tatty, for her to be employable. Because they believe you! They go out and buy blue-tinted contact lenses instead of getting treatment for cataracts; they plaster themselves with marine algae and waste the little money they have on expensive cures to rid themselves of cellulitis instead of buying the milk, fruit, and vegetables their bodies crave.

The South Americans are a handsome people, naturally attractive, but they have an inferiority complex that renders them especially vulnerable to expensive western decadence. You only have to stroll around the streets of Santa Cruz to convince yourself of this: thousands of beauty parlours, hair stylists, aerobic centres, gymnasiums, fashion shops and now, plastic surgeons.

As we lie on the rocks, waiting for the warmth of the sun to penetrate our goose-fleshed skin, we pay silent homage to the Boy Scouts whose naive hopes of adventure ended so tragically on the cliffs up ahead.

****

Five young men and one woman enter Amboró National Park on Christmas Day 1992. The men, whose ages ranged from 24-31 years, were Boy Scouts and included a pair of almost identical twins, Eduardo and Jorge Roca Urioste. They had decided to spend the Christmas break on an excursion to the the Rio Isama and to make the gruelling two-day climb to the summit of Cerro Amboró. Though we had cut an improved trail to the top of the cerro in 1990, by the time the party of Scouts made their fateful visit this trail had become overgrown and difficult to follow. Even so, the staff on duty at the Park office allowed the group to go unescorted, a serious breach of the Park's own rules.

On the 26th December, the three youngest men, Eduardo Roca (26), Rafael Muñoz (24) and Ernesto Quevedo (26), get up early and leave their camp site (the same place where Miriam and I had spent the night nearly 5-months ago) with the intention of looking for orchids and ferns in the surrounding hills, to reconnoitre the trail and return later in the morning. They take only a machete with them, no food, no spare clothing, no mosquito repellent. From the camp site the trail ascends steeply, dips slightly, and then, rises steadily in an exhausting climb of 2,500ft to the top of a ridge overlooking the hanging valley of a rocky stream, which we called the "Quebrada Cascada" ("Waterfall Stream"). When I first came here in 1982 I thought the "Quebrada Cascada" might lead me back to the Rio Isama. Later, when I was called in to help find the lost Scouts my memory of this minor point was to save two peoples' lives.

Standing at this same place, having already lost their way, and after an unpleasant night without food or water, the three Scouts are anxious to return to their camp beside the River Isama. Disoriented, they start their descent into that fatal quebrada. To reach it they have to hack their way through dense vegetation covering the steep, 45° degree slope of the hillside. Hungry and tired already, this was heavy work. Then, closer to the quebrada itself, the dense vegetation gives way to impenetrable thickets of Tacuara, the thorny bamboo that can only be cut with an axe. At this point the three men nearly turned back. But overwhelmed with fatigue and convinced they are close to the river, they somehow crawl their way through. How demoralised they must have felt when they reached the narrow gully cut by the quebrada, for now they knew this was not the Rio Isama. Here they had to spend another night, wet, cold and hungry.

Too late to turn back and with only one choice left to make, stay or go on, they choose to make their last mistake: go with the flow and follow the "Quebrada Cascada" in its fatal hanging valley.

Meanwhile, back at the camp site, after a fruitless attempt to find their friends, the other three decide to return to Buena Vista in search of help. Unfortunately, that night, New Year's Eve, the mountains are drenched by a tropical thunderstorm and they are unable to descend the swollen river. It is not until late on New Year's Day that they arrive at the Park office in Buena Vista. The following day three park guards and two volunteers return with Jorgé Roca to start the search for his twin brother and his two friends.

By the 3rd of January the Third Aerial Brigade had been called in to help and a helicopter with extra rescue teams is dispatched to the area. Sitting comfortably at home this is the first time I hear about the impending tragedy.

On the 4th of January I am visited by members of the missing men's' families who come to ask for my help. After giving myself a few hours to think about the possibilities, and remembering my own confusion in the same place, I convince myself that the three young men are not lost but are trapped in the Quebrada Cascada.

The three Boy Scouts start down the quebrada under a dark stormy sky which lashes them with heavy drops of rain. Running over soft sandstone, the quebrada has worn itself a deep, vertically sided ravine. Every now and again, where it passes over slightly harder rock, it drops away in a series of waterfalls. As the quebrada descends towards the river, gathering pace on the ever-steeper incline, these minor falls become more daunting, first a yard or two high, then five or six. With desperation their driving force, the men fashion crude ropes of jungle vines with which to descend the even larger falls. But with every drop, their return, and later their rescue, is made more difficult. After about another week without food or proper sleep, overlooked by hostile trees, and shivering in their wet clothes as the wind scours the narrow ravine, the Scouts collapse.

Whilst they lay in delirious sleep the number of would-be rescue teams had risen to eight, and eventually more than a hundred people equipped with radios and back-up helicopters were to be involved in the search for them.

I spend two frustrating days beside a military helicopter waiting for the weather to improve. We had tried to reach the Quebrada Cascada twice but were beaten back by strong winds and rain. Rescue teams descending the Quebrada reported signs of the Scouts but had turned back when they reached a place they judged to be impassable without special equipment. I insist they try again. They send back for longer, better ropes and more supplies.

On the 7th of January, having given up on the weather, Miriam and I are helicoptered to a base camp in the Park where we are left to make our way up the Isama as best we can. Here we were joined by Jorgé Roca and a park guard called Heriberto on their way to deliver the ropes and supplies required for the descent of the Quebrada

Cascada. By four in the afternoon, we arrive at the Scout's abandoned campsite.

Whilst all this has been going on the three delirious young men awake and make another attempt to reach the Isama River. Now, aware that they are fighting to survive they somehow manage to inch forwards. Late in the afternoon, circa the 4th of January, they arrive at a point where the water of the quebrada enters a sort of swallow hole.

After some desultory conversation Eduardo Roca decides to attempt the descent and brave man that he was enters the swallow hole. That was the last time he was seen alive. His companions continue calling to him and occasionally think they hear a response. By now night has fallen and the gathering clouds present a renewed menace, heavy rain which quickly fills the quebrada with the turbulent waters of a freshet.

By morning the weather has improved a little and the remaining two Scouts, Rafael and Ernesto, attempt to communicate with Eduardo but receive no response. Maybe they spend the day of the 5th of January anxiously awaiting the return of their companion. On the 6th, with the weather fast improving, they manage to squeeze down the swallow hole to emerge on the other side. They would not have been able to accomplish this feat if they had not lost nearly two stone on their starvation diet of leaves.

Then comes the final shock. They stand overlooking the Rio Isama, instantly recognisable by its broad deep canyon, but there is no way down. From where they stand the quebrada falls to the river below in two forty-yard-high waterfalls. An impossible vertical obstacle that even then they could not fully appreciate, standing as they had to, a little way back from the edge of the slippery rim. And there is no sign of their friend.

Here they spend a day and a half in a state of exhausted depression and nearly out of their minds with hunger. By the evening of the 7th they decide, suicidally, that on the morrow they will attempt the descent of the cliffs.

After another almost sleepless night, soaked by constant rain, Rafael and Ernesto make their way to the edge of the cliff. They approach it without any real idea of what faces them and without any clear plan. It is 10.00 a.m. on the 8th of January. They had been missing for thirteen days.

****

At 0530, on the morning that Rafael and Ernesto plan their suicide, I awake amongst a jumble of sleeping forms, Miriam's curls tickling my face and a friendly spider eyeing me from the end of its thread. When, in the middle of a rainstorm we had at last dropped off to sleep, sometime around nine o'clock in the evening, our little palm leaf shelter had been home to the two of us. Jorgé and Heriberto having bravely decided to deliver their precious cargo of ropes and provisions to the rescue teams far above. Whilst we slept two other rescue teams, together with a number of individual volunteers arrived to share the shelter with us.

By seven in the morning they have all left on their way up the trail to Cerro Amboró, leaving Miriam and myself in an agitated state. The night's unabated rain has put an end to our hopes of continuing upriver to the foot of the Quebrada Cascada's twin falls. Up ahead the Isama River's narrow ravine will be completely full of water and, as far as we know, there is no way around it.

We stand looking at one another with the resignation of those confronting failure. After eleven days we don't believe there can be any hope of finding the scouts alive and with the departure of the others any possible help has gone too. Then, quite unexpectedly, a little man completely swathed in yellow oilskins pops up from a dip in the path. He is he says, Señor Morejon, the leader of the YPFB's Exploration Team. We know he is our last chance. I launch into my most convincing explanation of why I know the Scouts are trapped, not lost and how it is that I know exactly where the missing men are to be found. How imprudent it has been that the River Isama, itself, had not been searched, and why they have to help me. I said "me" because Miriam foresaw that the impending scenario would be too much for her and wisely elects to remain at the camp site.

Señor Morejon, without wasting time on useless questions, agrees! We are back in business.

It is drizzling as we set off. We climb the right-hand side of the passage but my hope that we can immediately descend to the river again proves to be impossible. Once again defeat stares us in the face. I know that for the next half mile the cliffs rise from the river in an unbroken wall two hundred yards high. And, at this point, another vertically sided ravine enters from the right. I'm sure we're boxed in.

But my robust companions sweep aside my pessimistic comments with the ease of professionals and continue to hack their way towards the top of the cliff. After a further hour we manage to surmount a ridge overlooking a ravine in front of us. All around us we can see the dark depressions in the forest that betoken the curling passage of the Rio Isama and the many quebradas that enter it. But which is which?

I am an experienced explorer and know the Park well, but I had never in my life felt so confused. We had made many twists and turns during our ascent and back tracked several times. Now, standing on this ridge, I hadn't the slightest idea which was north or south, or which of the many dark depressions was the Rio Isama. Fortunately, Señor Morejon was not so fallible and with the help of his map and compass he eventually convinces me the ravine facing us is not the main river but the one that enters from the right. But how are we to descend?

My five companions set off again, this time at the pace of Olympic champions in the thirty-kilometre walk. Not only am I much slower but I am also hampered by one of my tennis shoes which has split in two. I had to borrow a pair as my own hadn't emerged from the helicopter with me.

After a further hour, during which we make a number of fruitless attempts to descend the ravine, I am told one of our party has finally been successful. When I catch up with the others, I find them hanging on to tree branches looking over the edge of a precipice more than one hundred yards high. At first, I am sure I have misunderstood the message because it was patently obvious that nobody has gone down there, rope or not.

But again, I'm mistaken, for clear as a bell we hear a "halloo" from somebody far below. Whilst I'm pondering on who this might be, whether or not we have at last made contact with the lost scouts, everybody else disappears over the edge with the spontaneity of mountain goats.

I am aghast. Okay...terrified! I don't want to be left behind and I certainly don't want to follow the others. I roll on to my belly and swing my legs over the edge. I find myself standing on a narrow ledge that dives towards the floor of the canyon below. It is covered with rotten leaves made slick by the passing of my companions. Every ten feet or so a tree trunk offers a hand hold, but for the rest I have to put my trust in ferns and flimsy tussocks of grass. At one point, where this frail path runs out, one of my companions is waiting with a rope and I have to descend a few yards to pick up another ledge below. It certainly is a nasty spot, slippery, and devoid of hand holds, but it offers a perfect bird's-eye view of the rocky stream fifty yards below.

The last bit of the descent is a nightmare. Encouraged by my companions I somehow or other manage to get to the bottom without breaking any bones or totally ruining my precious binoculars. But nothing will induce me to go back the same way.

The quebrada running through the canyon fulfils all our unexpressed hopes. It presents no obstacles or particular difficulty in its course towards the main river and within ten minutes we are sitting on a sand bank looking at a part of the Isama River I am very familiar with. Now I'm back in charge.

The waterfalls guarding the ascent of the Quebrada Cascada now lie only a few hundred yards upriver. The trouble is the Isama, it appears distinctly unfriendly, but we have to cross it or start cliff-scaling again. My companions did not waste time pondering on difficulties, filled as they are with faith in my prediction that we will find the missing men. Once again, they disappear, this time up the right-hand cliff of the Isama river, leaving me to keep up as best I can. The time is 10 a.m. on the 8th of January 1992.

Rafael and Ernesto sit on a narrow ledge above the falls. They know that what they are about to attempt will bring them death or salvation. At 10 a.m., determined to summon up the courage for the descent they edge towards the lip of the falls. At 10:15 they back off for a last-minute discussion back on their ledge.

At 10:20 a.m. our party reaches a point on the opposite cliff on a level with the top of the lower waterfall. By peering upwards through gaps in the trees we can just make out the top of the upper falls. How disappointed we are to see it, for no one stands there. We shout, we halloo, whistle and scream, but receive no response. Despair runs like passing shadows across our faces. Are they there? Could they hear us above the splatter of the rain that is now a bit heavier and the roar of the river and the falls themselves? Maybe they are there. We yell again and again, nothing!

****

Warmer now, Miriam and I put on our wet clothes and continue our journey upriver. We pick our way between the many boulders that have fallen from above and the muddy pools left by the recent rain. The Hawk-Eagle has gone, and the only sound is the crunch and clink of our passing. Even the animals have forsaken this stretch of the river, guarded as it is by the deep water in the passage. It's not until we reach the ravine on our right that we see the prints of Deer and Peccary.

Here we cross the river and sit for a moment on the sandy beach. The cliff marking the entrance to the ravine has been worn to a rounded smoothness by the grinding of rocks and pebbles brought down by the river. A Ringed Kingfisher, the largest of the New World species, sweeps past with a throaty rattle as if mocking the sound of our steps.

The way ahead is easier now, for the river is wider and free of rocks and deep pools. It is the approach to the falls of the Quebrada Cascada. The right-hand cliff is heavily wooded, the bare vertical left one stands back a little to leave a narrow band of trees between it and the river. As we pass, a mixed flock of foraging birds is busily exploring the potential of this wooded strip. Filled with the memory of my last time here we hurry on to the falls themselves.

The lower waterfall tumbles into a large circular pool filled with cold iron-stained water the colour of weak tea. It is a curiously uninteresting pool, devoid of any obvious signs of life — no animal smells, no plants, no fish, no tadpoles, only a few lethargic water beetles belie its absolute sterility. It's a cursed place, emanating a cold disdain for the exuberant life all around.

I try to fathom the reason for it. What is it that is different? Something to do with the water? The chill air? Evil spirits? Unfortunately, no stern looking Shaman is at hand to answer me.

[In addition to being a hanging valley, the valley of the Quebrada Cascada differs from all the others in another way. Near its source, a few miles away and three thousand feet up, stands a tilted slab of flat grey rock the size of several football fields. In the lower quadrant the rock has been gouged out into a circular depression about a yard deep and eighty yards in diameter. At the very centre is a nipple like eminence from which a large number of pale streaks radiate towards the circumference. It looks like a giant bicycle wheel, or to me, exactly like photographs I have seen of meteorite impacts. Could it be? And if so, is this related to the sterile nature of the quebrada? Could that meteorite have been made up of poisonous heavy metals? Could it have been radioactive? Someday, no doubt, someone will come along and explain it, but for now the cold cheerlessness of the place resides in the tragedy of the Boy Scouts.]

****

Nothing! We turn to go. The pool. That cold disdainful pool…of course! It was a natural deposit for anything washed down by the Quebrada Cascada. I suggest to Señor Morejon that we should look in the pool before we leave. If the scouts are trapped a little further up, maybe they have thrown something into the stream in the desperate hope of providing a clue to their whereabouts.

At 10.30 a.m. three of our group cross the surly river to the pool opposite. Almost immediately they find a pair of torn jeans and a pair of black running shoes. None of these items look like castoffs, they are quite new. Logic tells us that they must belong to the missing men.

At 10.30 a.m. Rafael and Ernesto return to the top of the falls, ready to do or die.

Me, with my befogged binoculars look up, squinting through the gaps in the leaves. For a moment I can't believe my eyes for standing on the top of the upper falls are two figures apparently contemplating the descent.

We all have a peek and start yelling and waving like a family of excited Spider Monkeys. There is absolutely no response from the figures above. I have to dismantle my binoculars to rid the objectives of the condensation that has collected in them. Then, when I have reassembled them, I see that the figures have noticed us at last. They appear to be calling, but feebly. We can't hear what they are shouting. We doubt, after all, that they are the missing scouts. There are only two of them and they appear peculiarly apathetic. They don't behave like men just saved from imminent death. They just stand there looking down. I examine the figures again. I only know what Eduardo looks like having come upriver with his twin brother. Both these men are dark. They certainly look emaciated but after many days so, too, do most of those involved in the rescue, and we can't discount the possibility that we have joined up with some of them. Curiously, one of the figures walks away from the lip of the falls and out of sight. I assumed that if they are the scouts, this one has gone back to talk to the third, Eduardo, to give him the good news. By then we are pretty sure they are the missing men.

How inadequate we are at communicating with each other. We shout. The single figure shouts back. We can't hear him. Can he hear us? Probably not. He keeps pointing down the cliff. I assume he is asking how he can get down. My companions think he is indicating that somebody is down below…fallen down below? I rapidly discounted this, though I'm afraid it is so. The other one comes back after five minutes and also points down the cliff.

Señor Morejon sends three of his men to try to find a way to the top of the lower falls. I don't think they can make it, it is all sheer cliff. They try. They fail. Maybe there is somebody down there, but even from our vantage point on the opposite cliff, the foot of the upper falls is shielded from our view by a rocky prominence.

All we can do is to return to the camp and let it be known with the radio that the missing men have been found. We all hold up our hands, palms out "Stay there". "Don't move". The two Scouts seem to understand. Then we move away, back the way we have come. Yes, back the same way.

I said I wouldn't go back that way, but nobody else wanted to swim the river through the narrow passage and I didn't want to do it alone. So back, that way, we went, and how easy it seemed, filled as we were with the elation of success and the urgency of our mission.

****

We weren't there to see the actual rescue but standing in the same place now it is easy to visualise the heroic efforts of those who carried it out. The "Grupo Pantera", the Special Air Rescue team, the stolidly dependable park guards, the youngsters of Miriam's Ecological Club and others worked together and lowered the two survivors down the first of the forty-yard-high waterfalls by rope. Sadly, tragically, at the foot of the upper falls they found the shattered body of Eduardo Roca Urioste lying on the rocks.

A Lama helicopter flew into the gorge. Struggling to survive, as well, the pilot and a young Austrian managed to pick up the two survivors and Eduardo's body. Everybody who witnessed the scene described it in the same manner. The tiny helicopter. The deep, narrow canyon. The buffeting wind. Silvio, the Austrian, hanging for grim death to the net in which he is being lowered from the helicopter, the trees lunging at the rope. The pilot frantically trying to stabilise his rocking machine. Silvio bundling the plastic-wrapped body on to the helicopter's skid and the two men into the cabin. The nerve-wracking ascent as the blades of the helicopter struggle to catch the air. The gasp of those watching as, once again, the trees lunge at the rope. Then, at last, the rise to safety and the swift flight to Santa Cruz.

Nobody will ever know exactly what happened to Eduardo Roca. Was he swept by floodwater from the swallow hole on to the rocks below? Did he slip at the top of the falls as he tried to grope his way forwards in the darkness? Neither Rafael nor Ernesto can tell us, for up to the last minute they thought that, somehow, Eduardo had managed to descend the falls and was going to come back to rescue them. This, in itself, can remain a fitting memory to the dead young man.

****

I brush the tears from Mriam's face andd pull her to her feet. We have yet to reach the cliff where the Military Macaws used to nest and, I haven't told her yet, I want to go further still, to sit on what I call Noel Kempff's rock and cogitate.

For the next mile the river runs between sheer-sided cliffs three hundred yards high, the deepest of the Isama's ravines. Being almost bare of vegetation they have little to offer the Macaws in the way of nesting places. Then the river widens a little, and the cliffs fall in a series of narrow ledges, some of which are wide enough to support tussocks of coarse grass. Behind these the macaws dig into the fibrous soil to make their nesting burrows. Even as we approach the spot a group of five fly off upriver. Unfortunately, they are the only ones we are to see. Even though we spend hours hidden in the bushes opposite the cliff, none return. This is disappointing and inconclusive. In past years we have seen as many as twenty pairs of birds at this spot. Those we have just seen suggest the cliff is still being used as a breeding site, but provides no information on whether or not the colony is still a thriving one. More than likely we have arrived too late to catch the main early-morning flight and that deep in their burrows unperturbed females are incubating their eggs. To gather useful information, we would have to spend a night below the cliff, as we had originally planned to do. Because we had to leave our packs behind, we can't do that now.

Kempff's rock, like that last tooth in the jaw of an old Irish whisky drinker now stands alone, loosely plugged into the shifting sands of the river's middle course. Scaled with centuries of mossy looking lichens, its scabrous skin offers us a purchase on its otherwise smooth grey surface.

But I must remind myself, I'm not here to give you a detailed account of a perfectly ordinary, albeit house-sized rock. I'm here, sitting on Kempff's rock, to make a decision: should I leave well alone, or should I, like a rotten molar, drag it out and expose the felonious root? And murder is felonious, and Noel's murder even more so.

I have been hesitant because the consequences for me would be no different than they would be for, say, Joc Schmuck, should he suggest that Martin Luther King's death was planned and executed by the Bishop of Boston. Scurrilous insinuations deserve short shrift and public rebuke. So, I must tread carefully.

The truth of the matter is Noel was my friend and deserves a better, albeit posthumous fate than has been his lot to date. I also feel a little guilty for in some measure I may have been responsible for what I believe really occurred.

Officially, Noel, his guide, the pilot of the plane they were in, and the Spanish biologist stumbled into a fatal situation little different from the way those poor boy scouts stumbled into the "Quebrada Cascada", accidentally. And, even though a martyr dies for a cause and not by accident (as Bolivia's ex-President Banzer described his demise), Noel was declared a martyr to science. Martyr or not, it remains true that, publicly, the inevitable consequence of Noel's blunder into the middle of a drug operation is viewed no differently than it would be had he trodden on a rattlesnake, just bad luck. The monsters of Huanchaca (rather different from those as imagined by Conan Doyle or reassembled in the film "Jurassic Park") eliminated the Kempff party so they couldn't snitch to the authorities; so, yes, a reprehensible, but perfectly logical, motive.

The official line of enquiry has been, and still is to find out whose Drug Centre, was it? because; the owner (for the sake of brevity I will assume only one) is ultimately responsible for the regretful incident.

Without doubt this question has to be resolved and the culprits made to answer on both counts, supplying chemicals for the cocaine business and by extrapolation, the murder of three members of the Kempff party.

Two employees from the Huanchaca Drug Centre are now languishing in gaol and have admitted that they were the men who shot Kempff and two members of his party. They claim that when they heard the plane land they panicked, ran towards the airstrip and opened fire as soon as they saw strangers coming up the track towards them. And, for the moment at least, this, too, is where the official investigation seems to have come to a stop even though not one of the chief perpetrators have been arrested. Why? I believe it is because the official line of enquiry has been barking up the wrong tree.

So, from now on I am going to try to think like a policeman. I make no direct accusations, nor do I shilly-shally; I will simply put down what I know of the affair and what I think happened. Whether or not my thoughts suggest a new line of enquiry for the authorities to think about does not really matter to me; they will certainly annoy some people, especially those shy of exhuming what is still considered a politically disruptive murder, and others who will take offence where I assure them no offence is intended.

There is something not right about the confessions of the two men who say they killed Kempff and his companions because they panicked. I believe this is unlikely for various reasons. They couldn't have been sure the passengers on the plane were not bona fide visitors to the Drug Centre. Panic is a reaction that occurs when one is faced with a sudden, threatening situation. As it was, the men must have had fair warning with the sound of the plane approaching the airstrip several minutes before it actually came down. With the advantage of surprise would two heavily armed guards, one with a machine-gun, the other with a rifle, have panicked because a light plane (with a maximum capacity of five people) lands in their territory? With more men at the Drug Centre, why didn't this panicky pair call for reinforcements? There must have been standing orders to deal with such emergencies; either they had orders to kill unexpected visitors, or not, but neither scenario would seem to warrant an element of panic to enter into the situation.

Was it panic, too, that led them to hunt for Castelló all morning, and into the night? And later, as Captain Añez made his first landing approach to the airstrip, were they still panicking when they ran towards Noel's burnt-out plane? Is it not apparent that they ran on to the airstrip with the intention of defending it, with more killing if necessary?

Then there is another major factor: all the world and his wife knew that the Kempff party was on its way to study the mesa; albeit they were probably taken by surprise when Noel, at the last minute, abandoned the plan to scale the cliffs in favour of flying up there. There were other airstrips on the mesa, but the one by the Drug Centre was the obvious choice since it was the only one in serviceable condition. So why didn't the owner of the Drug Centre warn off the

Kempff party long before they set out, or at least warn his own guards in time to prevent the disaster?

It should also be remembered that by this time, with the presence of the Drug Centre almost public knowledge, the owner had little to gain by having hapless visitors killed on his doorstep.

At first glance the pilot's behaviour would seem to have been negligent, for as any six-year-old would tell you landing your aeroplane on a remote airstrip would be a foolish action indeed. Pilots I have interviewed on the protocol involved in such cases all agree: "If you are planning to land on somebody's private airstrip, seek permission to do so long before you make your trip". Here I should explain, when passing over an unfamiliar airstrip you can't radio down to those on the ground unless you know their frequency. The pilots continue: "If you are forced to take a risk (because of bad weather or mechanical failure) make sure that you broadcast your position and intention to land to every possible listener; and delay your landing as long as possible so those on the ground have time to hide whatever it is they don't want you to see. At the same time, it is a good idea to let those on the ground know that your plane is in trouble by waggling the wings and making lots of engine noise during your approach. If you see buildings below it's advisable to taxi immediately to the most deserted end of the airstrip. Then cross your fingers".

Since I was not party to the arrangements made for the trip to Huanchaca, I can only speculate here. People I have interviewed, those who knew Juan Cochamanidis, tell me he was born in San Ignacio, the usual staging post for any visit to the Huanchaca area. That he knew every airstrip in northern Santa Cruz and the Beni like the back of his hand and was equally known by one and all. In other words, he must have been confident he would be welcome to land wherever he chose to.

However, that the pilot didn't live up to expectations became apparent at the last moment when he admitted he had never overflown the Huanchaca plateau and definetly had never landed on the Huanchaca airstrip. Even so, with his experience and reputation everything should have gone well, so what went wrong this time?

First of all, I think it is necessary to remember that the pilot was not selected for any explicit quality he had to offer the Huanchaca expedition, he happened to be one of two or three men available at the time. So selected as he was at relatively short notice, he would not have had much time to swot up on the flight. Also, in Bolivia pilots who come with chartered aeroplanes are not captains of their ship in the traditional sense, they are usually very respectful and defer to the opinions of their passengers. With Noel they wouldn't think of doing otherwise. Don't doubt it, Kempff was the boss.

At first sight it might appear that Noel had a guide who was careless or, just, naive. One who failed, under the circumstances, to take any of the necessary precautions. Why did the guide fail to do his job?

Setting aside my suspicions for the moment, we cannot really put blame on the guide. As a poor campesino he was not important enough to influence decisions unless asked. The only exception might be during an emergency, and as Cochamanidis prepared to land at the Huanchaca airstrip, there certainly was one. They all saw a suspicious looking encampment; and yet, according to Castelló's evidence, Kempff and Parada insisted they land. Now there's a funny thing.

All these doubts and riddles must be resolved before the enquiry into the death of the three people in Huanchaca can progress. The problem is that apart from Noel's immediate family I doubt anybody, except me, has spent so much time thinking about it. Castelló's evidence, given immediately after the withdrawal of the expedition from Huanchaca, is understandably brief. All the attempts of the police and the commission of enquiry to elicit further details from him have failed. Vicente Castelló Losada wants to forget.

****

Miriam nudges me to draw my attention to a Dusky-green Oropendola entering its nest. The nest, like a Weaver Bird's, is a deep pendulous grassy bag attached to a tip of a branch hanging over the river above us.

It reminds me of the Christmas stockings we used to hang up so expectantly when we were kids. How worried I was that my father would punish me for my annual catalogue of misdemeanours by filling mine with clothes, or sweets I didn't want, and nothing else. Everything I asked for was far too large to go into a stocking: an electric train set, that I never got; a steam engine, that I never expected to get; a real football, that was substituted for a smaller rubber one; a set of illustrated encyclopaedias that I believe my mother wanted us to have, but in the event was scaled down to one fat, child's version.

Somehow, Christmas was always so disappointing, or at least I made it so by the asthmatic expectation I invested in it.

The oropendola reappears at the mouth of its bag; at the base of it's ivory-coloured bill a bright blue eye examines us in a way that seems to ask, "Have you made up your mind yet?" Then the bird turns its head to watch us with its other eye. Actually, I'm never quite sure how birds' eyes work, maybe it's looking in the opposite direction.

I have made up my mind. Publish and be damned.

****

So, to return to the Kempff issue. I feel I must speak out and suggest that the answer is not as simple as those self-confessed assassins claim, the result of panic, or as the authorities think, or maybe, would have us believe. I have a responsibility to put my evidence forward because the owner of the Drug Centre, no matter how vile society regards his activities, may not be to blame for the murders among Kempff's party. Indeed, I believe they were perpetratcd without his knowledge or consent.

As I have told you, a few weeks before Noel was killed, I went to see him to grouse about some of the statements he had made during a TV interview. After we dealt amicably with these, we went on to review the progress I was making with Amboró, and his forthcoming expedition to Huanchaca, that he thought I might be interested in joining. Then we got around to talk about our old topics; the constant threat to Amboró and Huanchaca by loggers and the animal exporters.

As far as they were concerned, I wanted to know where we stood, as their previous opponents, and as threatened animals of sorts now that the Supreme Decree banning their activitics had been declared.

This is roughly how our conversation went:

Me: "Don Noel, I've heard they're still logging in Huanchaca and that they intend to force government to limit the boundaries of the Park to the mesa itself."

Noel: "Yes, unfortunately that is true. It'll take time to sort that out. But don't worry about it, I'm sure I'll get my way, threats or no threats."

Me: 'Threats?'

Noel: 'Yes. Many of the loggers don't want National Parks.'

Me: 'Have you received threats?'

Noel: 'Don't worry about it.'

Me: 'Okay. What about the animal exporters? They can't export legally now, but I've heard some of them are trying to find ways to continue their businesses. What do you think we need to do to stop them?'

Noel: 'We have to be very careful. I'm not very happy about the situation and I think it would be better if you leave it to me.'

Me: 'Okay, but we both know that I've been too closely involved for me to leave here without knowing where I stand.'

Noel: 'Yes. Maybe, you're right. To tell you the truth I'm very worried about the situation. I've reason to think they might kill us!'

Me: 'Who…exactly?'

Noel: 'Them. Be careful, Robin.'

And on that warning note Noel excused himself, a frowning, worried looking man, his pale blue eyes flicking from the papers on his desk to me as I left.

****

In the light of this conversation, I think it improbable that Noel blundered into a fatal situation as fortuitously as Goldilocks's stumble into the Bear family's home. So, could it have been the result of careful planning? I believe so, and fully aware that here in Bolivia the road to good intentions is paved with hell, I argue my case.

Here it is: it's important to establish that people who create protected areas, like parks, create enemies: local landowners, cattle ranchers, logging and mining companies, colonists, animal exporters, oil companies and in spite of what they pretend, hunters and, of course, investors and politicians. These, and others are the people who have the most obvious reason to sabotage the establishment of a protected area, and the easiest way is to eliminate the driving force behind this.

Historically, those conservationists and park guards who have been murdered (and there have been many of them) have being victimised by one or more of those mentioned above because; their interests are directly affected by the circumscription of large tracts of land. In the case of Huanchaca, it had been proposed to set aside more than two million acres, so most of these elements could have been involved. And we must look where there's motive.

On the other hand, the drug mafia, apart from a degree of inconvenience, is not much affected by the establishment of protected areas for their members can conduct their activities virtually anywhere.

They can move away. Indeed, their essential mobility is one of the main factors that make their capture so difficult.

It was public knowledge that Noel was busily employed in converting Huanchaca National Park from a "Paper park" into a real one, from a nearly unoccupied wilderness into a closely guarded sanctuary. Noel's plan to study the mesa, itself, was no secret. The newspapers had published interviews with both Noel and the Doñana biologists about their forthcoming venture, including last minute radio bulletins referring to their immediate itinerary. In other words, in addition to the public at large, every ne'er-do-well in the vicinity of Moira and Huanchaca would have known the Kempff party's plans and movements.

I don't know much about the drug barons and their activities in Huanchaca and, as I have said, I don't believe they were planning to eliminate the Kempff party on their airstrip, for the murder of a celebrity like Kempff and the national outrage sure to follow would not only jeopardise their freedom but would also harbinger the death of their very lucrative operation. Surely, they were not that stupid. Surely, they would have ceded the ground to Kempff and quietly moved out. Indeed, various informants and witnesses called by the commission of enquiry stated that the Drug Centre had been shut down two or three months before Kempff's visit and was in the process of being moved to another site, called "Murcielago" ("The Bat") on the Huanchaca mesa, and from where they continued to operate for two years after Noel's murder.

Others thought that the normal operation of the Drug Centre had been only temporarily run down; interrupted at a time (with the weather normally at its best) when it should have been most active. Whatever, the conclusion is that unwelcome visitors were expected and all activity at the Huanchaca airstrip suspended.

No doubt, they would not have wanted Narcotics Police nor Kempff and his team anywhere near their illicit Drug Centre. The fact remains that when Kempff and his team eventually got there most of their installation was still in position. Why?

Many testified to having heard an extraordinary burst of aerial activity at the Huanchaca airstrip immediately prior to, and into the night of the day before Kempff's arrival. The only interpretation that could support both theories that they had moved out; and they were temporarily shut down is neither of these. Quite clearly the answer is something in the middle: they were moving out, but appeared to be still operational, for the same reason that the Kempff team was still operational at Moira but had not yet moved to the mesa. We know that unseasonably bad weather had caused both parties unexpected delays. Whatever, somebody was clearly in the know.

But you might ask: "if they were in the know, why hadn't the drug factory been moved in plenty of time, months before Kempff's visit? Part of the answer is provided by copies of Kempff's correspondence with Doñana and one excerpt from Cabot's book (see below).

16th April 1986: Letter Kempff to José Cabot, keeper of the Scientific Collections at the Doñana Biological Station: 'How are the negotiations for the trip to Huanchaca going? Is it already financed?'

20th June 1986, ibid: "As time is flying, and we still have a lot to do before August, I beg you, as soon as possible, to confirm the trip to Huanchaca National Park. If it has to be postponed I still have time to take a different course, given that the month of August is always set aside for exploratory trips."

In Cabot's book: Nieves, Cabot, 1997. La expedición zoologica hispano-boiliviana a la Serrania de Huanchaca, p64: "...with June well advanced we Spanish still weren't certain that we could mount the [Huanchaca] expedition."

In other words, not even Kempff knew whether or not they would be going to Huanchaca until the last moment.

And talking about last moments, it was only at the beginning of September that the expedition abandoned attempts to force a way up Huanchaca's legendary rockface and took up the option of flying to the airstrip.

So, given the forced delays and last-minute changes to their plan, why didn't the drug barons warn Kempff off? Maybe they did; witnesses in the drug enforcement agencies testify that there was no plan to interview the Kempff party whilst they were at Moira, and in the event this never took place. So, who was it behind this last-minute ploy to scare off the Kempff party? [It would have been interesting to trace the origins of the radio message.]

Whatever the truth of this, I believe the drug barons had a last-minute understanding with Kempff. I believe they expected him to keep the biologists well away from the Drug Centre which lay seven hundred yards from the airstrip.

I believe Noel and maybe the guide as well, had been told that if it was necessary to use that airstrip, they had better be sure to do this.

This can be the only, and one and only, reason for the inexplicable reaction of Kempff, Cochamanidis and Parada, on seeing for themselves, an obvious, large, drug installation just before they touched down at the Huanchaca airstrip. I will remind you of their reaction to Vicente Castelló's immediate request to abort the mission: Cochamanidis, the pilot, said nothing; both Kempff and his guide insisted they continue as planned. Are we to believe, then, that a relatively naive Spanish visitor immediately recognised what an encampment of tents and dozens of oil drums implied, while the three Bolivians did not?

Of course not. In fact, they were so confident they didn't bother to make a reconnoitre of the encampment that posed a real danger, but made a quick one over the airstrip that didn't, before landing. Then, without further delay they opened the door and all of them scampered up the track towards the drug encampment, just happening to notice on the way what a busy little road it was, as if at the end of it they would find themselves welcomed to Bruce Lindenberg's stillborn camp for tourists.

I conjecture the precipitate rush for the encampment was to tell the guardians something like the following: "We're the ones you've been expecting. Sorry we're late. Don't worry about us. We'll stay well away from your camp."

Would this also explain why Noel insisted that he be on the first flight? Maybe he was expecting some sort of trouble and thought his celebrity presence would guarantee a rapid solution. Reading Cabot's book, I get the feeling that Noel didn't want any of his Spanish colleagues on the first flight. If he hadn't told them about the Drug Centre, he had good reason. Later, there is another little incident that supports this: during their dash up the path to the encampment Noel stops [Castelló implies that Noel was too tired to continue] and returns with Castelló to the airstrip.

And, Cabot informs us, Noel got very angry when told the narcotics police were planning to visit him. Why? Did he think they might put his arrangement at risk?

Maybe this is how it was. If so, why were they met with a hail of machine-gun fire?

The official report of the Commission of Enquiry apparently had no trouble explaining away all the inconsistencies I have referred to. In practice, not only are there a lot, but they come thick and fast.

Notwithstanding, their verdict was the death of Noel and two of his companions was the result of misadventure.

******

Misadventure? No. Noel was cut down by machine-gun fire because two assassins were waiting for him.

What nobody knew was that somebody, who I will refer to as Mr. X (equivalent to Noel's "them" during his conversation with me) had already taken steps to protect his logging interests in the country, almost certainly in Huanchaca and Amboró too.

Most loggers as I have already discussed, encourage close members of his family to provide meat for his workers, in return for this inconvenience his relations are allowed to take up the very lucrative business of exporting live birds and animals.

Reasons for Killing Noel are understandable. He had ambitious plans for expanding the number of National Parks in Bolivia and the size of existing ones, but why bother about me? I think it would be tedious for readers to repeat the answer covered in many chapters of the book.

Later, I would have included my successful efforts to triple the size of Amboró National Park as another motive, but at the time of Kempff's demise the Park was a relatively small area of only minor interest to the big time loggers.

However, what myself and Cheri were known for was our dedication to stop the wildlife exportation trade. Cheri and I were never assaulted or had our lives threatened because of our work in Amboró, but as you have seen, when we interfered with the Animal Mafia's interests, we became prime targets. With Cheri back in the USA the animal exporters, all of them, Noel's "they", would certainly like to see me out of the way.

It is necessary to mention that The Association of Animal Exporters still hoped their lucrative business would be allowed to continue for the 1985-86 period. On behalf of Antelo, Carillo, Claure, Roig, Romero, Aguilera, Minten and Onishi, Juan López signed a letter on the 11th of January 1985 to the Director of the CDF-SC asking for permission to export 300,000 birds, of which 30,000 were macaws with a value between $200-2,000 each, and $300,000 worth of monkeys, and other animals from the 1st March 1985 (twelve months after the initial one-year ban). That their request contained many protected species does not seem to have bothered them. That

Supreme Decree No 21774 killed their hopes of making millions, must have really bothered them! This Supreme Decree of the 27th of June 1986, announcing a permanent ban on all further exports of Bolivian wildlife raises another possibility. That this decree was signed barely more than two months before the tragedy in Huanchaca might suggest another motive for Kempff's assassination: revenge!

Mr X, or more likely his local henchman, almost certainly knew the pilot, Juan Cochamanidis, and maybe the guide, Franklin Parada, as well. He would have been party to all that was going on. Indeed, he might have been responsible for feeding Noel false information about the safety of the Huanchaca airstrip. It is even possible that the pilot or guide was Mr X's henchman. I think the pilot can be excluded if we remember that he did not choose to come, he was chosen, and with short notice at that.

On the other hand, Parada would have to be a prime suspect; a man who knew Kempff intimately, having accompanied Noel on numerous field trips he became a man in whom Kempff placed much confidence, and one in a strong position to influence him. But he was also a relatively poor man susceptible to bribery; a man who told two of the Spanish biologists that Noel had given him the opportunity to start his life again, a classic Judas profile; and (see Cabot, p. 113), more to the point, had rescued him from a very risky past that in Bolspeak would suggest he had been working in the drug business. Could this unexpected intimacy, proffered on the 1st September, the day before the scheduled flight to the mesa be a sort of confession on Franklin's part? Now that the time was so near, was he suffering from remorse? And, what of Parada's vehement support of Noel's decision when they saw the drug encampment below them? Of all those aboard the plane, Franklin with his risky past, should have been the first to state the obvious, protect his boss from this sudden emergency and suggest they get the hell out of there. And, later, at the fateful airstrip, is it not true that Franklin, standing quietly next to Castelló, was the first to be shot. To me, suggesting he was a prime target; and "dead men tell no tales".

And isn't it coincidence that the second to be shot was Noel, the other prime target? In fact, he and Franklin were gunned down at a moment when the most obvious target if they were all of equal standing as unexpected visitors would be, was Castelló, young, tall, lithe and rested, standing a few paces away forcefully pleading their case as innocent biologists.

Mr X's plan to have Noel killed was a very clever one, taking advantage of a situation that would be blamed on the drug mafia. It is almost certain that some of the men working at the Drug Centre were known to his henchman, since men who need work will quite happily switch from cutting logs to cutting cocaine base. This is probably why, after Kempff's death, the workers at Moira ran away from the sawmill; they had worked for the Drug Centre at some time or the other and could be sent to jail for doing so. A bit of money, then, placed in the right hands would have bought the hired assassins at the Huanchaca airstrip.

In other words, Mr X had a strong motive, and he had the information and plenty of time to organise an ambush and hire men to do the killing.

But who is Mr X? We've already suggested he has logging and wildlife export interests. He's Bolivian. He's clever and wealthy. He's greedy and powerful. He's a murderer. He's a "Nature Hater". He probably lives in La Paz or Cochabamba. But he's not a drug baron, at least not one of the Huanchaca cartel. He is, probably, as everybody believes a high-ranking politician.

So, what is the truth? We cannot have it both ways: either Kempff and his two companions died gratuitously, or their deaths were part of a carefully planned assassination. I have put forward my evidence. If Kempff's murder wasn't planned, his death, that occurred a scant five weeks after he said to me "they might kill me," was an extraordinary coincidence, the result of two panic-stricken guards; the selection of the Huanchaca airstrip was an awful mistake (which nobody, among the hundreds of people in the know, thought to correct while there was still time); the pilot and guide were insufferably incompetent; and Noel pitifully naive.

But if I'm right, then Noel was a true Martyr to Science. God bless him.

****

So that's it. I lied to you when I told you Miriam and I came to the Park to count parrots. I came to tell you this. I came to tell the world this.

And why have I waited till now to do so? And why have I chosen to do it this way? Why didn't I just step up to the Commission of Enquiry and say my piece?

You kidding? Like Salazar's, they might have been the last words I ever tried to say. Now the whole world knows. And if they come for me now, you'll know I was right.

****

Oh! I nearly forgot. Just one last thing to clear up: my suspicion regarding the Chief's what-you-ma'call-it. Amboró, I am reliably informed by a number of experts there is no such word as Amboró in Guaraní as almost none of their words begin with a vowel. 'Mboró, on the other hand, is a good Guaraní combination of two words that perfectly describe the rocky outcrop which gives the Park its name. So 'Mboró National Park it should be.

I have spent more than twenty years in Bolivia to establish 'Mboró National Park, and as "'Mbo" means prick, and "ró" means head, some people will say I was the right man to do the job.

"So you see, Chief, you're just a big prick after all."

****

"Miriam, shall we go home?"

9. The Way It Is

It is supposed that when man, *Homo sapiens*, first crawled out from his monkey skin he was little differentiated, except by potential from his forebears, Neanderthal Man. His brain was now better organized, he became well connected, God made him a member of the Old Boy's Club, putting into his reach the benefits of clothes, fire and weaponry. With a little organization he discovered that his needs could be acquired more easily, instead of gathering he planted, instead of hunting he herded. The price he had to pay was twofold. Division of labour was necessary and some had to accept a lower position than others; subservience and envy were born, equality lost. Somebody had to work every day, Weed the plants, Milk the cows and goats, Guard the horses and sheep. Man took on responsibility and lost his freedom. Chiefs were elected, rules were established and man developed power over his fellows. 'One for all and all for one' became the order of the day, and nonconformism emerged.

As anybody who has studied the process of evolution knows, the non-conformist is a paradox. He bears the seeds of destruction when he is wrong and the germ of salvation when he is right. Within a small community or organization the machinery is less structured and aberrations readily tolerated and, if they appear to be better absorbed and developed to their logical conclusion. Over the decades rapid advances are made; 'rocking the boat' can sink it or produce a more seaworthy design. It follows that the most successful renegade is he that can do both: point out the faults and produce an improved version. He may be called a 'Genius' or 'Jesus' for his efforts and become the new leader. New ideas, whether it be a more effective club or a better system of government, wield Power!

Large organizations by their very nature are less well adapted to change. A small change in one segment can trigger off an undesirable chain reaction and cause havoc to the system as a whole. Non-conformists are less welcome, they must dissimulate, be isolated or, maybe, killed.

In the past one way Society dealt with this problem was by territorial expansion, whether by war or colonization, and in doing so formed societies within which the nonconformists were welcome, at least at first. The 'Conquistadors' and the 'Children of the Mayflower' were largely non-conformists, not the contented 'hoi polloi', nor the wisest of men. But The 'Mayflower' carried in its hold a largely peaceful people, lured to the 'New World' by promises of land, whereas many of the Conquistadors were unsuccessful, dissatisfied men. Unable to improve their lot at home, they went looking for a new way of life to provide their needs.

So: The Spanish brigantines carried men of a different breed, men in search of treasure or the rapid fortunes to be made by the promise of free labour (i.e. the unsuspecting 'Indians'). They were 'freebooters', nonconformists of the most destructive kind. When they arrived in the new territories of South America these hopeful men discovered that nothing really changes, the golden opportunities they sought had already been exploited by the favoured few. Hardly any of them brought their wives and families since they had no intention of staying for long. Their independent streak, their intransigence remained and, faced by the frustration of all their hopes, many of them elected their own leaders and retired to the freedom of the backwoods.

But their inability to lead and reorganize themselves remained as dormant as it had been before, and the disheartening circumstances that confronted them stripped them of their resolve to improve their situation. After more time spent brooding about the injustices of life they took solace by adopting the pretty Indian girls as their wives ... and soon fell into the indolent ways of the vanquished natives.

In North America there was little talk of treasure and nobody expected the Indian to be his slave. Most brought their wives, even whole families, and they came prepared to work the land that was free for the asking. The North American was, in short, prepared for industry, the South American prepared for sloth. The North American was largely of English or of European extraction and came carrying the seeds of co-operation and government through mutual interest. The South Americans were largely Spaniards or Portuguese carrying the instincts of their forefathers, the urge to be independent, the will to disassemble.

Today, in South America, these fundamental characteristics are the root of its problems. Every man is pariah or king. Nobody wishes to work where everybody can live by only paying lip service to the idea, it is below one's dignity to work for others if you can manage without doing so. The labour available is mostly to be found among the Andean Indians, for they live in an inhospitable environment, where subsistence farming is difficult and requires more effort than working for somebody else. And let us recall that many centuries ago the 'Incas' taught them how to be servile. But, before you rush to open your shoe factory, farm, or whatever, you'd better be aware that the 'Altiplano Indians' are no better fitted for labour than the lowland Indians are, for by nature most of them have been found to be proud, intractable and (even) deceitful.

Even today a great part of South America's people cherish the concept of free-booting. In Bolivia, although many of the better classes, 'La Buena Gente' ('The Upper-classes') hold important positions in Government and Commerce but, they rarely manage to demonstrate much commitment to their work; and truly, it has to be said, they make very poor employees. It is as if they all consider their work an unfortunate circumstance, a temporary indisposition necessitated by the vicissitudes of the situation; a fall from grace (that) 'Ojala' ('With Luck') they will soon be able to forget. The inference being, that in some unstated way, they will be miraculously provided for.

As I have said, in Santa Cruz such miracles happened. During Fawcett's time, when the population of Santa Cruz was a mere 20,000, the First World War brought a huge demand for rubber and the town found itself well placed to take advantage of the circumstances. Thirty years ago they discovered oil and the town became a city overnight. What had been no more than a large village or a small market town, suddenly swelled into a sizable city with five times as many people as before. In the short space of ten years, where once stood a sort of Oriental Bazaar, up popped like mushrooms after a shower of rain, Hotels. Restaurants. Bars. Night clubs. Bordellos, Cinemas and Sports Clubs to cater for the influx of foreign technicians. Where once only the Bullock Cart with its solid wooden wheels, creaked and squeaked its way around the muddy streets. Now modern automobiles hoot and honk their way around paved roads. Where once shops were filled with Rope, Saddles, Hammocks, Pitchforks, Tallow, Candles and Seeds, they now sell Electrical goods, Computers, Car parts, Jewellery, Watches, Records and Tapes and Fashionable clothes. Supermarkets, Banks, Filling stations, Insurance companies, Travel Agents and Hamburger Stalls have swept the traditional commerce aside. Blacksmiths gave way to 'Gomistas' ('Puncture repairers') and Combustion Motor mechanics, Horse Trading to Car Parking Lots, Animal Hides to plastic sheets, Weaving to synthetic fibres, Hoes to Herbicides, Firewood to Propane, Farm produce to imported tins and packets, Wattle and Daub to cement and concrete. In a very short time the streets were no longer filled with 'Campesinos' selling their wares. Who needed them? Overnight the country people were dispossessed. Left with no alternative they turned around their horses and carts, and teams of oxen, to return to their rural villages where they still sit, unwanted, disillusioned and poorer than they had ever been before. Bolivia had a new class of indigents. In one short decade a whole way of life was eliminated. No longer were the country roads busy with the toing and froing of farm families as they made the long journey to town, voyages that sometimes required weeks to complete. Old friends never saw each other again, wayside businesses disappeared, highways and byways became overgrown and finally lost to the forest, that, always hungry for revenge, quickly reclaimed them.

Then the oil began to run out. The foreign technicians began to withdraw. Money began to dry up. Having been built brick by brick as money from the sale of drugs became available, new mansions in a derelict state were now a familiar scene. Garden walls only knee-high. Staircases leading to nowhere. Windows without shadows, feint footsteps echoing down the empty corridors. Gardens choked with weeds and the weeds overhung by trees. Supermarket shelves remained full, while restaurants stayed empty. Hotels hollow halls of passing shadows, faint footsteps echoing down the empty corridors.

The people of Santa Cruz waited for the next miracle that, like Spring, was sure to come. 'Oh ye of little faith', come it did. Not only a second miracle but, a miracle of miracles. Cocaine came to town. Where oil had been a blessing to the better-off townspeople and their dependents, cocaine promised riches for all. Now a new breed of strangers came to town: Colombians, Argentinians, Americans, Italians and Dutchmen to mention a few. Big men needing large supplies, little men seeking an ounce or two, Santa Cruz welcomed them all, for here was a new wave of freebooters, the 'New Conquistadors' ready to risk all for the sake of a quick fortune. Unlike oil, the exploitation and profits of which were largely in the hands of big foreign companies, cocaine dealers were largely Bolivians: 'Los Pichicateros' as they came to be known. Whereas oil is largely a technical business requiring huge investments of capital, cocaine

offered something for everybody, there was no-one too lowly or too
ignorant to profit from this magnificent new enterprise. Add to this,
the very essence of the business, its illicit nature, its secretive needs
(which favoured independent elements rather than a highly organized
machine) and it is easy to understand how it was, that more than half
the population of Santa Cruz became employed in its illegal machinations. Only a genius could have planned a business with such
potential. Like the explosion of an atom bomb, cocaine sucked
everybody into its core, peasant or politician, thief or judge, cat or
king, nobody could resist the inexorable corruption, few even gave it
a thought. Cocaine was a North American problem. Bolivia would
take advantage of it and among those who did were some officers and
generals in the military.

A Guardian newspaper report started with the headline: 'The town
that got high on cocaine' and went on to justify it with these words:
...'for Santa Cruz took the advantage and now is one of the World's
Cocaine Capitals. Ignored by President Hugo Banzer (1972-79) —
himself a Cruceñan — the trade sprouted and flourished, but the real
boom awaited the presidency of Luis Garcia Meza. With him, one US
official is quoted as saying, 'the drug mafia of Santa Cruz has bought
itself a Government'.

At the very top of this Mafia pyramid are the favoured few, the
moneyed brains. No more than half a dozen men, these 'Padrinos'
control ninety percent of a trade said to be worth more than $1 billion
a year. Each one has an organization of several hundred persons. I
have not used the noun 'employee' or the verb 'employ' because
being a hireling, as the words suggest does not apply to a proud
Bolivian's concept. For them, the notion of 'personal contacts, loyal
to him and eventually receiving from him their share of the profits'
would be more respectful.

The 'Mafia pyramid' are a varied lot, Judges, Generals,
Politicians, Bankers, Policemen, Pilots, Chemists, Gunmen, Para-
militaries, Mechanics, Drivers and a lot of odd-job-men and ne'er-
do-wells. Where possible they are recruited from within the family,
and families here are very extensive indeed, given that the main
occupation of the menfolk is to wander around impregnating as many
of the local females as is possible. Each family member will have his
own mini-organization drawn from his own circle, thereby greatly
extending his Padrino's own central family. Happy to be paid to
answer the phone; run and get the boss (often drunk) from his

favourite restaurant. A chauffeur to clean the Mercedes. The gate to be opened and closed by the 'Sereno'. The family to be kept clean and fed by the maids. The Dobermans (of which there are many in Santa Cruz) to be exercised by the Garden Boy.*

Only a short step removed are yet others, members of the 'Respectable Family' occupied in laundering the money. Among their ranks are many that do not know that ultimately,

they are working for the Cocaine Mafia. They set up the businesses whose social legitimacy is best described by a joke born from the *Cruceñan*'s wry sense of humour.

First man: *"I see you have built yourself a new drug parlour."*

Second man: *"Yes, but at the back we've got a bakery!"*

A joke, that not only refers to the socially acceptable status of the drug barons, but is also a jibe at the Money Launderers in a small town where competition is stiff and new ideas at a premium.

The intricate web of kinfolk and friends loyal to the 'Pichicatero' extends from the heart of the city to the main rural towns and from there, to the depths of the countryside where the coca bushes are cultivated. Here a separate organization having little connection to the main family takes over. These people are the contractors and their agents, working busily to fulfil their drug quotas to avoid finding themselves chucked out of their business.

Author's note: The detailed account of the drug trade was given to me by two of my friends. One of them, a Bolivian pilot making a living by working for the head of the family, flying shipments of 'base'(the illegal precursor bought by the contractor to produce Cocaine) from the boondocks to the city of Santa Cruz.

My other informant an American who came to Santa Cruz to make his fortune by smuggling 'base' back to the USA but, as naive as he was he fell into the hands of the law. He had supposed that his Bolivian contact was helping him to do this but he was wrong, his contact was an informer employed by the Bolivian equivalent to the DEA. And being so my informant was freed from prison pending his return to the States. It was during his rather long wait that, in great detail, he told my wife and myself about his situation as a hostage, he would be allowed to return to the USA as soon as his father transferred thousands of dollars to someone's anonymous account.

The contractors deal with those at the bottom of the cocaine pyramid, the Campesino farmers who produce the raw material, who plant, tend and, after three or four years, harvest the crop of leaves. In 1984 it was said that 500,000 Bolivians were occupied in growing, processing and transporting coca.

Apart from the traditional farmers (many of whom also sell part of their harvest directly to the drug trade), there are those that have been forced to farm coca for the contractors. Most of these Lost Souls are a shiftless lot, repaying their debts or working themselves back into favour, or occasionally through blackmail being forced into this new way of life. They may also act as agents for the contractor, keeping an eye on those farmers producing the quota for him, some of whom will have been paid in advance to do so. An average coca farmer will have between four and ten acres of bushes and will produce from one to two tons of leaf each year: worth about $600/ton when dry and ready for sale. [Incidentally, a ton of leaf will produce about one kilogram of pure cocaine.]

Coca will grow quite well under forest shade and many farmers do so to avoid detection. But, most prefer to fell the forest because the plants produce more leaf and the bushes are smaller, making harvesting easier, and some of the farmers say that mosquitos and other biting insects, not to mention snakes, are less of a problem.

The bushes are usually grown from cuttings and planted in spring before the onset of autumn. The bushes are perennials and are pruned to keep them to manageable proportions and seldom exceed two metres in height. Only some of the leaves are removed from each plant, the apical ones being left to stimulate more rapid growth. The major problem for the farmer is drying the leaves after harvest since they must not be allowed to go mouldy. Drying is done on sunny days when the leaves are spread out on concrete or hard mud aprons and turned every few hours. As soon as they are dry enough, crispy, they are placed on raised platforms under a palm leaf shelter. Every once in a while all the leaves in this crude store will be taken outside for re-drying. The women take charge of this important part of the operation and for convenience the store is situated close to the house.

[Flying over a coca growing area, a friend of mine took the drying leaves to be duck ponds, and mentioned their prevalence in an official report!]

These leaves are destined for the market places of the altiplano where the people chew the leaves mixed with a pinch of baking soda or lime called 'yibte' to release the stimulating alkaloids. Legal growers tend to sell a sackful of leaf as it becomes dry. Illegal growers, that is unregistered producers within the two recognized areas, and growers in other parts of Bolivia, may store a ton or more of leaves with the consequent problem of keeping them dry. As time goes on this takes up more and more of the family's energy and those who have not pre-sold their crop may end up throwing the whole store- full away for want of a buyer. This is one reason why most coca farmers are only too willing to grow on contract; the other big incentive is that the drug Mafia pay higher prices. As soon as they have a sufficient quantity, the contractor's agent is summoned to inspect the stored leaf. If he is satisfied he will contact his superiors and return with the money.

The leaf is now taken to the place where it is to be processed: 'the kitchen'. As far as the contractors are concerned the risks start with the removal of the leaf from the farm and its transport to the kitchen. For the sake of secrecy this is done in his own truck at night and for obvious reasons the kitchen is usually in some well-hidden place that can be approached without recourse to any of the main roads. The more inaccessible the better. Nowadays, with the increase of anti-coca operations the leaf is rarely processed on the farm and may be transported by boat to distant locations hidden in the depths of the forest.

The kitchen is where the mass of coca leaves are reduced to manageable proportions. It may not be more than a thatched roof shed, a few troughs made from thick plastic sheet and a stack of drums containing sulphuric acid and kerosene. The whole set up can be easily removed or abandoned without much loss of investment should the need arise.

The leaves are placed in the troughs and soaked with kerosene, then macerated by trampling them with bare feet, like grapes, until they lose their colour. This messy solution is then washed with the acid until a thick paste called 'Pasta' is formed. After more washing and when dry, this is the paste commonly smoked in cigarettes called 'Pitillos'.

More uncommonly, the kitchen is converted into a more soph-isticated and fully mobile set up by putting everything into the back of the Contractor's truck. Electricity is provided by throwing cables over some unwitting person's private power supply. Instead of trampling leaves they're macerated by fixing outboard motors into the drums or, in a state-of-the-art set up the drums are discarded and replaced by washing machines.

Irrespective of the set up used, and after the chemical processing of the leaves has been reduced to 'Pasta', this paste can be further refined with kerosene and sulphuric acid to produce the infamous 'base', the illegal precursor sold by the Contractor to the Mafia to produce Cocaine.

Each contractor may have as many as 200-400 farmers supplying him with leaf, and from their total production he will produce 4-6 tons of base each year, worth about $15 million on the streets of Santa Cruz.

In the past only base was produced in Bolivia, but with the increase in law enforcement in other countries (such as the moves against the Cartel de Medellin in Colombia) more and more Bolivians are supplying the finished product. To produce the crystal (the 'Sugar' of users) the dark brown coloured base has to be washed and re-washed in ether, and the resultant liquid distilled and dried out to make the familiar looking crystals of pure cocaine. The problem is that the process requires a properly appointed laboratory and chemicals hard to obtain, factors that greatly complicate the operation, the size of the investment and the risk of detection.

But here I am getting ahead of myself. Let us first look at the situation as it largely remains, Bolivian 'Pichicateros' with base to sell. The Colombian, before taking this hard paste back to his country needs it to be more concentrated, by how much dependent on his logistical problems. Gone are the days when any old stuff was acceptable, the really crude base of yesteryear is now being further refined in Bolivia. As we have seen Kitchens have become increasingly sophisticated. The base now purer, prepared to acceptable International standards. Any reduction of quality between one ton and another could affect profits of many millions of dollars.

Mr Big in Colombia wants a ton of refined base of sufficient purity to make 900 kilograms of crystal cocaine, worth $50 million delivered to the United States. Mr Big tells his agent in Santa Cruz to scout around and find the people who can supply, as a single lot, what he requires. He doesn't want to buy from more than one source, because of security and because he requires as homogeneous a product as possible to avoid any reduction of quality that could affect his analysis.

When Mr Big's agent lets him know he has lined up a potential supply, Mr Big will dispatch his courier with the cash, has to be cash nowadays. On several occasions I have heard people discussing how big a million dollars is, all in new $100 notes: Eight two-kilogram dried milk tins?. A medium-sized suitcase?. A rucksack full?. The auxiliary petrol tank of a Landrover?, and so on. In fact one million dollars in crisp $100 notes will fit neatly into an average sized attaché case, occupying about three-quarters of a cubic foot.

I know because, believe it or not, an absent-minded friend of ours once left such a case in our house. The point is the courier cannot just take the first plane out. A million dollars is too much to lose. So, the courier will have to make his preparations carefully and take his time. He is an expert and will be paid (maybe, as much as 4%) for his time when he delivers, and when he doesn't, well ...

As soon as Mr Big gets the message the money has arrived safely he will dispatch his Chemist on the first flight out.(I give him a capital C because he is a very important man). The Chemist will in fact play a double role, analyst and hostage.

Meanwhile, the local 'Contractor', waiting in Santa Cruz has been told to get the stuff ready for analysis. Nobody is going to make unnecessary trips out to the campo so the base has to be brought to the city. The problem is a ton of base is bulky, heavy, and what's more ... it stinks!, literally. It has the unpleasant strong odour of tobacco-filled old socks. These factors present difficulties but with careful organization, timing and adequate wrapping, the consignment is moved to a favoured location, a very private residence near to the centre of town.

The day the Chemist arrives is a day of sheer farce. Telephones start ringing all over the city. *"Has he arrived yet? Is he the same chap as last time?"*

"Does he look nervous? When does he want to start work?"

"What hotel is he booked at?"

"Is he awake yet? Is he in a good mood? Does he need anything? Food? A driver? Ah, a girl. No?, is he feeling sick?"

Dozens of people await the moment the Chemist decides to start work. When he does so he will be locked in with the base until he has finished his analysis, work that might occupy him for several days. He will have brought with him his personal equipment: balances, reagents, standards to calibrate his tests, and so on. The gross equipment will have been furnished by the local drug dealer. The chemist's job is to ensure the goods are up-to-scratch and any variation is minimal among the samples he takes. He cannot leave his temporary goal until his verdict is made and the money is handed over.

The moment the Chemist gets busy, the money men, those who will be counting the cash, will be got together. Thirteen thousand one-hundred-dollar bills (the price of a ton of base) will be tested for forgeries using a small calculator-sized machine which emits a sharp "beep" when passed over the surface of a genuine note. Silence indicates a fake. Nobody counting the money or guarding Mr Big's interests can leave the room until the Chemist's decision is known. Actually, the men counting the money are third parties, contracted for that purpose and they are not taken to the money until the very last moment. Since they cannot leave, no one person knows where both money and Chemist are.

When the Chemist is satisfied with the analysis of his samples he telephones Mr Big in Colombia. Mr Big now telephones the okay to his courier and men guarding the money. The third parties counting the money now ring the 'Contractor' to tell him the chemist is satisfied and the money is correct and ready for transfer. The 'Contractor' now phones his men guarding the Chemist and the base, to say that the sale can go ahead. They are given a telephone number to ring where Mr Big's men are waiting to take charge of the shipment. When everybody is in place, Mr Big's men telephone the courier that they are now in possession of the drug, and the 'Contractor's agent telephones the guards with the Chemist that they now have charge of the money. All three parties: Mr Big's men with the drug, the 'Contractor's personal men, and the third parties who counted the money can now leave on their separate ways. The money counters, together with the 'Contractor's agent, will deliver the money to him, less $100.000 to cover their expenses. Honesty doesn't come cheap. But who cares! In Santa Cruz its pay day for everybody!

No wonder the three horizontal bands, green, white, green of the *Cruceñan* flag are said to represent the coca leaf, cocaine and the dollar.

By the time the family has been paid the base will be well on its way. Taken by truck to a waiting aircraft, hidden somewhere in the campo, within a few short hours the shipment will have cleared Bolivian airspace during the night and will arrive in some isolated jungle strip with the Colombian dawn. Here the drug will be carefully refined, washed in acetone and ether, crystallized and re-crystallized to produce the white sugar of pure cocaine. Now the most risky part of the whole operation takes place, getting the drug safely to the United States. Once there, it will be diluted with inactives and often cut with 'speed' to about 20% of its original purity. Then it will be sold on the streets for about $100 a gram. It is said that amongst the younger generation one in ten persons will experiment with cocaine at one time or another. On one occasion — I was reliably informed by a journalist, it was put out in bowls during a reception at the White House for anybody who felt so inclined. Indeed, cocaine has been called the 'Intellectuals Drug' and many medics and scientists are regular users.

It may be too naive of me to say that the Bolivian drug mafia undertook the production of pure cocaine reluctantly, but I believe this was so. My argument is based on the character of the *Cruceñans*, the free-booting mentality with its basic nonconformism and independence, and the evidence of the past. To begin with the latter. Since the early part of this century cocaine has been smuggled into the USA. Since 1970 the popularity of the drug increased enormously and now amounts to 30 tons a year, worth $25 billion sold on the American streets. The Colombians with, so it is said, their technical know-how and criminal contacts within the States, commanded almost total control of the drug business, in spite of the fact that the 'Cocaine' was purified far to the south in Peru and Bolivia.

Only now, when the Colombians find themselves with increasing difficulties, are things changing, with Bolivia producing about 25% of the pure cocaine arriving in the USA. Given that a lot more profit is made from the pure drug, "Why?" we can ask ourselves did the Bolivians not do this before?" To say that they did not have the technical capability is simple-mindedness. After all the process of distillation is not very complicated, and Bolivia with many of its own technicians employed in the Government oil industry (the 'Yacimiento Petrolifero Ficales Bolivianos') could have supplied the demand. And, even had they not been able to, there was enough money around to import the expertise had they wanted to. The same argument can be applied to the marketing of the drug in the States. Miami is almost second home to many *Cruceñans* and it is difficult, for me, to accept that whatever contacts were lacking could not have been quickly established. Had we been considering a newly born venture the argument might remain 'quod erat demonstratum', but we are talking about a lapse of more than a decade.

No, I believe, as I have said, the Bolivian Mafia did not want to involve themselves in supplying cocaine, as there may have been a tacit agreement with the Bolivian Government to this effect. Now they did so because they had been forced into it. Their free-booting mentality demands perfect freedom and still remains a basic tenet of the Bolivian character, allowed as it is to develop in the, still, primitive conditions of Bolivia's society and government. The unsophisticated process of base production, its rural backdrop, its simple necessities, its buccaneering dominance of the unworldly coca growers, its independence from International overseers and its reliance on the extended family, is ideally suited to the idolatrous, easy-going nature of the people; clever people who know their country well and whose motto could be 'keep things simple'. (If only all their countrymen would bear this in mind!) Why? then would they wish to forego this relatively trouble-free, pastoral vocation for something quite different? Full as cocaine production is with added risks at all levels of the operation. More profit? Probably not, since much of it would be swallowed up importing the necessary chemicals, the provision of laboratories, larger payrolls and larger pay-offs etc, and lack of motive, for, after all, the 'base' business already made them very rich men indeed. Also, the pure drug attracts other problems, the focus of International Police, especially the inappositely named 'US Drugs Enforcement Agency'. The interest of the World's top criminals, the big Mafia, with the consequent loss of autonomy and the easy life style. The pressure of International Politics and, subsequently, pressure from National politicians. This is not to say that production of precursors is ignored by these forces, but put simply, nobody here pays too much attention to a dead dog in the street, it's the stink that demands attention. Presidents Reagan and Bush could scream foul play over the exportation of cocaine to the States, but 'base' being one step removed, is a less controversial substance.

So what really happened to force the hand of the Bolivian entrepreneur?

One significant event involving Roberto Suarez, the so-called 'Rey de la Coca,' almost certainly provoked a change. The Cocaine King had been accused by the Colombians of supplying sub-standard base, that led to a serious argument during which it is suspected five

Colombians were shot dead by Suarez's cohorts. After this episode Suarez set up his own connections with the States and proceeded to supply cocaine directly to them, proof of his capacity to do so and proof, maybe, of his reluctance 'a priori'. [Incidentally, this incident was adopted by a Hollywood movie in which Roberto Suarez was called Mr Sosa.]

This event, significant as it was, could not account for the changes that were to come to the Colombian organization. Times changed; the Colombians found themselves operating under increasing difficulty. Moving large quantities of illegal substances around had become more difficult with the introduction of more effective law enforcement. And maybe more important than all those referred to above was a swing in public opinion at home, those that had been apathetic to the trade in drugs now disliked the reputation that Bolivia and the Bolivians were getting what they deserved provoked their outrage.

This change in public opinion, in spite of the Suarez setback, occasioned the Colombians to transfer part of their operation to Bolivia where the drug Mafia had bought the Government and armed forces. Where large tracts of relatively unpopulated territory favoured the clandestine nature of the Colombian's enterprise; and where, perhaps more importantly, the raw product was in abundance. With such obvious advantages in union one may wonder why it was that the Colombians waited so long.

I would maintain it was the Bolivians who opposed it. Whatever, events overtook them. Bolivia got itself a democratically elected Government that pandered to public opinion. Had this not happened, one might assume that the Colombians would have moved their entire operation to Santa Cruz. But as it was, Bolivians were tired of the coca coups of the past and came to realize that the wealth generated by the drug trade was not in their long term interests.

Even the campesino farmers stood to lose in the long run. The illicit activities of the coca barons are better served by maintaining the peasant population in a state of subjugation. Their basic strategy has to be one of anti-development. The campesinos are living in a state of perpetual subsistence. Illegal coca cultivation, in the absence of other cash crops, may be their only source of income. While this remains so, they are forced to take the risks involved. 'Hobson's Choice'. Given alternatives through cheap loans and agricultural development schemes, they are likely to turn upon their present masters; especially now that Government is, for the first time, trying to tackle the problem in earnest: to meet the conditions laid down by International agencies who will not supply development loans to countries that don't demonstrate effective drug control. Although the drug business as a money earner is far more lucrative than any other Bolivian enterprise, estimated to be worth a billion dollars/year (more than the Gross National Product), the money earned goes to the drug families, or disappears into foreign bank accounts. The sale of drugs does not supply the funds to invest in schools, industry, build roads, hospitals, nor, as we have seen, pay for development projects.

The profits of the drug trade can't benefit everyone. But what a pity!, for Bolivia is sorely in need of Industrialisation and markets for exportable commodities like good cash crops which are labour intensive. 'Pacha Mama's' gift has turned out to be a two-edged sword that only a change of International Law could alter.

Should cocaine and other drugs be legalized? *"No!"* say the politicians, the Church, the Police, the Mafia, the old, the mothers and fathers. *"Maybe"* say the legislators, the medics, the social workers, the singles. *"Yes"* say the liberals, the mystics, the philosophers, poets, hippies and the young. But, *"I don't know"* say most of us.

But what is disturbing, is much closer to home. A drug is by definition a substance that in small doses produces alterations in the body and mind, or both. Cigarettes, alcohol and a variety of common household items: coffee, tea, nutmeg, salt, lettuce, to mention only a few, could also be defined as drugs.

In 1982 the United States registered: 260,000 deaths due to alcohol-induced road accidents, more than 30.000 deaths from lung cancer and 96 deaths attributable to cocaine overdose. In 1985 the American Lung Association estimated that there were 250,000 premature deaths due to smoking in the USA/year. Supported by this data, apparently supplied by people with vested interests, the Legislature knew they couldn't do much about road accidents, nor bother with the latter but; yes they should campaign for the abolition of tobacco.

Recently, we have seen more and more restrictions placed on tobacco-smokers. Now there are new laws which prohibit lighting-up in many places: public transport, public offices, theatres, restaurants etc. How far down the road are these increasing restrictions taking us? How long before smokers can't smoke in the streets or before cigarettes are banned altogether? Is it not logical that a ban on alcohol would follow? After all, if tobacco and alcohol were new products they would be immediately circumscribed as very dangerous drugs indeed, and with sound reason.

What, then, would a smoker do when he/she can't get a packet of cigarettes every day? Will they have to turn to something else? Whatever's available? Cocaine? Are they to be made criminals? Society and commerce has made them drug addicts. Tobacco companies, their parents and friends encouraged them to smoke and Harold Wilson taxed them for his bounteousness while he blew pipe smoke into their eyes, blinding them to their true status as social menaces.

Legalize drugs? I would have to say, "YES! Bugger the snakes and lizards!"

Univision News, a Washington based TV network made the statement that 'on the occasion of the Pope's visit to Bolivia five million drug dealers knelt down in front of Pope John Paul II when he arrived in the Bolivian altiplano'.

Whatever is done drugs are here to stay. The law of demand and supply. Law enforcement can only make it more difficult for those involved in the industry. The only outcome is to drive it underground and raise the price of the product. The drug mafia are experts singularly organized to take advantage of the black market. The profits to be made from drug trafficking are so immense there will always be a supply of criminals and would-be criminals to run the industry. Increased law enforcement can create a violent backdrop to Society. More people are killed during internecine drug wars than die from drug overdose. The inflated prices paid for drugs on the black market leads to further violence: muggings, rapes, robberies and kidnappings directly affecting innocent people.

Pharmaceutically pure cocaine sells for $50 an ounce. By the time an ounce of pure cocaine has been cut and sold on the streets it will have made the underworld a profit of $10.000. The cocaine industry in the United States is worth $30 billion in gross revenues and is a growing industry. Most of this money is sent to tax havens abroad that contributes to a negative balance of payments. Some of it is used to corrupt government officials, police and young people. Law enforcement officials acknowledge these realities and emphasize the need for controlling cocaine at its source. Any Bolivian would scoff at the proposition that the coca bush be eradicated or the Campesinos persuaded to cultivate alternative crops.

Only cynics, it is said, claim that the drug issue is being manipulated to the benefit of politicians, prosecution and defence
lawyers, customs officers and narcotics police.

Presidents and most of the Senate, essentially old men lacking any
first-hand experience of drugs, see in their campaign to extirpate the
problem with the same God-given right that earlier potentates saw in
the Crusades. Maybe we will have to wait for the next generation to
put things right. But now, the majority of the legislature, the millions
of older people taught to believe in the possibility of eliminating
drugs, and a number of well-intentioned young people will support
the President in his beliefs.

But Presidents have had their fingers burnt napalming other
nations (even friendly ones like Cambodia). So, why make war against
Korea or Vietnam if the war against drugs will win the next election?

The police, like any organization, includes many ambitious individuals in search of promotion and job security. Because of its manifest
failures drug enforcement offers plenty of scope for those who claim
that an increase of narcotic agents will correct the shortcomings. They
have vested interests in exaggerating the issue, anaesthetizing Society's
concern for the real killers alcohol, poverty, war, pollution and most
threatening of all, climate change.

I can just visualise the latest news story, 'Mr X was arrested last
Tuesday morning in connection with the fraud at the Holdsafe Banking Trust. A search of Mr X's house produced a substance that on
analysis proved to be hashish. Mr X admitted to being a sometime
smoker and told the Judge, Mr Somebody that he believed the habit
had temporarily impaired his judgement and had led to his fingers
being caught in the till'. How often does one read that a bottle of Old
Crow Bourbon and an open tin of Budweiser beer were found in the
arrested man's house? The inference being that we are all perfectly
aware of alcohol's seamy side and tailor our behaviour accordingly.
The poor misbegotten son is no longer on trial, it's drugs themselves
that stand accused. The muddle comes from a common failure to
realize that drugs do not contain our way of behaving, they can only
help us express our inner selves.

Of course the police are justified in fighting crime and as long as
drugs are illegal part of that fight must be directed against the Drug
Mafia. But the Mafia itself is, paradoxically, the Police Department's
most ardent supporter. Yes, the Mafia says drugs must be stamped out
not legalized. It's common knowledge that the underworld made

more money bootlegging alcohol during Prohibition, that 'Noble Experiment', than at any other time. Like mould, the Underworld establishes itself on putrid fruit. If there is something rotten about Society, something dishonest in the way it treats its citizens, then, be sure, the Mafia will exploit the situation and thrive.

Drug users themselves are not criminally minded, but if the Law makes the use of drugs an offence, then they are criminals. If a police-man can raid your house, beat you on the head, smash your furniture, simply because you were having a quiet joint with a friend, one could well develop deep down resentments that may explode into violent behaviour. Most people consider their private habits their own bus-iness and a law that states otherwise an infringement of their 'Free-dom of Rights.' Drug users are no exception. Many innocuous drug users, like the 'Potheads' forced by the Law to act in a furtive manner develop anti-authoritarian attitudes. One only has to recall the success of films like those of 'Cheech and Chon' and the very popular Bert Reynolds movies suggest that this is so.

Authority has often stated that the 'Casus belli' of their campaign against private drug users is to save those same people from them-selves. Drugs, they say, have been shown to be physically and mentally damaging, and by association people get led into a life of crime. Yes, as long as drugs remain illegal the innocent become involved with the criminal pushers who may lead their clients to harder, more costly drugs (mescalin, opium, heroin and cocaine), and by doing so provoke more serious crimes to pay for them. Old ladies don't get mugged because somebody has taken a snort, they get mugged for the money in their purse to pay for the next gram.

I seem to remember that some time ago the activities offered by Drug Clinics were frowned upon by many politicians and some social workers. I have to ask myself, how would driving the drug business underground help? How does it help to separate the help of a Wise Counsellor from the innocent youngster wishing to know what it's all about? Why take away his/her opportunity to experiment under prof-essional guidance? I believe it is true that there is nothing so attractive as something forbidden, something slightly mysterious and people, the young especially, have a sneaking admiration for those who buck the system whether they be pirates, astronauts, rock singers or smokers of grass. Educating people about drugs to the full, letting them try them will remove the mystery; making the lesson a bit clinical, explaining the advantages and disadvantages, will not produce a nation of drug addicts as opponents of this scheme claim. People are just not that stupid. The real danger lies in ignorance, not knowing exactly what one is letting oneself in for. Keeping skeletons in the cupboard whether it be our descent from the apes, grandma's cancer or the sexual experience is now recognized as absurd. Society gets what it deserves; handing the real trouble makers, the Mafia, the profits of the drug business on a plate is also absurd. So why do we do it?

Driving drugs underground, as we have seen, creates more problems than it solves. More tellingly, perhaps, is that drug control is just not effective. The DEA's 'Coca Substitution Programme' claims to be making headway, the truth is they are not. Their policy is to offer $2000 for every hectare that a Campesino voluntarily eradicates. The total area of coca eliminated in this way is proffered as verification of the scheme's merit. But they don't count any new coca fields, whether they are planted by new immigrants or by those farmers who, having relinquished their Coca fields in exchange for the cash, soon make up the loss by planting new ones. Nevertheless, the Substitution Programme remains in place.

In 1981 the Drug Enforcement Agency made only 4379 cocaine arrests. Florida State made a further 3559 arrests most of which were for simple possession. Cocaine traffickers make many millions of transactions every year and they will continue to do so because they are making very good money and, as the figures indicate, the chances of getting caught are small.

Hard drugs, as we have already discussed, are difficult to detect because they occupy very little space. Most of us have probably seen on the TV harrowing episodes in which some desperate 'Mula' ('Drug trafficker') has been arrested at the Airport. The poor man/woman knew before setting off there was a good chance they would end up in prison.

And not only that but their humiliation and disgrace when arrested filmed for all to see. When they have served their sentence the real punishment may be just beginning. Once they are home, in whatever small town or village they come from they need to find work.

[I have assumed that most of these unfortunates come from rural areas where life for everybody can be hard. For most 'City people', with populations running into the millions, there is always work on offer, and if you are arrested on TV what does it matter, hardly anybody will recognize you.]

Sometimes TV will televise even sadder cases, people who have placed 800 grams of pure cocaine in small capsules and swallowed them (but) having been detained the capsules have sprung a leak, disabling the ability to behave normally he/she has had to confess and now will plead for a life-saving operation. Even worse are those cases when ingested capsules have broken open during the flight and the 'Mula' died.

In 1980 US Customs seizures of cocaine totalled more than two tons, about 5% of shipments entering the country. Without a complete change in attitudes we face an endless cycle of arrests, seizures, convictions, violence and problems of addiction.

And, to me, as a Conservationist, a problem that is overlooked by the foreign press: the continued destruction by the growers of Coca in the Western hemisphere's tropical forests, or, to mention just another example, the usurpation of natural grasslands by the Far East's Opium growers.

The coca bush is a perennial shrub and will go on yielding leaves *'ad infinitum'*. Because it is relatively labour intensive the average Campesino family cannot manage more than two hectares. The value of the crop, even for the legal leaf-chewing trade, is enough for them to make a living from these five acres. When Coca is produced for the illicit drug trade the profits are much higher and with more profit the practice attracts a lot more people, people who are not genuine Campesino farmers. They set about cutting down even more forest. With the essentially illegal nature of the business there is no endeavour to improve plant variety or techniques of cultivation that could reduce the area farmed by increasing coca yields per acre. The fundamentally lazy nature of the people may discourage them from taking advantage of the increase, satisfied as they would be to sacrifice a decent living in exchange for less work.

Also, there have been strong moves from the Americans to obtain permission from the Bolivian Government to spray coca fields from the air with environmentally damaging herbicides. Some campesino syndicates claim that the US administration has already carried out extensive field testing of a defoliant called 'Spike'. Even the company, Chevron Chemical, who supply the product has pointed out the ecological implications and has tried to block drug enforcement agencies from obtaining supplies from them. What a scenario!

Another factor of major concern to Bolivians is the effect of the coca war on the campesinos themselves. Large profits from illegal coca farming have made them very aggressive in defence of their 'rights'. Led by agents provocateurs working for the drug Mafia the Coca farmers have banded together and reverted to a state of anarchy.

Take for example this Bolivian newspaper heading (El Mundo 24th May 1988): 'Coca producers give 48-hour notice to the [American] Drug Enforcement Agency to get out of the Chaparé' and the statement made by Campesino spokesman Julio Rocha, "Only the gringos are responsible for what is happening. We are not prepared to let the abuse continue. We demand respect and our human rights which are being violated under the pretext of drug control".

And what about the insurrection at Bulo-Bulo?, a small settlement fifty miles west of Buena Vista. Following reports of illegal coca cultivation and base production, thirteen anti-drug agents and a Public Prosecutor went to Bulo-Bulo to investigate. Dozens of angry peasants armed with machetes surrounded the group of officials and prevented them from leaving. Their message was quite plain, "No queremos perros intrusos en nuestro pueblo, vayanse, déjennos tranquilos porque sin el narcotrafico nosotros no vivimos". ("We don't want intruding dogs in our town, go away, leave us in peace because without the drug business we can't make a living".) The crowd of peasants, which had by now grown to three-thousand, started to throw sticks and stones and turned over an official jeep. Then they took the thirteen government agents hostage and brutally beat them. The Public Prosecutor managed to get away to sound the alarm.

Now, other campesino organizations, heartened by the firm stand of their Bulo-Bulo colleagues, are making demands that are completely unreasonable: titles to lands they have invaded and occupied illegally; expensive access roads complete with tarmac and bridges to service only twenty or thirty families; schools, clinics (and staff to run them) in tiny out-of-the-way hamlets; credit facilities to clear protected forests (some even inside National Parks); and, coming full circle, their right to grow coca outside the officially recognized areas.

And of all the counter-arguments made by the authorities, their refusal to allow the last is the weakest. The Bolivian people, including many who claim to represent the poor coca growers, are not aware of what we can call 'The Great Cocaine Debate', so we can't blame them if they take the nonconformist view. Coca and all its derivatives, they say, could be legalized; the money the country would earn could benefit everybody; the move to prevent this is a Gringo plot to disinherit the poor of Bolivia: 'Yanqui sour grapes'.

Coca leaves are rich in essential vitamins and chewing them taps the locked up energy stores of the body enabling a tired, listless person accomplish his menial tasks or complete long hard journeys over mountains weighed down by produce for distant market places. Coca leaves or cocaine, International law has made them both Class II drugs; forgive the Campesino if he fails to see the subtleties of the legislation; what's good for him could be good for everybody else. And truly, why not? Products containing extracts of coca were once highly esteemed for their invigorating properties by people all over the world. Why could they not again? There is no evidence whatsoever, neither physiological nor psychological, to justify the ban on coca based pick-me-ups, cordials, teas and other beverages. The fact that they are banned, classed together with dangerous drugs, does look like protectionism. Clearly, the present doctrinaires would not allow the export of leaves, lest they be processed for their alkaloid content, but the ready-made products could be manufactured in South America and exported to the benefit of the poverty-stricken Campesinos. Lucozade would go bankrupt, Coca-Cola made redundant. 'No Sir! We cannot have that.'

'Better we buy the whole 80-million kilogram crop.' Yes, believe it or not, this alternative has been seriously debated and, in spite of the manifest weaknesses of the proposal, still has powerful advocates today. The programme, they say, would *only* cost $2 billion a year.

There is really only one option available to us if we are to free drugs from the monopoly held by the Mafia and ensure that drugs are not over-abused. Drugs should be sold with prescriptions at Government controlled clinics, at a price sufficiently high to deter the casual user, but low enough to eradicate the influence of the Black Market. All sorts of refinements can be made to the basic idea: registered users could receive rebates for attending rehabilitation programmes; less harmful drugs could be sold more cheaply than those known to be highly addictive; financial awards could be given to those who became completely independent of drugs and, with their own story to tell, a selected few could be gainfully employed as creditable advocates of a drug-free way of life.

Surely it's time we act as grown-ups. After all, the horror is not really there, addiction and madness takes possession of a minute proportion of men and women, statistically inevitable in any modern

Society. In 1980 the 10,000 cocaine abusers who entered treatment programmes in the USA represented only 0.1% of the ten million users who carried out their day-to-day lives in perfect harmony, evidence that happy secure persons are much less likely to turn to drugs.

The horrors appear when people feel oppressed by the law and its enforcement agencies. They turn to drugs as a way out of the hell they find around them. Coming to terms with the problem by reducing a conflict exacerbated by the unsophisticated ease with which drugs are blamed for the breakdown of our uncompromising institutions, our morals, our religious beliefs, our system of education and our families. Do you remember that documentary 'Reefer Madness'? Is that the only way out?

'Ultimately', as Professor of Law, Steven Wisotsky declared, 'the Crusade against [drugs] is not a question of public health but an assertion of power to set the standards of proper behaviour.'

Earlier we saw how the biggest operatives in Santa Cruz make a deal. There are also shoals of small fry making minor deals: the 'Pitilleros' sell their base-filled joints on the pathways of the poorer barrios or from kiosks on the main streets of the city, where customers ask: 'Do you have cigarettes for sale, nudge-nudge the good ones?' And, don't forget the 'Cocaine Parlours' where an addict can spend the whole day being entertained by his supplier friend, and the many salesmen operating from taxis in the avenues of the city centre. Then we have the worst, the small-time dealers who supply an ounce to a few kilograms of the pure drug to the Gringo traffickers who, like ants to the sugar bowl, moths to the candle, fly in to town.

It is easy to pick these Gringos out when they first arrive: strangers in town, of slightly offbeat appearance, carrying none of the usual paraphernalia of *'bona fide'* tourists. Content to while away their time in the many bars available, they stick out like sore thumbs. Some are here today, gone tomorrow, but the majority spend a month or two and a few, the more professional ones playing for higher stakes, a year or more. Many do not speak Spanish and prefer to deal with one of the resident gringos. To give you an idea of these strange visitors we can briefly examine four case histories.

CASE 1. Harry is a slim handsome Canadian, well dressed in a casual way and full of nervous charm. In his mid-twenties and a graduate student in business, he has tired of working in his father's used-car lot. He is always the first in the bar and quick to buy you a drink (in the first few days) and chat about nothing in particular. He says he had heard so much about Bolivia he thought it was time he came down to take a look. He carries a very expensive camera and a large custom designed purse —well-ordered with lots of little pockets and separate compartments in which he has neatly stowed away all the papers he will need and many he won't. He takes a little too much interest in each person entering the bar and doesn't really concentrate on the conversation once he realizes you aren't the company he is looking for. He smokes long American cigarettes, putting them out as soon as they are lit, quickly filling an ashtray with long, wormy looking stubs. At first he only drinks Coca-Cola but as the days pass these become 'Cuba Libres' and eventually straight brandies. He still does not state the real purpose for his visit but one senses in his nervous glances his wish for the subject to be brought up. Some of the regulars think he is D.E.A., but the professionals have sized him up immediately: small time, leave him for someone else. Then, one day you enter the bar and he isn't there, but in he comes, later than usual, wearing the slightly glassy stare, with its dilated pupils of a 'Cocaine trip'. He drinks a lot and is louder than normal — much more relaxed. Now you know that last night he has made his deal, tried out the merchandise and will, probably, be gone in a day or two.

Being the weekend, and feeling a little sorry for him, Cheri and I had invited him back to the house for lunch when we saw him the day before. By now he has a fair idea of 'Who's Who' and no doubt he knows I am not in the trade but am interested in those who are. He knows I am considered a bit of an academic, a trustworthy oddball if you like.

"I guess you know what I've been doing" he remarks, almost before I've managed to fix us a drink. *"Yes"* I say, and nothing more. We settle back into the garden chairs placed in the shade of a Cupesi tree.

Then out tumble his justifications to lie like incriminating pieces of broken glass at our feet. *"Yea, man you're lucky. You don't know what it's like working for my old man. He pays me next to nothing. Yea, I'm tired of him. I just want out and start up my own business with my girl friend,"* Then pulling out photos of very attractive blonde, *"Yea, Barbara and me we want to start up a photographic business"*. This explains the expensive camera with which he, now, takes a picture of me, the house, the garden and Cyrana (who has flown to the back of my chair). *"My old man won't lend me the dough so Barbara and me thought up this little scheme. Yea, I've never done anything like this before"* and to end the awkward hiatus while he appears to consider his conscience he jumps up from his chair and says: *"Hey, you got somewhere I could sit and mail off some stuff?"*

I show him to my writing desk where he proceeds to pull out a dozen copies of today's local newspaper, a packet of tiny plastic bags, a pen, some plain sheets of paper and a sizable bag of cocaine. *"Three-hundred grams,"* he volunteers. *"Say, do you mind if I do this here? It won't take me a moment and I would sort of like to get this over with so I can relax."*

I fetch him another Cuba Libre that he drinks while dividing the cocaine into equal amounts that he puts into the plastic bags, every now and again helping himself to a pinch which he snorts into each nostril with a rolled-up hundred-dollar note. *"Say, would you like some?"* he absent-mindedly asks.

"No", I shake my head, *"I don't like cocaine"*.

Harry stares at me: his bright blue eyes almost sensual-looking now that the pupils are two or three times their normal size.

"Not your mustard, eh?" he mutters in parody of my Englishness. Harry continues his work; rolling up each newspaper with its little packet of Cocaine tucked inside; writing a name, not his, and an address on separate pieces of paper. Then he asks: *"Say, you got any Scotch tape?"*, I give him some. When he has finished with the first newspaper roll he passes it to me ... *"What do you think? Will it pass muster?"* he asks, putting on the English accent again.

"Well, Harry," I begin, *"I suppose it's not too complicated. As long as it doesn't fall out of it's wrapping"* I add. *"But that's not your name and is that your address?"* We laugh together for a moment. *"Yes and no. But I can always claim it's not my mail,"* he assures me with a confiding wink. I can't help mentioning it sounds pretty amateurish. Harry is not offended but goes on to explain that it's his girlfriend's address. With my chin withdrawn and my mouth puckered I send a disapproving message to him. Realizing the implied censure he quickly adds: *"Don't worry, the name's fictitious, there will be no problem."* Sounds even more amateurish to me but I don't say so.

"Yea". Harry starts a lot of his sentences with *"yea"*. "Yea I reckon to clear about sixty-thousand on this one", spoken like a man who has done many a deal before, belying his earlier statement that for him this was a first, a one off. Harry is getting a little drunk and a little too cocky to be interesting anymore.

By the time he has finished his preparations—all the packages now ready, the addresses written with capital letters —Cheri calls to come in and eat. When lunch is over I am ready to get rid of him. *"Well, so long. I'm due out Tuesday but I'll be back"*, he assures me. Back?, I'm not sure of the implication. *"So long, Harry. Nice knowing you"*.

We don't see Harry again, having turned down the invitation to his farewell party to be held in the the bar the following evening.

About three months later, talking to a friend, I hear Harry was busted: criminal use of the mail services, or something like that. Nobody knew any of the details or how long he got. So long, Harry.

CASE 2. Peter's tale reminds me of Harry (but) Peter is a very different kettle of fish: thirty-two, self-confident and extremely arrogant. A man who likes to be respected. An American, he speaks fluent Spanish and years ago he spent some time in Bolivia. Now he lives in Spain where he has a healthy business on the Costa Brava.

Peter has contacts, oozes bonhomie, likes to stay in the most expensive hotel in town, does a lot of coke, and hides his thoughts behind the most expensive sunglasses that money can buy.

His ambition is not *only* to be respected, *but* worshipped, a fact that Peter justifies by two singular characteristics: the first his wealth, accumulated by smuggling Cocaine from Bolivia to Spain without detection, while most of his smuggler friends spent time in jail and, the second his ability to convert his illicit gains into licit ones by multiplying his fortune in the Spanish tourist trade, while his circle of friends took a few years off to indulge in booze and girls. In other words he was brainy, while his friends, unable to recover their senses lapsed into poverty, that led to some of them killing themselves.

He always arrives from Spain with a bagful of funky gifts: expensive electric gadgets, flashy cigarette lighters, cheap camera equipment, obscene playing cards and mind boggling doodads. Peter likes razzmatazz, is generous to a fault, hates to be anything but the centre of attention and takes quick and deep umbrage to having his leg pulled or being shown up for his lack of sincerity. He is also an academic snob: "If I had wanted, I could have been a 'Distinguished Anthropologist'. He has violent tendencies and behind his superficial show of being a 'Laid-back Man,' sits a very confused person wracked by feelings of inadequacy.

Peter likes best the company of those who will be profiting from his presence; he refers to them as: 'Those I am helping out'. For him a trip to Bolivia is also a month's vacation and the moment he arrives the party has to start. His time is spent in mad dashes around town from one bar to another (always ones he knows and to his frequent embarrassment ones that no longer exist), or late night visits to any one of his old friends. The constant eating and drinking —reinforced by repeated trips to the bathroom for a 'snort', often together with his friends (just like a group of giggling girls). The debauchery offset by flurries on the squash court or hectic short outings to the countryside. Peter doesn't like to be unfit and takes a certain amount of justifiable pride in his sinewy strength. When he gets upset (should anyone have dared a *too* acute observation) his eyes bulge, he pulls back his scalp with his hands and contorts his face into a maniacal grin; in fun, but the joke only drawing attention to his real psychosis.

"Hey man. Hey man, what do you want, man? We're just having a good time, yeah. Now you wanna start something. Right? Well, man, I don't need this shit," then rising from his seat he says: "Didn't I just give you a present? So I don't go for this shit. Right? So fuck you," and turning to his camp followers: "Let's go. We didn't come here to be insulted", and off he storms.

The next thing you hear is: how he got drunk that night and shot a man dead he didn't even know. He's decided to cut his trip short and will be leaving the day after tomorrow. This puts his friends in a panic because, now, they have to get everything ready at short notice and hurried operations mean mistakes. But Peter's party is over and everybody had better jump to it —or else! "Or else I will go right back to Spain and not bother to help you out anymore—ever! What do y'all think I'm doing this for, for Christ's sake! Doesn't everybody know I don't need the cash. So, for Christ's sake!, pull your fingers out willya."

The main burden of the unexpected turn of events falls on the shoulders of Sid. Sid is, or so I am told, a genius. Peter tells me he taught Sid everything there was to know. Sid is an expert at packaging. He will take the guts out of anything, rebuild it then fill it with cocaine. He will then make sure that it weighs the same as before and that it works perfectly, thereby passing muster should it be examined by Customs.

Sid's greatest problem is that he is lazy and likes to drink. Sid, especially, hates to be hurried and what he does takes time. His *'forte'*

is packing half a kilo of coke into radios and tape-recorders, extracting superfluous bits and replacing working parts: motors, trans-
formers, speakers by equally efficient ones from smaller machines.
As far as I know, nobody who has used Sid has ever been caught and
a number of them have made the trip several times. Sid, by making
his friends rich, has worked himself out of a job. At heart a timid man,
Sid does not have what it takes: nerves of steel to entrust himself with
his own genius. He's like many of us, who for no reason at all, start
shaking at the knees the moment we have to pass through Customs.
Sid, having proved once again his ability, has received a call from
Peter in Spain to say all went well. "See you. Maybe next year. Right?"

CASE 3. The story of James the Englishman is rather different.
James is older than most of the traffickers, about forty; a thin
cadaverous man with a potbelly and balding blond hair. With his
extreme pallor of face and his sunken cheeks, James could have
stepped straight out of Gormenghast —Steerpike, perhaps?. He likes
to project the image of an English Gentleman brought low by sudden
misfortune. Actually, James is an idiot, holding himself firmly in with
the aid of a tight waistcoat, black velvet smoking jacket complete
with buttonhole flower and a spotted bow tie. James confided to me
how he used to be the manager of the rock group Pink Floyd, was
escaping a disastrous affair with a millionairess in Miami and was not
really interested in remaking his fortune by smuggling dope (but) was
researching a book.

James spent eighteen months in the Santa Cruz gaol. It was here I
got to know him as, occasionally, I visited the gringo inmates, a duty
that members of the British Mission were encouraged to do. We took
them something to read, matches and cigarettes for the smokers,
medicines that they had asked for, and so on.

When I first saw James he was asleep, coiled in a foetal position
on a low bunk bed, he was wearing only a pair of white underpants.
His room —for in no sense could it have been called a cell —
measured about six by ten feet and had a low roof of corrugated
asbestos. The crudely plastered walls covered by calendars and
pictures of pretty girls with very little on. The first thing he said, when
his roommate, a young Bolivian, woke him, was: 'Those are not
mine,' waving an arm towards one of the more lurid pinups. Everything James said was in the negative. "This isn't pretty is it? My God,
you don't know how bad the food is here. 'Those others,' flapping his

hand in the direction of a group of prisoners who were taking a momentary interest in my arrival, 'aren't humans. If it were not for my constant vigilance they would have cut my throat long ago. They wouldn't hesitate cutting anybody's throat for a few dollars.' Then he would go off into long explanations why it was that he was still here. 'Not that I haven't tried to get out. Believe me!'

'I'm not going to be milked of every penny I own, but they won't let me go until there's nothing left. No, by God, I'm not that stupid, not now, not after my first mistake. Cost me two-thousand quid and still I'm not out; the bastards just pocketed it.' Meaning that he had paid the Prison Governor and a lawyer the customary fee to obtain a rapid discharge, but these unworthy gentlemen smelt a sucker and wanted more. James would tap the side of his nose on enunciating each fact and then heave his skinny white chest to expose his even skinnier ribcage, as if to confirm his courageous resolution. I had to admit, for a man who had bought the prison restaurant (as he claimed) he certainly looked underfed when I first saw him. It was later he developed that unsightly paunch that sits so badly on thin men with nowhere to exercise.

When James spoke his bottom lip would begin to tremble as if strained by the effort and his small pale green eyes would flit from object to object, as if constantly checking their loyalty. He reminded me of a highly nervous monkey. His cell mate was not to be trusted. His lawyer was no good. The British Embassy had not tried to do anything or him. This last apart, James tended to be a bit of a liar.

But fact is often stranger than fiction and what happened to James the day he was arrested, the day the President came to town, was an affair of such high farce it could have been torn from the script of a comic opera. Just as the jubilant crowds were settling down to listen to the President's address, with the brassy chords of the National Anthem still ringing in their ears, just as Siles Zuazo's first words began to silence the restive people, at the precise moment when all of Bolivia's attention belonged to this one man, the video cameras swung away to film a commotion in the main airport concourse. While the sound track obediently recorded the President's words, the images were dedicated to James: held by two policemen, half naked, his white belly hidden by bags of cocaine strapped around his middle. James had just managed to upstage the President of Bolivia.

James had probably been the victim of a setup in which unsuspecting traffickers are supplied with cocaine by men working in cahoots with the police. When the sucker tries to leave the country these men not only retrieve the drug but also offer a deal, usually $1000-2000, to hush things up. James says that he was offered such a deal. Although he was prepared to pay the bribe, indeed came prepared with the cash, he turned it down he said, because they wouldn't allow him to leave with the cocaine.

As it happened, things did not turn out in his favour and he soon found himself in gaol. It is my guess, it was here he first discovered he had been cheated.

Life in the Santa Cruz prison need not be the terrible experience of the imagination, not, that is, *if one has money*. For twenty dollars prisoners are allowed out for the evening when they wish. Girls can be brought to them in their rooms and food can be delivered to them from outside restaurants. Arrangements can be made to provide TV, or even carry on one's business activities.

Not so long ago the local newspapers stated that a ring of drug dealers was operating from within the prison and, an even greater scandal came to light, in which it was claimed (probably correctly) that a group of prisoners were sent out each night to engage in burglary. This affair, that I have related before was brought to the public's attention when an American shot several of them dead in his house.

I did not see James again for some time after my last visit. When I did, there he was, complete with flower and spotted bow tie, sitting at a bar in town sipping gin. Apparently, he had been released. He was now more reticent about his experiences, but since he hung around for some time I supposed, that wiser now, he felt confident of fulfilling his original mission. What eventually happened to him I don't know because I didn't see or hear tell of him again.

CASE 4. Tooey's story is, maybe, the strangest and saddest of all. I remember meeting this young American not more than a few minutes after he had got off the plane in Santa Cruz. Cheri and I were sitting drinking beer on the patio of a reasonably cheap restaurant called La Fonda de Ariel. He came rubbernecking down the palm-lined sidewalk and turned into the restaurant's gate. Tooey looked very handsome with his fair hair, muscular body and tanned freckled face: the tan was Los Angeles, his winning smile Steve McQueen. My wife, never one to miss the opportunity of being friendly, invited him to join us; a move which apparently pleased Tooey immensely. I don't remember for sure but I think he said he was originally from Carolina, and he spoke with the patois of a country boy. He said he was keen to hear about Santa Cruz and promised to give us a ring later in the week. He paid his bill (I think I only ever heard of him doing so twice) and left.

He did come round to the house shortly after that and I saw him a couple of times in the bar, or met him in the street. Some six or seven months later I met him again and I remember, when I asked him why he had overrun his vacation, he said: 'I guiss I feyal like sticking 'roun a liddle longa.' I guessed that he had decided to set himself up as one of the many resident small-time dealers.

After this incident I didn't see him for, perhaps, a year. He was hardly recognizable. He must have lost a good three inches in height and now stood well under six feet. His muscular body had collapsed like a dried puffball; his skin was burnt to the same colour, a deep leathery brown; his arms were like sticks, protruding from a dirty torn T-shirt; and he had lost a lot of teeth: 'I haid to sell 'em to kep going. Yo'all recall I 'ad sevral gold ones. Weil, I 'ad to knack em out.' His distorteded jaw had trouble mouthing the words. He was living in a grass hut some twenty miles outside the city, and I remember him telling me how he loved the countryside. 'Thucking in the thenery an gitting high.' Tooey, I was told later, was spending all his time smoking Pitillos. If he had any resemblance to Steve McQueen before, he now looked more like Mahatma Ghandi.

The next time I heard something about him he was said to be hiding out, having cheated somebody who took it personally. The heat on, Tooey tried to leave the country with the usual extra and, like James, got caught. He spent a month or two in the Santa Cruz gaol before he was seen again, once more openly living in town. I know from a friend that at least part of his activities was selling marijuana on the streets. He approached a friend of ours in the central plaza, mentioned he knew me well and offered my friend a small amount of grass for nothing, but as a favour could my friend lend him twenty dollars? Tooey was paid and told my friend to wait a few minutes - he would be back. My friend, suspecting a frame-up, took fright and hid in a travel agent's from where he could keep an eye on the street. When Tooey returned in a taxi he was in the company of several men. My wily friend felt sure they were narcotic's police. Indeed, Tooey had turned narco-informer— a fact corroborated by my acquaintances — and was thought to be mixed up with the crowd that had caused

James' demise. Not long after the marijuana incident I saw Tooey for the last time, looking more like his old self and with money enough (for the second time) to pay his bill. When I approached him he acted in a guilty manner, hedged my enquiries, and soon fled the bar.

About two months later he was found stabbed to death. Like most informers here, his life was a short one.

Harry, Peter, James, Tooey, there were many such gringos living in and visiting Santa Cruz. At one time there would be three or four new faces a month. Like Harry, fly-by-nights, here today gone tomorrow. From what I have seen, unlike Peter who uses Sid's peculiar genius, most come to a poor end and many now spending time in the less convivial prisons of North America and Europe. I often wonder if this is not part of the plan: throw a few sprats to the system for the sake of demonstrating Bolivia's avowal to stamp out the trade; and all the better that those caught should be gringos, although, to be fair, more and more nationals are being arrested within the country. Also, I have never quite understood why the Bolivians should tolerate these undesirable visitors if not for the aforementioned reason. The art of being a dope head is not to look like one (James knew this) but many (like Tooey) made no effort at deception. Maybe, it's a throwback to the spirit of free-booting, maybe, they are liked. Certainly, the majority of them could be called real characters and, when they are not down and out, spongers can be very amusing.

Some were frequent visitors to our home and good-naturedly discussed my premise that life could be one long holiday, if that's what they wanted, without resorting to chancy illegalities. Cheri and I gave ourselves a pat on the back for having saved more than a few of them, probably the less determined ones from a stretch behind bars.

One case was very close to me, because it involved my British Mission (BTAM) student. She had become infatuated with her Italian boyfriend. We didn't know him very well since he was not one of our close circle of friends. Cheri had never met him and I only knew him as one of the young men taking part in our local pub's Dart Competition. The general opinion of my colleagues at the pub was disconcerting, that he was probably another of those mixed up in drug trafficking. When my student told me that they were planning to elope, but she was worried because he told her that the expense of their plan to elope would be mitigated by his intention to take a few ounces of cocaine with them. Cheri and I spent a whole day inveigling her to drop the idea. Fortunately for her, with a torrent of tears she did. A few days later we found out (that) her lover was arrested on his arrival in Rome.

BUT, wolves run together (or whatever the cliché is), and it was inevitable that our enemies insisted on tarring us with the same brush.

I had to consider my employer's (HMG's) position (that I should have done in the first place). My problem was that I really believed we were achieving something worthwhile, fighting my own private war against the foolishness of drug abuse and, more importantly, the misery that awaited all those fresh young people who thought a bit of trafficking was not a serious offence. I thought the discrepancy would be covered by the few allies I had at work. To a certain extent this mechanism was operating, my friends did defend me, my failure was to recognize that it had to be a two-way process—I should have made my position clear from the start. Whatever, the whole issue came to a head when I was told that one young man in the British Mission, whom everyone knew had a predilection for Cocaine, sneaked on me in return for official absolution.

More pity for him for he was one of the young persons we were trying to help. It's not what you are doing that matters, it's what people think you are doing, and when those same persons are spiteful they can do a great deal of damage.

I definitely had enemies among my colleagues at work, and who hasn't?, and their intention to destroy me (maybe for no better reason than they saw me as a threat to their own reputations) was, I realized (a bit late) coming between me and my 'straight' friends who supp-orted me. When these same friends began to avoid me the faith-keeping mechanism broke down and I found myself out on a limb. By the time I took steps to correct the situation it was (for other reasons you will see) too late. *'Monsieur Guillotine'* was already on his way.

In discouraging our friends from falling from grace the main argument we employed was basically a simple one and, maybe, in their reluctance to face the terrifying realities for themselves, was sometimes an effective one. I believe they thought we knew what we were talking about because of our wide circle of 'dubious' acquaint-ances and contacts and the inside knowledge we gained from them.

Author's note: See Chapter 3 page 99, how my tenuous position with BTAM was resolved by P. W. Little's letter to me.

As I mentioned before, we only tried to convince the real amateurs and in their case we sounded persuasive. Why, I wonder, cannot some more eloquent advocate prepare an advisory pamphlet that could be given to young people as they board the plane going to 'Problem Countries?' Has the scheme been tried before and found wanting? Does somebody think it would encourage the innocent? 'S'truth is, it may be a case of *Struthio camelus*, 'head burying'.

Our simple argument went like this: amongst the many people we knew who had tried a bit of amateur drug trafficking it was the smaller operators who frequently failed. I suppose the reason was partly political. Bolivia wanted to demonstrate lip service to the idea of drug control. The 'Narcos' were probably better organized than was generally assumed, after all they had American advisors working with them. It was likely that they knew about nearly every little deal since there were many informers (like Tooey) and hidden motives that we (as 'Extranjeros') were not aware of. Big deals were covered by pay-offs, big deals were profitable to the Mafia (the 'Coca Nostra') and when they failed caused considerable ripples within the organization. The main dangers did not lie, as the amateurs thought: in getting through Customs on their return home, but in avoiding the duplicity of the dealers, police and narcotic agents in Santa Cruz; the pratfalls intensified by the political ramifications within Bolivia itself. After all, Miami, Amsterdam, Rome, the Canary Islands are but a short phone call away and a favour is a favour. Unlike the professionals, little people rarely bothered to check out those pratfalls: they bought their drugs from one person, often knowing nothing about him, and left. Since these amateur efforts were more often than not (as in the case of Harry) 'one-offs', the reliability of a particular supplier was not based on past experience that is so important to the serious dealer. *'Viceversa'*, no genuine supplier likes to see his customer busted, it's bad for his reputation, and he will steer clear of amateurs as: 'small time, leave him for somebody else'. In any crime prevention network there is always an element of agreement between felon (pay-offs, tip-offs) and police (looking the other way, reduced pleas). Handouts only go so far, they might keep money in the pockets of an individual policeman but they don't keep him in his job; every policeman must show an element of success. Tip-offs provide the easiest way of fulfilling this necessity and, for his part, any criminal has vested interests in the 'devil he knows.' New faces can be expensive.

So, if you want to risk a venture in drug running, better make it big. Of course, even in Bolivia a couple of kilograms of cocaine requires a considerable investment and is usually beyond the resources of the small operators. Furthermore, bigger deals require a considerable knowledge of the system and the people operating it. We have seen how complicated the system is in the case of Mr Big, but even smaller deals necessitate inside knowledge and pay-offs: something James did not take into account.

As you can imagine, this down to earth logic, unadorned with moral issues, made its point among the casual operators we got to know and was sufficient to put some of them off. As you have seen, those who didn't heed the argument frequently came unstuck. I can think of quite a few who are serving their time and ... only know two who have got away with it. So please, if you are thinking about it, do it right. To do this you will have to know with whom you are dealing. To know them you will have to spend considerable time and effort in Santa Cruz itself — up to a year. The days of easy money are past; there are many snakes and rotten ladders waiting for the unwary and even the wary cannot predict the fall of the dice.

Through my contacts I knew a modest dealer living in Santa Cruz. His name is Rolando, he is a thirty-five-year-old Bolivian, speaks almost perfect English and, most of the time (except when somebody puts on the heat) asks a favour: *"be trustworthy."*

He has had a lot of dealings with passing 'Extranjeros' and knows what he is talking about. So, let's listen to what he has to say.

> "I wouldn't bother with an ounce or two, why get put away for three years when a really worthwhile deal may only cost four. What with parole and first offender status, the time difference you get if things go wrong will probably be measured in months. If you want to be rich, do the whole thing; why play for small stakes when the risk is about the same? Better, get a couple of friends. One of you will have to come down here, get to know how things work. Later, when everything's fixed, another of your friends can come to Santa Cruz and take the stuff to Brazil or Argentina. Your third friend can take the stuff back to the States or wherever. In that way he won't be singled out for special attention because his passport will be clean: no Bolivian stamps in it, he just went to Rio for a vacation like lots of other people. The first guy has got to do his homework, he's got to fix it so that he can get the dope out of here without any trouble. Sure, it needs a bit of capital investment, say fifteen grand. If three guys

really want to be rich they can put that together in no time. Reckon on a minimum of one kilo, the rest will cover the costs. Get to know the people here, find someone reliable, a friend, offer him part of the profit when it's all finished. Nobody is going to fuck you up if they stand to make a few grand out of it, and they can fix the fellows that have to be fixed. No problem if you follow the rules and take your time."

Here Rolando paused to think for a moment but soon continues:

"I never like dealing with anybody who is in a hurry. I don't know, it's sort of fishy. You don't know anything about the guy and I like to, at least, know a fellow's name. Wouldn't you? Even if it is false."

For a moment we laugh quietly together.

"A lot of these guys think only about themselves. Look, I never approach anybody. They have to come to me. You know, people talk, the fellow gets the idea, tries to get to know me. 'Waddaya drinking? Wanna smoke?' You know. Well, I'm as suspicious as hell, maybe this guys DEA or something. They forget we got to be careful too. Which reminds me, no Hippie types, I never deal with them, too obvious. So, it's no use just busting in here and expecting to do a deal; hell it's a lot more complicated than that. That's the way it is."

After a slight pause Rolando continues:

"If you think you have got a new angle, forget it, you haven't. A lot of people spend a lot of time and money thinking up ways to hide coke and get away with it. Chess sets, ashtrays, figurines made out of base. Hollow planks. False bottoms, tops and sides to boxes, suitcases, bags, packets of sugar, coffee, flour. You name it, they have all been tried. The chances are, if they suspect you are carrying something they will find it. Are you the type? Could you keep calm as the Custom's officer raises his eyes to yours and asks: 'Have you anything to declare?' Will your throat dry up? Your knees give way? Your lips tremble? Do you want your vagina probed with plastic covered fingers and impersonal peering eyes? Would you like to have your anus explored by big burly men at five in the morning? That's the way it is."

And so it is. Let's consider some of the unlikely horror stories of recent years. Teenage girls arrested on their arrival in Miami, each was said to have concealed 250 grams of cocaine in their vaginas. Santa Cruz surgeon accused of complicity in drug trafficking, the authorities claim he helped two men hide a total of three kilograms of pure cocaine in their abdominal cavities. Foolishly they had caught the same flight. Couple accused of hiding cocaine in their baby's rectum.

The pity of it is (that) these nasty things can happen to the innocent.

'Would you mind coming this way, Sir? I understand you have just been to Bolivia. Just routine, Madam. No need to get upset. No, Madam, I am not sure that won't be necessary. If you will just step behind this curtain. And if you wouldn't mind removing your clothes. I'll send somebody in to see you in a minute. No, Sir. I am afraid that it's no good calling a lawyer. We have a job to do, Sir. If you wouldn't mind bending over, Sir. If you would just lie down on that table, Madam. That's it, open your legs, this won't take a moment. Yes, Sir. I am afraid we will have to. Yes, we know she is only fourteen, Sir, but we have to make sure, Sir. There are so many cases now from Colombia. Yes, I know she's your daughter, Madam. Please don't upset yourself.'

This particularly harrowing story happened to an acquaintance of mine now living in London. He had to return to Bolivia to deal with some land he owned. After a short stay he left and on his way home went to see some wealthy friends of his in Buenos Aires. When they were claiming their baggage after the flight an Argentinian official came forward and started to systematically search each and every suitcase. Next to my friend stood a completely unmarked one which nobody wished to lay claim to. Not surprisingly, for when it was forced open it contained a large quantity of cocaine barely concealed under some clothing. For no other reason than his proximity to the suitcase, my friend was suspected of being the owner and was promptly arrested. Flustered by the turn of events, the poor fellow (unwisely) revealed the address of his wealthy friends and in the company of several narcotic agents he was escorted to their house. Unfortunately for everyone there, a small amount of marijuana was found and they too were arrested. This ill-fated discovery placed them all in real jeopardy; the simple deduction being that a man with friends who smoked grass is a man guilty of trafficking cocaine.

Notwithstanding their failure to provide a single shred of evidence against my friend: the forensic analysis failed to link the case to him, the clothes were obviously not his. In spite of this evidence the authorities withdrew his passport and placed him under open arrest. Six months later the poor man was eventually allowed to leave. His friends fared equally badly: the couple who owned the house were released after a week. When they got home the house had been ransacked and $30.000 worth of jewellery and household items were missing. Their two visitors were held as accomplices in cocaine trafficking and were only released after two months in gaol.

We all believed that the owner of the case was a member of the flight crew (in conformity with several cases in the past), and one of the Argentinian Custom's officials their accomplice (also in line with past events). Who else but a member of the plane's crew could get an unmarked suitcase aboard? Who else but somebody in cahoots with Customs would neglect hiding the cocaine? Why didn't it's discovery conform to official procedure? The case was opened before the owner was identified. The only possible conclusion is that the owner of the suitcase was absolutely confident it would not be opened but, as happens so often, somebody who had not been paid turned informer. So, at the last minute an order was made that all the cases on the flight from Santa Cruz be examined. The owner of the suitcase, probably a lowly courier, a 'Mula' received a signal (maybe from a Custom's official who knew of this order) not to claim the case. Maybe, had the courier made any attempt to conceal the drug the case would have passed inspection and the courier would have stepped forward to claim it. But with everybody standing around the discovery was too blatant to overlook. The game was up. My innocent friend, standing handily by, framed. More than likely somebody has paid for the fiasco. I wonder if that courier, who bungled the plan is still alive? I wonder if the man who failed to pay the necessary bribes is well? I wonder if the informer is still in the land of the living? Because heads roll when the drug barons are annoyed. *"That's the way it is."*...

"That's the way it is." A lot of weak people become criminals overnight. Some innocent people have to suffer and others die whilst Presidents, Mafia Kings, Narcotic's Agents, Police, traffickers and drug pedlars play out their secret games. When the chips are down the roll of the dice will decide the outcome.

'Would you mind stepping this way, Madam?'

"That's the way it is."

THE END

Glossary of Spanish and
Guarani Words and Acronyms.

AASANA: Administración de Aeropuertos y Servicios Auxiliares a la Navegación Area. The Airport and Airline WorkerÕs Union.

Academia Nacional de Ciencias: The National Academy of Sciences, among whose members one name sticks out, that of Oscar Wilde.

Acción Democratica Nacionalista (ADN): General BanzerÕs political party. They were widely expected to win the 1985 general elections, which in the event was won by the MNR. In the 1994 general elections the ADN suffered a crushing defeat (again at the hands of the MNR), which led to Banzer's abdication from politics, and his retirement to his home in Miami. [Until 1997 when he made a surprising return to win the 1998 elections.]

ADN: see above.

Alcaldia: The office of the mayor.

Almendro: probably *Dipterix alata*, but used rather loosely for a variety of similar trees.

Alojamiento: A boarding house equivalent to a Spanish pensi—n.

Altiplano: The high valley (mostly over 3,000 metres) that lies between the eastern and western cordilleras of the Andes.

Ambaibo: A softwood tree, *Cercropia leucocoma*, with finger-like fruit much liked by birds, especially toucans and parrots, and the local children.

ANP: correctly with an accent: Amboró National Park.

Araçari: a middle weight toucan of the genus *Pteroglossus*.

Asesor legal: lit. a Legal advisor.

Autóctono: equivalent to our loose use of the word 'Indian'.

Bagré: a common small scavenging catfish, probably of the genus *Rhamdia*.

Barrio: the suburb of a town.

Beni: the Beni is one of nine departments into which Bolivia is divided. It lies to the north-west of the department of Santa Cruz.

Boga: a small elongate fish of the genus *Leporinus*.

Cachasa: also spelt cachaza, is crude alcohol distilled during the production of sugar. Since it contains methyl alcohol it is not recommended for human consumption.

Cachuela: the rocky rapids of a river.

Cacique: village or tribal headman.

Camba: a native of the department of Santa Cruz.

Cambista: a street money-changer.

Camion: a lorry or truck.

Campesino: noun, a peasant farmer; adj., as in campesino housing.

Campo: countryside or rural areas.

Caña: sugar cane, or the alcohol produced from it (see cachasa).

Cantina: a cheap bar and restaurant.

Carachupa: the Common Opossum, *Didelphis marsupialis*.

Casa de Cultura: lit. the House of Culture.

CDF: Centro de Desarollo Forestal, the Forestry Department; also responsible for National Parks and Wildlife.

Cedro: not a cedar, as the name implies, but *Cedrela tubiflora*.

CEPROINCT: Centro de Promoción e Investigación Científica y Tecnológica. Centre for the Promotion and Investigation of Science and Technology.

Cercropia: see ambaibo.

Cerrado: a mainly Brazilian ecological zone comprising semihumid deciduous forest.

Cerro: a large rocky outcrop or mountain peak.

Chagas: a trypanosome disease transmitted by triatome bugs (vinchucas). Humans become infected directly (from bites) or indirectly (through excreta). The parasites lodge in the heart muscles and cause progressive debilitation and eventually death. It has been stated that 40% of the population of Bolivia are infected with this disease; unless treated early it is incurable.

Chapare: the lowland part of Cochabamba Department where coca is officially cultivated.

Cheque de gerenencia: cheap bank notes issued by the Bolivian Government as temporary replacements for the valueless peso notes.

Chofer: a driver or motorist.

Cholo/Chola: derogative word for male or female Colla.

Chonta: a palm tree of the genus *Bactris*, notable for its (up to) twenty-inch-long spines.

CIAT: Centro de Investigación en Agricultura Tropical, the counterpart organisation working with the British Mission. Ciento-diez: literally the telephone number, 110, of the mobile police force.

Coatimundi (locally known as the tejon): *Nasua nasua*.

Coca: the bushy shrub *Erythroxylon coca*, which produces the alkaloids used in the manufacture of cocaine; and was previously used in the manufacture of Coca Cola.

Colectivo: a bus; the cheapest form of interurban transport.

Colla: originally anyone of the Collasuyo tribe, which included the Aymara and Quechua Indians of the altiplano. Often used derogatively by the Cambas.

Colonos: colonists, normally peasant Collas.

Comunicaciones: lit. a report. Used by Reg Hardy as the title for his PRODENA newsletter.

CORDECRUZ: Corporación de Desarollo de Santa Cruz.

Cruceñan: a native of the town of Santa Cruz.

Curichi: an overgrown pond or ox-bow lake.

Curucusí: or curucuci, pronounced *coo roo coo see*, any click beetle of the genus *Pyrophorus*.

Damnificados: lit. those who have suffered loss, or who have been made homeless.

Decreto Supremo: a Supreme Decree; carrying less weight than a Decreto Ley.

Denuncia: a denunciation.

Departamento: one of Bolivia's nine federal regions; equivalent to a State or County.

Departamento de Vida Silvestre y Parques Nacionales: the Wildlife and Parks Department; part of the CDF.

Doñana: Doñana National Park in southern Spain.

Ducal: brand of beer brewed in Santa Cruz.

El Niño: the unusual warming of the Pacific Ocean current, which causes severe perturbations of the weather from Australia to California. It was named 'the child' by Peruvian fisherman who noticed that the sea got warmer around Christmas time every five years or so, and when it did, there were no fish.

Espundia: Delhi Boil or White Leprosy, medically *Leishmaniasis*.

Estancia: a country estate or cattle ranch.

Excombatiente: an exserviceman or war veteran.

Fajas sub-andinas: the eastern Andean foothills, approx. 300-2,000 metres in altitude.

Falangista: a member of the falange fascist organisation established in Spain in 1934; later adopted by Franco as his party.

Fiesta: a party or celebration.

Flota: an inter-city bus.

Gallo fino: a fighting cock.

Galpon: a large shed or storehouse.

GEA: Grupo Ecológico Amboró, Miriam's youth group in Buena Vista.

Grabadora: a tape recorder.

Gringo: a foreigner or fair-haired person; more especially a North American or north European, when it is often used disparagingly.

Grupo Pantera: a paramilitary exploration and rescue team.

Guajojo: the Common Potoo, *Nyctibius griseus*.

Guapomo: the tree, *Salacia impressifolia*, or its fruit.

Guaraca: the Speckled Chachalaca, *Ortalis motmot*.

Guaraní: the original lowland Indians of Santa Cruz.

Hectare: measure of land equivalent to 2.4 acres.

Huelga: a worker's strike or walkout.

Huevo de gato: acc. to Kempff this is *Covelia nodosa*.

Ingeniero (Ing.): a title that stands before a name to indicate that the person holds a college degree in agriculture - an agronomist. The Spanish use many such titles: Profesor for a teacher (as Noel Kempff); Licenciado for a university degree; lawyers and vets are Doctors, Arquitecto an architect, etc.

Instituto de Colonización: the Government Institute responsible for the colonisation programme, including the granting of land titles. The office was closed down in 1993 because of institutionalized corruption; more especially the granting of illegal land titles. In 1994 it, and the equally corrupt office of the Agrarian Reform, were replaced by the new Instituto Nacional de Reforma de Tierras (INRA).

Islas Malvinas: the Falkland Islands.

Jaguarundi: *Felis yagouaroundi*.

Jararaca: the South American equivalent of the Bushmaster.

Jochi: usually the Brown Agouti, *Dasyprocta variegata*, but may also refer to the Paca, *Agouti paca*.

Joichi: a nuisance.

Kiosko: lit. a kiosk, or small roadside bar.

LAB: Lloyd Aereo Boliviano, the state-owned airline company.

La Linea Roja: the Red Line, introduced into the Bolivian language by the author, to mean a boundary between illegally settled lands and fully protected lands. Later the term was used to mean a temporary boundary setting off a protected area.

Libreta: lit. a licence; also libreta de servicio militar: certificate of military service.

Lloyd Aero Boliviano: see LAB.

Loteodores: a squatter or illegal settler.

MACA: see Ministerio de Asuntos Campesinos y Agropecuarios.

Machetero: a Beni tribe known for the large feathered headdresses they use in their traditional dances. The feathers are taken from macaws, principally *Ara araruana*, *Ara macao* and *Ara chloroptera* - all protected species.

Mañana: lit. tomorrow; 'the spirit of mañana' means laissez faire.

Marihuí: a species of biting sandfly, especially numerous near rocky rivers; these flies, of the genus *Phlebotomus*, are vectors for a leprosy-like disease, a type of *Leishmaniasis*. See Espundia.

Mataral(es): an overgrown thicket(s) or dense woodland, often marshy.

Maúri: the Smooth-billed Ani, *Crotophaga ani*; gregarious non-parasitic species of cuckoo common in Santa Cruz.

Metiche: lit. an interfering busybody.

Micro: a minibus.

Ministerio de Asuntos Campesinos y Agropecuarios: Ministry of Peasant Affairs and Agriculture.

MNR: Movimiento Nacional Revolucionario, the political party that won the 1985 elections; and later the 1993 elections.

Mono leon: in Bolivia the Saddleback Tamarin, *Saguinus fuscicollis weddelli*, not to be confused with the Golden Lion-Tamarin, *Leontopithecus rosalia*.

Morpho: any one of large blue butterflies of the genus *Morpho*. Their wings are used for art-deco pictures.

Motacú: the palm (*Scheelea princeps*), the leaves of which are used for thatching. The nuts are an important food source for many wild animals (squirrels, peccaries, agoutis) and large parrots like macaws; unfortunately trees exploited for thatching material do not produce nuts for several years.

Movimiento Nacional Revolucionario: see MNR.

Narcos: short for narcoticos.

Narcoticos: Narcotic's Police; often working together with the US

DEA: The Drug Enforcement Agency.

Nota de remision: an airways bill.

Notario: a public notary or solicitor; often an honorary position in Bolivia's rural areas.

Oropendola: any bird of the genus *Psarocolius*.

Paisano: lit. a fellow countryman; used derogatively by the Cambas when referring to the Collas.

Paquió: the tree known as the Stinking Toe, *Hymenea stilbocarpa*.

Patujú: pronounced pah two who, any plant of the genus *Heliconia*.

Peligro: dangerous.

Peni: a large lizard of the genus *Tupinambis*; called tegu in other South American countries.

Peons (ref Fawcett): unskilled workman or porter.

Peso: the Bolivian unit of money, which later gave way to the Boliviano.

Piazza: lit. a covered walkway or pavement.

Picaflor: a hummingbird.

Pica-pica: a tree-sized stinging nettle of the genus *Urera*.

Pichicata: the crude paste or base used for the production of cocaine. Also known as sulfato.

Pichicatero: from pichicata - cocaine, drug dealer or producer.

PL-480: lit. Public Law-480; a US-AID sponsored scheme whereby the funds used by the Bolivian Government to buy US wheat is returned to Bolivia for use in development work.

Plaza: the town square.

Politico: a politician.

Poliza de exportación: An export certificate.

PRODENA: acronym for Pro Defensa de la Naturaleza; originally a Peruvian natural history society for the protection of the environment; later used by Reg Hardy and others to found the equivalent Bolivian society in 1979.

Pueblo: lit. a village, or the people.

Quebrada: from quebrar, to break; thereby a stream in a deep cleft, or a mountain torrent.

Quinta: lit. a small estate; a weekend farm.

Reserva Fiscal: a protected area set aside for commercial reasons.

Resolución Ministerial: a Ministerial Resolution; a temporary law, usually prior to the preparation of a Supreme Decree.

Sabalo: a migratory fish, Prochilodus scrofa, that is common in most lowland rivers.

Salteña: a chicken or meat pie, or pasty.

Sapo: lit. a toad; also a Bolivian and Peruvian game in which the contestants throw lead or bronze disks into the mouth of a metal toad.

Sardinas: any one of a number of small species of fish belonging to the genus *Astanyax*.

Sayubú: the Sayaca Tanager, *Thraupis sayaca*.

SEARPI: Servicio Encausamiento del Rio Piray; a European Community financed project to protect the city of Santa Cruz from the flood waters of the River Piray.

Sereno: a watchman or guard.

Seriema: the Red-legged Seriema, *Cariama cristata*; similar to, but smaller than the African Secretary Bird.

Serrania: a range of hills or mountains.

Siesta: an afternoon nap.

Sindicato: a syndicate; sic, more especially a cooperative of peasant farmers.

Subprefecto: a deputy sheriff.

Sumurucucu: the Tropical Screech-Owl, *Otus choliba*.

Sur or surasu: a southern wind that blows from Argentina, bringing with it a sharp drop in temperature, and often rain.

Suri: the Lesser Rhea, *Pterocnemia pennata*.

Tacuara: a large species of spiny bamboo of the genus *Guadua*.

Tajibo: a hardwood tree, *Tabebuia caraiba*, which puts out a spectacular yellow bloom; and is much prized for furniture making, and by the building trade.

Tamandua: the Southern Tamandua, *Tamandua tetradactyla*.

Tinamou: any representative of the family *Tinamidae*, partridge-like birds known locally as perdizes.

Toborochi: the tree, *Chorisia speciosa*, the emblem of Santa Cruz.

Trago: a glass of alcohol, rum, gin, whisky, etc.

Ucureños: originally people from the town of Ucure–a near Cochabamba, whose revolt in 1952 lead to the Agrarian

Reform. Later used specifically for a peasant army sent by the Bolivian Government to quell an uprising in Santa Cruz.

UGRM: its correct initials should have been UBGRM, Universidad Boliviana Gabriel Rene Moreno, but not even the University used this form; and later it became the UAGRM, Universidad Aut—noma Gabriel Rene Moreno.

UMOPAR: Unidad Móbil de Patrullaje Rural, the Rural Mobile Police Force.

Uni—n: sic., see below.

Unión Juvenil Cruceñista: a Santa Cruz association for the protection of young peoples interests; considered right-wing.

Verdolago: probably *Buchenavia oxycarpa*.

YPFB: Yacemiento Petrolifero Fiscales Boliviano: the state-owned oil company.

Yoperobobo: the S. American 'Fer-de-Lance', a viper of the genus *Bothrops*.

Yuca: manioc or cassava, cultivated plants of the genus *Manihot*; not to be confused with the household Yucca.

Yuki: a small nomadic tribe of primitive Indians (comprising about 180 known individuals) whose ancestral lands lay just to the north of Amboró National Park. They had no fire, and kept slaves drawn from their own people, who on the death of their masters were killed to accompany him in his afterlife. Because the Yukis were hunted down and killed by loggers and colonists, they were herded together and put under the protection of American Missionaries.

About the Author

Robin Clarke, English Naturalist, Scientist, Taxonomist, Ornithologist and explorer, was born in Tanzania. Since he was very young, he was a passionate nature lover. His collection of British beetles is housed in the Tervuren Museum, Belgium and the British Museum in London. In 1983 he secretly investigated the illegal multimillion trade in wildlife in Bolivia organized by the Mafia out of San Francisco, California, and with the help of The Unión Juvenil Cruceñista, the export trade was close down by popular demand. Together with Bolivia's famous Noel Kempff Mercado, he founded the Amboró National Park. He was awarded the PREMIO JOVEN due to more than 20 years of his self-financed commitments to the conservation of our Planet and for being an inspiration for young people to love nature.

Other books published by
Harrow and Heston
All available **FREE** on the
Harrow and Heston web site

FICTION

9/11/TWO

by Colin Heston

This gripping novel offers a glimpse into the real world of counter terrorism, hints at why 9/11 was allowed to happen and warns us that it could easily happen again. It's politics as usual in New York City when Larry MacIver, world renowned criminologist, is tapped by NYC Mayor Ruth Newberg to save NYC from a second 9/11 attack. Will it be nuclear? Will it be bio? MacIver and his geeky assistant Manish Das must overcome FBI ineptitude, CIA intrigue and, most of all, the evil and ruthless Iranian terrorist Shalah Muhammud, to save the city. In the underground of this story, so suspenseful, so frustratingly funny at times, are the crucial questions of counter terrorism that worry anyone within a couple of hours drive from New York City: Will it be a drone next time? Will New York politics doom the city's defences? Written before drones were widely in use, the novel seems prescient of much that has happened (should and should not have happened) in the world of counter terrorism.

Miscarriages

by Colin Heston

Teen Chooka grows up in the weird world of 1950s Aussie pub life. When his alcoholic dad dies, he searches for his identity, and that of his shadowy underage girlfriend, Iris. Captivated by the pub's many crazy customers and their raucous stories, Chooka becomes a boozer just like them. But Iris, after a miscarriage, disappears and Chooka sets out on a search that takes him to foreign places including Melbourne university and Vietnam. The search ends in a Melbourne pub, where they start over, but this time there's a different ending.

"…a brilliant, unforgettable book about real people…a sensitive, touching and poignant story." - *Reader's Favourite.*

Ferry to Williamstown

by Colin Heston

In this raucous Aussie story, corpses pop up in the Yarra river while Lizzie entertains her powerful and kinky clients in her Winnebago, parked on the ferry to Williamstown. Tightly bound Detective Striker, confronted by the mob of Catholics, wharfies and communists who rule Williamstown, struggles to solve the mystery. Lizzie gets engaged to her uncle Bobby, the lame ferry driver, and her mum, Babs, spellbound by the strange Father Zappia, tries to solve her own mystery of St. Robert's toe. She throws a raucous send-off party for Lizzie, and out of the chaos emerge many truths.

Holy Water

by Colin Heston

In this very naughty, hilariously irreverent farce, Alphonso, a Mexican drug lord, captures the market in Holy Water, acquires a university for his LGBTQ daughter, and makes an Australian cardinal the pope. "The worlds of gender, religion, and university life will never be the same again. Whoever wrote this book should be locked up with the outrageous characters s/he invented!" (*Chronicle of Lower Education*).

NONFICTION

The Art of Punishment. 2 Volumes. by Graeme R. newman
A Primer in Private Security by Mahesh Nalla and Graeme Newman.
A Primer in the Psychology of Crime by Mark Seis and Shlomo Shoham.
A Primer in the Sociology of Crime by John P. Hoffmann and Shlomo Shoham.
Close Control: Managing a Maximum Security Prison by Nathan Kantrowitz.
Corporate Crime, Corporate Violence by Michael J. Lynch.
Crime and Social Deviation by Shlomo Shoham.
Discovering Criminology from W. Byron Groves edited by Graeme R. Newman, Michael J. Lynch.
From Gangs to Gangsters by Marylee Reynolds.
*God as the Shadow of Man b*y S. Giora Shoham
Justice with Prejudice by .Michael J. Lynch

Just and Painful 2ⁿᵈ Edition. by Graeme R. Newman
Migration, Culture Conflict, and Crime edited by Joshua D.
    Freilich, Graeme R. Newman, S. Giora Shoham, Moshe Addad.
Personality and Deviance by S.Giora Shoham.
Punishment and Privilege 2ⁿᵈ edition edited by Graeme R. Newman.
Race and Criminal Justice edited by Michael J. Lynch and E. Britt
    Patterson
Representing O.J.- Murder, Criminal Justice and Mass Culture by
    Gregg Barak
Salvation through the Gutters by S. Giora Shoham
Sex as Bait by S.Giora Shoham
The Mark of Cain by S. Giora Shoham
Valhalla, Calvary and Auschwitz by S. Giora Shoham
Vendetta (Italian) by Graeme R. Newman and Pietro Marongiu.
Vengeance: The Fight against Injustice by Pietro Marongiu and
    Graeme R. Newman. 2ⁿᵈ edition .
Who Pays? Casino Gambling and Organized Crime by Craig A.
    Zendzian.

PLUS

**Just about every book written by Sir Arthur Conan Doyle,
Charles Dickens, and many hundreds of other classics
many recent, of fiction and nonfiction.**

HARROW AND HESTON
Open Access Publishers
https://www.harrowandheston.com/

(HH)

AUSTRALIA, NEW YORK & PHILADELPHIA

CPSIA information can be obtained
at www.ICGtesting.com
Printed in the USA
LVHW061344030623
748753LV00002B/214

9 780911 577648